Foundations of
Contract Law

Interdisciplinary Readers In Law
ROBERTA ROMANO, *General Editor*

Foundations of Corporate Law
ROBERTA ROMANO

Foundations of Tort Law
SAUL LEVMORE

Foundations of Administrative Law
PETER SCHUCK

Foundations of Contract Law
RICHARD CRASWELL AND **ALAN SCHWARTZ**

Foundations of Contract Law

RICHARD CRASWELL
ALAN SCHWARTZ

New York   Oxford
Oxford University Press
1994

Oxford University Press

Oxford   New York   Toronto
Delhi   Bombay   Calcutta   Madras   Karachi
Kuala Lumpur   Singapore   Hong Kong   Tokyo
Nairobi   Dar es Salaam   Cape Town
Melbourne   Auckland   Madrid

and associated companies in
Berlin   Ibadan

Copyright © 1994 by Oxford University Press, Inc.

Published by Oxford University Press, Inc.
200 Madison Avenue, New York, NY 10016

Library of Congress Cataloging-in-Publication Data
Foundations of contract law / [edited by] Richard Craswell, Alan
Schwartz.
p.   cm.—(Interdisciplinary readers in law)
ISBN 0-19-509035-7
ISBN 0-19-507940-X (pbk.)
1. Contracts—United States.   I. Craswell, Richard.
II. Schwartz, Alan, 1940–   .   III. Series.
KF801.A7F65   1994
346.73'02—dc20
[347.3062]   93-43175

9 8 7 6 5 4 3 2 1

Printed in the United States of America
on acid-free paper

For Laura
and
For Leah and Seth

Preface

The selections in this book emphasize economics and, to a lesser extent, moral philosophy. Obviously, this emphasis is partly a reflection of our own biases (biases for which we make no apology). In ways that are not so obvious, however—at any rate, they were not obvious to us when we began this project—the emphasis on economics and philosophy also results from three other goals we consciously adopted from the outset.

First, we wanted readings that took explicit stands on the normative question of what the law *ought* to be. In particular, we have not included any excerpts whose only aim is to describe current patterns in the case law. However, this selection criterion also led to the underrepresentation of other important disciplines that traditionally do not speak directly to normative issues. For example, there is little historical analysis in this volume (although the excerpt by Clare Dalton is an exception). For better or worse, in modern legal scholarship it is the economists and philosophers (or the lawyers most influenced by those disciplines) who have taken the most explicit normative stands.

Second, we wanted readings that help explain why parties make the contracts they do make. Economists recently have created a rich literature devoted to explaining contracting behavior. Much of this literature is quite mathematical, and the empirical work often uses rigorous econometric techniques; we have excluded those articles as well. However, the technical economic literature has already influenced legal articles, and we have included some of that work here.

Third, we wanted readings that addressed particular legal issues that arise in a first-year course in contract law. Our goal was a set of readings that could be used throughout the semester or throughout the year, rather than being relegated to a "theoretical overview" at the beginning or end of the course. Thus, there are very few readings in this volume that address the system of contract law as a whole. Most of the readings address specific legal topics— the limits on the recovery of consequential damages under *Hadley v. Baxendale,* for example, or the scope of liability during preliminary negotiations under decisions such as *Hoffman v. Red Owl.* In effect, we take the pragmatic position that normative perspectives are best understood by examining what they have to say about specific issues, rather than by trying to analyze them in the abstract. But this criterion, too, may have excluded perspectives whose principal contribution is to provide a theory of contract law as a whole without having much to say about particular contract issues.

The organization of this volume reflects our aim of integrating these readings with the first-year contract course. The core of the readings are in Chapters 2 through 5, which are organized in the "backwards" or remedies-first manner originally introduced by Lon Fuller. Thus, Chapter 2 discusses legal remedies for breach of contract, on the assumption that a valid contract has been formed and that one party has failed to do what the contract requires (whatever that might be). Chapter 3, "Defining the Performance Obligation," discusses how to determine what it is that the contract requires, while retaining the assumption that a valid contract has in fact been formed. This is the chapter that deals with implied excuses (impracticability, frustration, mistake) and implied warranties; it also includes the recent scholarship on long-term "relational" contracts. Finally, Chapters 4 and 5 discuss the formation of a valid contract. Chapter 4 deals with consideration and traditional offer-and-acceptance issues; Chapter 5 addresses unconscionability and other defenses to formation.

Obviously, this division is somewhat arbitrary. For example, the readings on rescission and restitution damages (infra at pages 115–26) also relate to the interpretation of conditions and the assessment of whether one party's performance was "substantial," those being two of the legal tests for rescission and restitution. These readings could therefore have been included in Chapter 3 ("Defining the Performance Obligation") rather than in Chapter 2 ("Remedies for Breach") where we actually put them. The readings on nondisclosure have been placed in Chapter 3 (infra at pages 160–74) in connection with mutual and unilateral mistake, but they could just as easily have been placed with the defenses to contract formation discussed in Chapter 5. Examples like these merely confirm the observation that the law is indeed a seamless web. Responding on a practical level, we have tried to make each section of the book self-contained (with cross-references to related sections) so that the selections can be assigned in any order.

This self-contained or "modular" principle is also reflected in Chapter 1, which contains readings potentially relevant to every part of the course. The three topics discussed in Chapter 1—the enforcement of promises, the selection of default rules, and the distributional effects of contract law—are as relevant to contract formation as they are to the remedies for breach; thus, these readings could be assigned at any point in the course. Our placement of these selections at the beginning of the book does not reflect any judgment that a first-year contract course ought to begin with these readings. (Neither of us teaches the readings in that order ourself.) Instead, these selections have been placed in a separate chapter precisely so that they can be used *wherever* they best fit the organization of any given course. If these selections had instead been parceled out to the various parts of the reader dealing with particular legal doctrines, this flexibility would only have been hampered.

Though our organizational scheme derives from the first-year law course, this book is intended to be useful to a wider audience. The readings, taken as a whole, reflect much of the contract theory that normatively minded lawyers, economists, and philosophers have produced in the last fifteen years. These have been rich years for contract theory. The book can therefore be used in advanced law school seminars (we have so used it), and it can also be part or all of the reading for an undergraduate or graduate law and economics course or a course in contract theory.

Many people assisted us in the preparation of this collection. Particular thanks are due to our series editor, Roberta Romano; and to Ian Ayres, David Carroll, Jason Johnston, and David Slawson for helpful comments and suggestions. We have also benefited from the research assistance of Stacey Cole, Hanoch Dagan, and Terrence Gallagher. Deletions from the original texts are indicated by ellipses, but footnotes and subheadings have been deleted without indication.

Los Angeles, Calif.　　　　　　　　　　　　　　　　　　　　R. C.
New Haven, Conn.　　　　　　　　　　　　　　　　　　　　A. S.
December 1993

Contents

2. *Remedies for Breach,* 41

Foundations of
Contract Law

Normative Underpinnings

The issues discussed in this chapter are fundamental to all of contract law and contract theory. The readings in the first section ask why promises should be binding in either a moral or a legal sense. Patrick Atiyah's essay questions whether promises as such should always bind, especially in cases where the promisee has neither relied on the promise nor conferred any benefit on the promisor in exchange for the promise. Charles Fried defends the obligation to keep promises—even promises that have not yet been relied upon—by grounding that obligation in the promisor's voluntary choice to bind himself or herself, and in a theory of individual autonomy that demands that such choices be respected.

The readings in the second section deal with the related obligations that make up much of contract law. To say that a promise is binding raises two ancillary questions about the scope of the obligation: (1) What legal consequences flow from its breach? (2) Under what circumstances is an otherwise binding promise excused? There is also a further question concerning the rules of remedy and excuse: should these ancillary rules override the parties' own agreement, or should they be "default rules," which govern only to the extent that the parties have not agreed otherwise. Specific default rules will be discussed in Chapters 2 through 4, but the excerpts by Richard Craswell and by Ian Ayres and Robert Gertner provide a general framework for analyzing default rules. The Ayres and Gertner article presents various economic perspectives on default rules; the Craswell article discusses the possible contributions of noneconomic moral theories.

The third section's topic—whether contract law can or should have redistributional effects—will also be relevant to other readings, especially those in Chapter 5. The essay in this section provides an introduction to these distributional effects, and also discusses the relationship between distributional and efficiency concerns.

§1.1 **Enforcing Promises**

The Rise and Fall of Freedom of Contract
P. S. ATIYAH

[I]t is necessary at the outset to distinguish three situations in which contracts may be held legally binding, or promises may be found morally binding:

1. In the first situation a contract or a promise may be found binding after a price has been paid for it. For example, a person may borrow §100 from a friend and may simultaneously promise to repay it. In this situation there would, both legally and morally, be a liability to repay even if there were no promise. The promise may, indeed, be said to be "implied" but my contention is that in this situation the primary justification for imposing a legal or moral obligation on the party borrowing the money is that he has received a benefit at the expense of the other party, and that is, in a property-owning society, usually sufficient to establish a liability. I express this conclusion by saying that the liability is benefit-based in this type of case; it could also be said to arise from broad notions of unjust enrichment. If it should be asked what is the function of the promise in such circumstances, one answer might be that the promise has evidentiary value. It is evidence that the promisor has received a benefit (for if it was not a benefit would he have promised to pay for it?) and it may be evidence of many ancillary matters such as the precise terms of the arrangement, the date of repayment and so on.

2. In the second situation, a contract or promise may be enforced where the promisee has acted in reliance on the promise, or on the promisor's conduct, and would in consequence be in a worse situation than if no promise had ever been made. The case of a simple loan discussed above may, of course, also be a case of such action in reliance, for the lender may only have

lent his money in reliance on the borrower's promise to repay. But cases of action in reliance may arise without any element of benefit or unjust enrichment. In the law, a common example is to be found in the typical contract of guarantee, as where A promises to guarantee repayment of a loan to be made by B to C. In this situation if B acts in reliance on the guarantee, he will lend money to C and may (if C is himself not good for the money) thus make his position worse than it would be if there had been no promise. In my terminology I would refer to liability in such a case as reliance-based. As in the previous case, I suggest that many forms of reliance-based liability arise, or would arise even in the absence of a promise. The party relying may be relying, not on a promise, but on other words, or mere conduct. Such reliance is a commonplace in modern societies and often gives rise to liabilities even in the absence of a promise. For instance, a person who buys a new house, in reliance on the proper performance by the local authority of its duties of ensuring compliance with the Building Regulations, may have a remedy against the authority for malperformance even though they give no promise. Here again, as with benefit-based liability, the result may be justified or explained by saying that there is an "implied" promise. But it will be observed that in such circumstances the liability comes first, and the implication is made subsequently to justify the decision already arrived at. Once the liability itself is well established (whether in law or in social custom) it is easy to make the implication. But in the first instance, it is the conduct of one party, followed by the action in reliance of the other, which creates the liability. As with the case of benefit-based liability, it is likely that an actual, express promise, will serve a useful evidentiary role in reliance-based liability. Whether the party acting did in fact rely on the other (or, for example, on his own judgment) and if so, whether in so acting, he acted reasonably by the standards of the society in question, are questions whose answer may be greatly assisted by the presence of an express promise. But again, it does not follow that it is the promise which creates the liability.

3. The third situation concerns a promise or a contract which has not been paid for, and which has not yet been relied upon. In the law such a contract or promise would be called "wholly executory." If such a promise or contract generates any liability, the liability must be promise-based, since it cannot be benefit-based or reliance-based. In the first two cases, distinct grounds exist for imposing the liability, apart altogether from the promise. In this case, no such distinct grounds exist. If the promise is held to be "binding" or to create some liability, it must be for some reason which is inherent in the promise itself. The principal grounds which (it is suggested) can be found for imposing such liability in this case are these. First, it may be said that a promise, even while executory, creates expectations, and that these expectations will be disappointed if the promise is not performed. In this sense, there is a similarity between a promise-based and a reliance-based liability. The promisee whose expectations are disappointed may feel he is worse off than he would have been if no promise had been made at all. Psychologically this may be true; but in a pecuniary sense, it is not. The party who acts in reliance may spend

money which he would lose if he could not claim recompense from the party on whose conduct he relied. But the promisee who has not yet acted in reliance on a promise, and not yet paid any price for it, will not be worse off in a pecuniary sense merely because his expectations are disappointed.

Secondly, it may be said that contracts and promises are essentially risk-allocation devices, like simple bets. The nature of this device is such that the transaction must generally remain executory prior to the occurrence of the risk, and the whole point of the transaction would be lost if the arrangement could not be made binding for the future.

The third possible ground for the enforcement of executory promises or contracts is that it may be desirable to uphold the *principle* of promissory liability, even in cases where the non-performance of the promise has little practical effect. The argument here comes to this, that if executory promises are held binding (whether in law or in social custom and morality) then people are more likely to perform promises which have been paid for, or relied upon.

Now it will be seen that many promises and contracts are likely to be wholly executory at the outset, but may quickly pass into one or other of the first two situations discussed above. A promise may be given which is at first executory, and only subsequently is it acted upon by the promisee, or paid for by the promisee. In this book I suggest that once this happens, the ground for impos-ing a liability shifts. The liability becomes benefit-based or reliance-based, where it was previously promise-based. This may seem strange, and indeed, it is precisely because this seems so strange that it has not generally been recog-nized, either in law or in general discussion of the nature of promissory liability. This, I suggest, is because promise-based liability is seen as the paradigm case for discussion both in law and among philosophers, and perhaps in ordinary discourse. One of the purposes of this book is to suggest that this is itself part of our cultural and legal heritage, and that an alternative perspective may be possible and even preferable. If benefit-based and reliance-based liabilities are taken as the paradigm cases of obligation, whether legal or moral, it may be suggested that promise-based liabilities are neither paradigmatic nor of central importance. Far from being the typical case of obligation, a promise-based liability may be a projection of liabilities normally based on benefit or reliance. Because these are normally found such powerful grounds for imposing obliga-tions, it has been thought that the element of promise (express or implied) which is often combined with benefit-based and reliance-based liability, is itself the ground for the obligation. And from this, it has been an easy move to the inference that promise-based liability, even without any element of benefit or reliance, carries its own justification.

Much of this book is based on the conviction that this traditional attitude to promise-based obligations is misconceived, and that the grounds for the imposition of such liabilities are, by the standards of modern values, very weak compared with the grounds for the creation of benefit-based and reliance-based obligations. The protection of mere expectations cannot (it is suggested) rank equally with the protection of restitution interests (arising from benefit-based liability) or reliance interests (arising from reliance-based liability). A person whose expectations are disappointed, but who suffers no pecuniary or other loss from the failure to perform a promise, has surely a relatively weak claim for complaint or redress. No doubt if there is *no* excuse or justification at all for the failure to perform the promise or contract, the promisee may be felt entitled to some redress, but even then it does not follow that he should be entitled to demand full performance of the promise, or redress based on such an entitlement. Frequently, a promise-based claim is based on relatively short-lived expectations; for it is where the promisor has (for instance) made some mistake, or overlooked some fact, that he is most likely to attempt to withdraw a promise. Where the promisor does not do this, the probability is that some action in reliance (or some payment) will soon be performed by the promisee, and he can then claim the much greater protection due to reliance interests or restitution interests. Adoption of this alternative approach would, of course, have a profound effect on the conceptual pattern of moral and legal obligation, but to argue, as I do, that the justification for creating promise-based obligations is usually weak, does not mean that this approach would involve a serious undercutting of typical moral or legal obligations. For in practice, even liabilities which are usually perceived as promise-based in law or social custom, are, in my terminology not exclusively promise-based at all. In fact most such liabilities are, or rapidly become, reliance-based or benefit-based, and the period during which they remain promise-based is usually relatively short. Indeed, in practical terms, the approach I advocate would tend chiefly to affect those relatively marginal cases in which promises are revoked shortly after they are given, and before they have been paid for or relied upon.

The second ground for maintaining the binding force of an executory contract is, as I have suggested, that such a contract is essentially a way of allocating a risk, or perhaps of transferring a risk from one party to another in advance. Where this is indeed the case (as for example, with bets, or some forms of insurance) it seems that the arrangement must, if it is to have any point at all, be binding on the parties at the outset. But this argument is open to two possible answers. One is that most contracts are not in fact entered into for the purpose of transferring or allocating risks. If contracts are construed as being risk-allocation mechanisms, this is because they are often seen as such in the eye of the beholder. Pure risk-allocation contracts are relatively rare, and it may be that special considerations do apply to them. The second possible answer is that even in contracts of this nature an element of reliance is still needed before it becomes essential to maintain the integrity of the transaction. Even an executory insurance arrangement, for instance, could be made cancellable so long as the insured still has time to find alternative cover. A person might, in principle, be given the right to withdraw from a bet on a race before the race is run, so long as the other party has time to place his own bet elsewhere at similar odds. There would be nothing logically impossible about such a possibility though it might be inconvenient.

The third ground for the creation of promise-based liabilities is also, I suggest, very weak in comparison with the grounds for the creation of benefit-based and reliance-based obligations. For this ground is, in effect, nothing more than an argument for the use of promise-based liability as a subsidiary method of ensuring compliance with benefit-based and reliance-based obligations. There are, of course, great difficulties in arguing that promise-based liabilities should be observed even though there is no independent justification for their observance, in order that reliance-based and benefit-based obligations should be better observed. Now it cannot be claimed that the case for the enforcement of promise-based liabilities is entirely vitiated by these difficulties, because the fact that my approach is so unorthodox itself testifies to the practical strength of this argument. Both morally and legally, promise-based liabilities *have* traditionally been thought worthy of protection, even where there has been no element of reliance or reciprocal benefit. It seems certain, therefore, that where there is some element of reliance or benefit, the case for redress has been felt a fortiori to be the more powerful. This too, I suggest, is part of the cultural heritage which I explore further in this book, and it is enough to say here that the whole trend of modern times is against arguments of principle of this character. It appears more in accord with contemporary beliefs to reject the argument of principle, and to insist upon the difference (for example) between maintaining the sanctity of a promise or contract because it has been relied upon, and because it might have been (but was not) relied upon. What is lacking is a theoretical or conceptual (and perhaps even linguistic) recognition of these differences.

In suggesting that these ideas are, at least intuitively or implicitly, gaining much ground today, and in advocating open recognition of these facts, it does not follow that I approve or disapprove of them. The nature of the conflict of values which underlies this question will become clear during the course of this book, but in its essentials the conflict is perfectly plain. Promise-based liability rests upon a belief in the traditional liberal values of free choice. Many still admire these values but they bring with them, inescapably, many other consequences which are today less admired, especially in England. They bring, in particular, the recognition that some individuals are better equipped to exercise free choice than others, through natural aptitude, education, or the possession of wealth. And the greater is the for the exercise of free choice, the stronger is the tendency for these original inequalities to perpetuate themselves by maintaining or even increasing economic inequalities. For example, in contracts which really are risk-allocation arrangements, to hold the contract binding must, in general, favour the party who has the better skill and knowledge for assessing future risks.

By contrast, other forms of liability rest on different values. Even benefit-based liability, though it may tend to perpetuate existing inequalities of wealth, does at least militate against increasing those inequalities in the way in which promise-based liabilities may do. For where liabilities are benefit-based, the law (or the moral norms) strive for a reasonable or just balance in the reciprocity of benefit; where liabilities are promise-based the free choice of the parties determines this balance, and it is inevitable that this will tend to favour those better able to exercise free choice.

Reliance-based liabilities are still more hostile to the values of free choice. As soon as liabilities come to be placed upon a person in whom another has reposed trust or reliance, even though there is no explicit promise or agreement to bear that liability, the door is opened to a species of liability which does not depend upon a belief in individual responsibility and free choice. Not only is the party relied upon held liable without his promise, but the party relying is relieved from the consequences of his own actions. The values involved in this type of liability are therefore closely associated with a paternalist social philosophy, and a redistributive economic system.

Contract as Promise

CHARLES FRIED

What is a promise, that by my words I should make wrong what before was morally indifferent? A promise is a communication—usually verbal; it says something. But how can my saying something put a moral charge on a choice that before was morally neutral? Well, by my misleading you, or by lying. Is lying not the very paradigm of doing wrong by speaking? But this won't do, for a promise puts the moral charge on a *potential* act—the wrong is done later, when the promise is not kept—while a lie is a wrong committed at the time of its utterance. Both wrongs abuse trust, but in different ways. When I speak I commit myself to the truth of my utterance, but when I promise I commit myself to *act,* later. Though these two wrongs are thus quite distinct there has been a persistent tendency to run them together by treating a promise as a lie after all, but a particular kind of lie: a lie about one's intentions. Consider this case:

> I. I sell you a house, retaining an adjacent vacant lot. At the time of our negotiations, I state that I intend to build a home for myself on that lot. What if several years later I sell the lot to a person who builds a gas station on it? What if I sell it only one month later? What if I am already negotiating for its sale as a gas station at the time I sell the house to you?

If I was already negotiating to sell the lot for a gas station at the time of my statement to you, I have wronged you. I have lied to you about the state of my intentions, and this is as much a lie as a lie about the state of the plumbing. If, however, I sell the lot many years later, I do you no wrong. There are no grounds for saying I lied about my intentions; I have just changed my mind.

Now if I had *promised* to use the lot only as a residence, the situation would be different. Promising is more than just truthfully reporting my present intentions, for I may be free to change my mind, as I am not free to break my promise.

Let us take it as given here that lying is wrong and so that it is wrong to obtain benefits or cause harm by lying (including lying about one's intentions). It does not at all follow that to obtain a benefit or cause harm by breaking a promise is also wrong. That my act procures me a benefit or causes harm all by itself proves nothing. If I open a restaurant near your hotel and prosper as I draw your guests away from the standard hotel fare you offer, this benefit I draw from you places me under no obligation to you. I should make restitution only if I benefit *unjustly,* which I do if I deceive you—as when I lie to you about my intentions in example I. But where is the injustice if I honestly intend to keep my promise at the time of making it, and later change my mind? If we feel I owe you recompense in that case too, it cannot be because of the benefit I have obtained through my promise: We have seen that benefit even at another's expense is not alone sufficient to require compensation. If I owe you a duty to return that benefit it must be because of the promise. It is the promise that makes my enrichment at your expense unjust, and not the enrichment that makes the promise binding. And thus neither the statement of intention nor the benefit explains why, if at all, a promise does any moral work.

A more common attempt to reduce the force of a promise to some other moral category invokes the harm you suffer in relying on my promise. My statement is like a pit I have dug in the road, into which you fall. I have harmed you and should make you whole. Thus the tort principle might be urged to bridge the gap in the argument between a statement of intention and a promise: I have a duty just because I could have foreseen (indeed it was my intention) that you would rely on my promise and that you would suffer harm when I broke it. And this wrong then not only sets the stage for compensation of the harm caused by the misplaced reliance, but also supplies the moral predicate for restitution of any benefits I may have extracted from you on the strength of my promise. But we still beg the question. If the promise is no more than a truthful statement of my intention, why am *I* responsible for harm that befalls you as a result of my change of heart? To be sure, it is not like a change in the weather—I might have kept to my original intention—but how does this distinguish the broken promise from any other statement of intention (or habit or prediction of future conduct) of mine of which you know and on which you choose to rely? Should your expectations of me limit my freedom of choice? If you rent the apartment next to mine because I play chamber music there, do I owe you more than an expression of regret when my friends and I decide to meet instead at the cellist's home? And in general, why should my liberty be constrained by the harm you would suffer from the disappointment of the expectations you choose to entertain about my choices?

Does it make a difference that when I promise you do not just happen to rely on me, that I communicate my intention to you and therefore can be taken to know that changing my mind may put you at risk? But then I might be aware that you would count on my keeping to my intentions even if I myself had not communicated those intentions to you. (*You* might have told me you were relying on me, or you might have overheard me telling some third person of my intentions.) It might be said that I become the agent of your reliance by telling you, and that this makes my responsibility clearer: After all, I can scarcely control all the ways in which you might learn of my intentions, but I *can* control whether or not I tell you of them. But we are still begging the question. If promising is no more than my telling you of my intentions, why do we both not know that I may yet change my mind? Perhaps, then, promising is like telling you of my intention and telling you that I don't intend to change my mind. But why can't I change my mind about the latter intention?

Perhaps the statement of intention in promising is binding because we not only foresee reliance, we invite it: We intend the promisee to rely on the promise. Yet even this will not do. If I invite reliance on my stated intention, then that is all I invite. Certainly I may hope and intend, in example I, that you buy my house on the basis of what I have told you, but why does that hope bind me to do more than state my intention honestly? And that intention and invitation are quite compatible with my later changing my mind. In every case, of course, I should weigh the harm I will do if I do change my mind. If I am a doctor and I know you will rely on me to be part of an outing on which someone may fall ill, I should certainly weigh the harm that may come about if that reliance is disappointed. Indeed I should weigh that harm even if you do not rely on me, but are foolish enough not to have made a provision for a doctor. Yet in none of these instances am I bound as I would be had I promised.

A promise invokes trust in my future actions, not merely in my present sincerity. We need to isolate an additional element, over and above benefit, reliance, and the communication of intention. That additional element must *commit* me, and commit me to more than the truth of some statement. That additional element has so far eluded our analysis.

It has eluded us, I believe, because there is a real puzzle about how we can commit ourselves to a course of conduct that absent our commitment is morally neutral. The invocation of benefit and reliance are attempts to explain the force of a promise in terms of two of its most usual effects, but the attempts fail because these effects depend on the prior assumption of the force of the commitment. The way out of the puzzle is to recognize the bootstrap quality of the argument: To have force in *a particular case* promises must be assumed to have force generally. Once that general assumption is made, the effects we intentionally produce by a particular promise may be morally attributed to us. This recognition is not as paradoxical as its abstract statement here may make it seem. It lies, after all, behind every conventional structure: games, institutions and practices, and most important, language. . . .

. . . The conventional nature of language is too obvious to belabor. It is worth pointing out, however, that the various things we do with language— informing, reporting, promising, insulting, cheating, lying—all depend on the conventional structure's being firmly in place. You could not lie if there were not both understanding of the language you lied in and a general convention of using that language truthfully. This point holds irrespective of whether the institution of language has advanced the situation of mankind and of whether lying is sometimes, always, or never wrong.

Promising too is a very general convention—though less general than language, of course, since promising is itself a use of language. The convention of promising (like that of language) has a very general purpose under which we may bring an infinite set of particular purposes. In order that I be as free as possible, that my will have the greatest possible range consistent with the similar will of others, it is necessary that there be a way in which I may commit myself. It is necessary that I be able to make nonoptional a course of conduct that would otherwise be optional for me. By doing this I can facilitate the projects of others, because I can make it possible for those others to count on my future conduct, and thus those others can pursue more intricate, more far-reaching projects. If it is my purpose, my will that others be able to count on me in the pursuit of their endeavor, it is essential that I be able to deliver myself into their hands more firmly than where they simply predict my future course. Thus the possibility of commitment permits an act of generosity on my part, permits me to pursue a project whose content is that *you* be permitted to pursue *your* project. But of course this purely altruistic motive is not the only motive worth facilitating. More central to our concern is the situation where we facilitate each other's projects, where the gain is reciprocal. Schematically the situation looks like this:

> You want to accomplish purpose A and I want to accomplish purpose B. Neither of us can succeed without the cooperation of the other. Thus I want to be able to commit myself to help you achieve A so that you will commit yourself to help me achieve B.

Now if A and B are objects or actions that can be transferred simultaneously there is no need for commitment. As I hand over A you hand over B, and we are both satisfied. But very few things are like that. We need a device to permit a trade over time: to allow me to do A for you when you need it, in the confident belief that you will do B for me when I need it. Your commitment puts your future performance into my hands in the present just as my commitment puts my future performance into your hands. A future exchange is transformed into a present exchange. And in order to accomplish this all we need is a conventional device which we both invoke, which you know I am invoking when I invoke it, which I know that you know I am invoking, and so on. . . .

The Moral Obligation of Promise

Once I have invoked the institution of promising, why exactly is it wrong for me then to break my promise?

My argument so far does not answer that question. The institution of promising is a way for me to bind myself to another so that the other may expect a future performance, and binding myself in this way is something that I may want to be able to do. But this by itself does not show that I am morally obligated to perform my promise at a later time if to do so proves inconvenient or costly. That there should be a system of currency also increases my options and is useful to me, but this does not show why I should not use counterfeit money if I can get away with it. In just the same way the usefulness of promising in general does not show why I should not take advantage of it in a particular case and yet fail to keep my promise. That the convention would cease to function in the long run, would cease to provide benefits if everyone felt free to violate it, is hardly an answer to the question of why I should keep a particular promise on a particular occasion. . . .

Considerations of self-interest cannot supply the moral basis of my obligation to keep a promise. By an analogous argument neither can considerations of utility. For however sincerely and impartially I may apply the utilitarian injunction to consider at each step how I might increase the sum of happiness or utility in the world, it will allow me to break my promise whenever the balance of advantage (including, of course, my own advantage) tips in that direction. The possible damage to the institution of promising is only one factor in the calculation. Other factors are the alternative good I might do by breaking my promise, whether and by how many people the breach might be discovered, what the actual effect on confidence of such a breach would be. There is no a priori reason for believing that an individual's calculations will come out in favor of keeping the promise always, sometimes, or most of the time.

Rule-utilitarianism seeks to offer a way out of this conundrum. The individual's moral obligation is determined not by what the best action at a particular moment would be, but by the rule it would be best for him to follow. It has, I believe, been demonstrated that this position is incoherent: Either rule-utilitarianism requires that rules be followed in a particular case even where the result would not be best all things considered, and so the utilitarian aspect of rule-utilitarianism is abandoned; or the obligation to follow the rule is so qualified as to collapse into act-utilitarianism after all. There is, however, a version of rule-utilitarianism that makes a great deal of sense. In this version the utilitarian does not instruct us what our individual moral obligations are but rather instructs legislators what the best rules are. If legislation is our focus, then the contradictions of rule-utilitarianism do not arise, since we are instructing those whose decisions can *only* take the form of issuing rules. From that perspective there is obvious utility to rules establishing and enforcing promissory obligations. Since I am concerned now with the question of individual obligation, that is, moral obligation, this legislative perspective on the argument is not available to me.

The obligation to keep a promise is grounded not in arguments of utility but in respect for individual autonomy and in trust. Autonomy and trust are grounds for the institution of promising as well, but the argument for *individual* obligation is not the same. Individual obligation is only a step away, but that step must be taken. An individual is morally bound to keep his promises because he has intentionally invoked a convention whose function it is to give grounds—moral grounds—for another to expect the promised performance. To renege is to abuse a confidence he was free to invite or not, and which he intentionally did invite. To abuse that confidence now is like (but only *like*) lying: the abuse of a shared social institution that is intended to invoke the bonds of trust. A liar and a promise-breaker each *use* another person. In both speech and promising there is an invitation to the other to trust, to make himself vulnerable; the liar and the promise-breaker then abuse that trust. The obligation to keep a promise is thus similar to but more constraining than the obligation to tell the truth. To avoid lying you need only believe in the truth of what you say when you say it, but a promise binds into the future, well past the moment when the promise is made. There will, of course, be great social utility to a general regime of trust and confidence in promises and truthfulness. But this just shows that a regime of mutual respect allows men and women to accomplish what in a jungle of unrestrained self-interest could not be accomplished. If this advantage is to be firmly established, there must exist a ground for mutual confidence deeper than and independent of the social utility it permits.

The utilitarian counting the advantages affirms the general importance of enforcing *contracts*. The moralist of duty, however, sees *promising* as a device that free, moral individuals have fashioned on the premise of mutual trust, and which gathers its moral force from that premise. The moralist of duty thus posits a general obligation to keep promises, of which the obligation of contract will be only a special case—that special case in which certain promises have attained legal as well as moral force. But since a contract is first of all a promise, the contract must be kept because a promise must be kept.

To summarize: There exists a convention that defines the practice of promising and its entailments. This convention provides a way that a person may create expectations in others. By virtue of the basic Kantian principles of trust and respect, it is wrong to invoke that convention in order to make a promise, and then to break it.

Notes on Enforcing Promises

1. There is an extensive philosophical literature—too extensive to cite here—on the possible moral bases for the obligation to keep one's promise. For a recent discussion, with citations to the literature, see Thomas Scanlon, Promises and Practices, 19 Philosophy & Public Affairs 199 (1990). Other discussions in the legal journals include Randy E. Barnett, A Consent Theory of Contract, 86 Columbia Law Review 269 (1986); F. H. Buckley, Paradox Lost, 72 Minnesota Law Review 775 (1988); Richard Craswell, Contract Law, Default Rules, and the Philosophy of Promising, 88 Michigan Law Review 489 (1989) [excerpted infra at pages 16–22]; and the symposium published at 17 Valparaiso University Law Review 613 (1983).

2. Atiyah suggests that enforcing executory contracts is most justifiable when those contracts allocate risk, as in the case of bets or insurance. He adds that "most contracts are not in fact entered into for the purpose of . . . allocating risk," and that pure risk-allocation contracts are "relatively rare" (supra at

page 7). Is this empirical claim correct? More important, what if an executory contract both allocates risk and performs some other function—would Atiyah's theory imply that only part of the contract should be enforced?

3. In his concluding paragraph, Atiyah suggests that reliance-based liability is "hostile to the values of free choice," and instead is "closely associated with a paternalist social philosophy" (supra at page 9). Is this correct? Consider Fried's argument that (1) if people choose not to make a promise, the bare reliance of another party should not give the relying party any moral claim on the first party; and (2) when people do choose to make a promise, imposing liability is not at all inconsistent with "the values of free choice."

In a subsequent book, Atiyah acknowledged the argument that one person's reliance on another cannot, by itself, explain why the second party should bear moral responsibility for the first party's reliance. P. S. Atiyah, Promises, Morals, and Law (Oxford: Clarendon Press, 1981). Atiyah concluded that this issue could only be resolved by recourse to "some socially accepted values which determine when expectations and/or reliance are sufficiently justifiable to be given some measure of protection" (id. at page 68). Atiyah went on to suggest that a person's promise might be interpreted as the promisor's admission as to the existence of some other social policy that would justify the promisee's reliance—and that such admissions might in some cases be regarded as conclusive on that issue of social policy (id. at pages 184–202). If Atiyah is willing to attach this much weight to an individual's voluntary promise, does that eliminate most of the differences between Atiyah and Fried?

4. Most articles in the law-and-economics tradition address the desirability of particular rules of contract law without addressing the more basic question of whether or why promises ought to be binding. However, economists have identified conditions under which one party would want to give another the assurance represented by a promise in order to encourage the second party to rely in some way that would increase the value of a transaction. For example, a factory owner may be unwilling to spend money customizing her machinery for a particular transaction unless the potential buyer promises in advance that he will not back out of the deal, or threaten to back out in exchange for a more favorable price, once the factory owner has spent the money to customize her machines. For further discussion of this reliance-based argument for enforcing voluntary commitments, see Charles J. Goetz and Robert E. Scott, Enforcing Promises: An Examination of the Basis of Contract, 89 Yale Law Journal 1261 (1980); Oliver E. Williamson, The Economic Institutions of Capitalism (New York: Free Press, 1985), chapters. 1 and 7–8.

Economic theories of the benefits of enforceable promises are also implicit in economists' analyses of remedies for breach of contract. As the readings in Chapter 2 will illustrate, most economists do not believe that promises ought always to be *performed*—although they would add that in many cases where the promise should not be performed the promise should still be regarded as enforceable, in the sense that the promisor should have to compensate the promisee for nonperformance. From an economic perspective, making promises enforceable in this sense may lead to more efficient decisions between performing and not performing, or more efficient precautions against accidents that might prevent performance, or more efficient levels of reliance by promisees, or more efficient levels of insurance against risks (to list just a few of the possible effects). The readings in Chapter 3, discussing the conditions under which nonperformance ought to be excused without liability for damages, will also bear on these issues.

5. What if an individual has no intention of making a binding promise, but (either deliberately or carelessly) uses words that give others the impression of a binding promise. In portions of his book not excerpted here, Fried argues that such an individual may properly be held liable, but on principles more akin to tort liability than to contract, because this individual has not voluntarily assumed any liability. Fried, Contract as Promise, chapters 5–6. For an argument that this liability is really contractual rather than tort-like because the only "consent" relevant to contract law is the objective *appearance* of consent, see Randy E. Barnett, A Consent Theory of Contract, 86 Columbia Law Review 269, 300–309 (1986).

If both authors agree that such an individual ought to be held liable, does it matter whether the liability is classified as contractual or tort-like? The relationship between contract and tort liability will be discussed in a number of other readings, especially those in Chapter 4 dealing with offer and acceptance.

6. The Atiyah excerpt suggests that contract law might systematically favor some classes of people at the expense of others. This issue is explored in more depth in the final section of this chapter, infra at pages 30–39.

§ 1.2 Selecting Default Rules

Contract Law, Default Rules, and the Philosophy of Promising

RICHARD CRASWELL

The rules of contract law can be divided into two categories: "background rules" and "agreement rules." . . . [B]ackground rules define the exact sub-

Richard Craswell, Contract Law, Default Rules, and the Philosophy of Promising, 88 Michigan Law Review 489, 503–11, 514–15, 517–23 (1989). Copyright © 1989 by Richard Craswell. Reprinted by permission.

stance of a party's obligation, by specifying (among other things) the conditions under which her nonperformance will be excused, and the sanctions which will be applied to any unexcused nonperformance. By contrast, agreement rules specify the conditions and procedures the parties must satisfy in order to change an otherwise applicable background rule. Agreement rules thus include most of the rules governing offer and acceptance, as well as such doctrines as fraud or undue influence, which define the conditions necessary for a party's apparent consent to be counted as truly valid. . . .

[Philosophical theories explaining the moral force of promises] may well have some relevance for contract law's agreement rules. For example, theories that ground the enforceability of promises on individual liberty might argue that the parties should be allowed to overturn nearly all of contract law's background rules by an appropriate agreement, thereby affording a much wider scope for the operation of whatever agreement rules the law adopts. Different philosophical theories might even have different implications for the content of those agreement rules—for example, the degree of force needed to make an individual's consent no longer voluntary, or the amount of information needed to make an individual's consent sufficiently informed. In this article, I am not concerned with the content of contract law's agreement rules, so I will not explore these implications further.

It is less clear that the philosophical literature discussed above has any implications for the content of contract law's background rules. . . . [W]hen a background rule concerns a topic that everyone agrees the parties should be allowed to vary—say, the extent of the warranty in the sale of a used automobile by a private individual—many parties simply will not address that topic in their agreement, so there will be nothing in the agreement's explicit content to resolve this issue. As a result, some method must be found to *interpret* the parties' agreement, to provide rules governing any topic not explicitly settled by the parties. Indeed, creative interpretation is often needed to determine whether there has been any binding promise at all, for even this fundamental question is not always explicitly settled by the parties. While it is perhaps more common to speak of "interpretation" in cases where parties attempt to resolve an issue but do so with insufficient clarity, and to speak of applying default rules in cases where the parties made no attempt to address an issue, the principle is much the same in either case. Both require an outside agency, such as a court, to choose the exact rules defining the parties' obligations where the parties have not unmistakably chosen some rule of their own. . . .

Existing Expectations

The frequent references in the philosophical literature to the "practice" or "institution" of promising could be taken to suggest that the exact scope of any promissory obligation is a matter of sociological fact, to be discovered by careful investigation into the practice of promising as it exists in the relevant community. For instance, an inquiry into the use of promises in late twentieth-century America might show that promisors were regularly excused whenever performance became commercially impracticable, or it might show that promisors were never excused no matter how difficult performance had become. In either case, the results of that inquiry would define the exact scope of the obligation that any late twentieth-century American had accepted when she made a promise.

Of course, any serious sociological inquiry would very likely identify several different forms of promising, each with different background rules and assumptions, even within a single community. At the very least, it would certainly be *possible* for a society to recognize several different kinds of promises, each with a different set of rules defining the exact scope of the obligation. For example, a society could have one kind of promise that imposes an absolute obligation to perform (in legal terms, one that exposes the promisor to a suit for "specific performance"), another that imposes an obligation to perform or to pay the equivalent in money ("expectation damages"), and a third that imposes an obligation to perform or to make good any losses the promisee may have suffered by relying on the promise ("reliance damages"). The community could also use promises that impose an obligation to perform unless performance became extremely difficult in some unexpected way (in legal terms, promises subject to the defense of commercial impracticability), and promises that permit no such excuse.

The possibility of more than one kind of promise greatly complicates the difficulties involved in interpreting the sociological data about a society's practices. For example, obvious questions arise concerning the number of people who must follow any set of rules for those rules to be accepted as a legally relevant practice. Must an institution be recognized in the community prior to its invocation in any particular transaction, or can any two parties create a custom-made form of promising on the spur of the moment? A related question involves the way we conceive of an individual who appears to be violating the rules of an existing practice: Is she merely an ordinary rule-breaker, or a pathbreaking pioneer in the creation of a new, perhaps more desirable form of promising? Other difficulties include the problem of conflicting expectations at different levels of generality—for example, if people expect written contracts to be binding, but they also expect goods to be sold at a fair price, what is their expectation regarding the force of a written contract that sets an unfair price? And what of the potential for circularity that arises when people's expectations are themselves affected by existing legal rules?

These problems in inferring morally relevant categories from purely empirical data are well-known, so I will not pursue them here. Instead, for the remainder of this section I will assume that sociologists can identify the set of promises—call them $promise_1$, $promise_2$, . . . $promise_n$—available to members of any particular society. However, this identification of the relevant choice set does not exhaust the possible uses of sociology. In order to reach a decision about any particular case, the courts must have some method of determining which kind of promise was actually made by the parties to any given transaction.

One can imagine societies in which this second question would be easy to answer—e.g., societies with a system of formal devices by which individuals could signal their choice of institutions. For example, the society might require all binding promises to be signed with a seal (ignoring for the moment the possibility of different kinds of binding promises), while treating any promises not made under seal as nonbinding. In such a society, the problem of interpreting the parties' utterances would deserve the lack of attention it received in the philosophical writings about promises, for it would be, quite literally, nothing but a formality.

The difficulty, of course, is that most societies do not use this method of interpreting parties' utterances, and for good reason. Even when there are only two kinds of promises from which to choose, many writers have commented on the difficulties of expecting all lay people to understand the use of the seal, and the apparent harshness of enforcing one set of rules against parties who clearly intended a different set to apply but who forgot to use the appropriate formality. These difficulties multiply rapidly if there are more than two kinds of promises—that is, more than two permissible sets of background rules—from which to choose. It would be very difficult to design a different seal for each of one hundred possible sets of promissory rules—and even more difficult to expect everybody to remember which seal they should use for each purpose. . . .

Substantive Moral Values

A more serious objection to this total reliance on sociological data is that it provides no perspective from which to criticize existing promissory practices, or to propose reforms in those practices. One might criticize particular legal rules for not properly conforming to those practices, but there would be no way to criticize the practices themselves. However, sociology is not the only possible source of content for contract law's background rules. An alternative is to look to the substantive values which justify the binding force of promises in the first place . . . to see if those values have implications for contract law's background rules.

Economic analysis is the most familiar instance of this method of determining the content of contract law's background rules. From an economic perspective, if society is justified in giving individuals the power to make morally binding promises, it is because such promises will, under certain conditions, lead to the most efficient satisfaction of human wants. This notion of efficiency (or some variant of it) can then be used to choose among various possible background rules, by identifying the rule that would contribute most to efficiency. For example, there is an extensive body of literature analyzing different contract remedies to determine which remedies are most efficient in which situations. Economists have also addressed the question of the most efficient rule for excusing promisors who fail to perform because of unexpected difficulties in performance, and the conditions under which individuals' promises should not be treated as binding because of "market failures" that distort the promisor's incentives.

John Rawls provides another example of how background rules might be chosen in order best to serve the substantive values that justify the binding force of promises in the first place. Rawls argued that the binding force of promises is justified if and only if the rules of promising—the background rules, in the terminology used here—are themselves consistent with principles of justice. For Rawls, this meant that they must lead to an equal distribution of all "primary social goods" (liberty, wealth, etc.), except to the extent that an unequal distribution would benefit every member of society. This provides a slightly different criterion for judging possible background rules—although one that will overlap with economic analysis when assessing the ways in which different rules benefit the contracting parties. . . .

It is important to realize that the selection of background rules designed to promote the substantive values that justify making promises binding is not necessarily inconsistent with freedom of contract. Under a strong version of this approach—that is, a version arguing that the selected background rules should be *mandatory*—freedom of contract would indeed be restricted. But this approach can also be used in a milder version, endorsing the preferred background rules merely as default rules or methods of interpretation for those cases where the parties have not specified a preference for some other rule. Any system of law, however committed it may be to the idea of freedom of contract, must have some way of resolving those issues on which the parties' contract is silent or ambiguous. A rebuttable presumption in favor of the rule that best serves some substantive moral value is one way to resolve such cases. . . .

Individual Autonomy

Another view of promising justifies the moral force of a promise as a necessary corollary of individual liberty or autonomy. If promises were not binding, it is argued, individual freedom would be unjustifiably restricted, as individuals would be deprived of the freedom to place themselves under a moral obligation respecting their future conduct. While there may be a slight paradox in the notion that freedom must include the freedom to limit one's freedom in the future, advocates of this theory resolve that paradox in favor of allowing individuals to make binding promises.

Autonomy-based theories may well have implications for what I have called "agreement rules," or rules concerning the conditions under which individuals will be allowed to vary the background rules that would otherwise govern their relations. For example, these theorists generally oppose the restrictions on freedom of contract represented by the rule denying enforceability to promises unsupported by consideration, or to promises that are deemed unconscionable. More precisely, they oppose restrictions on the enforceability of promises unless true consent is lacking (e.g., cases of duress), or unless the subject of the promise is not the promisor's to give away. Thus, one necessary part of these theories is a specification of the conditions under which a party's apparent consent will be recognized as valid. As long as these conditions are satisfied, autonomy-based theories hold that any rule or obligation agreed to by the parties should be allowed to govern their relationship.

In cases where the parties have not specified the rule they prefer, however, autonomy-based theories have much less to tell us. . . . Just as any default rule would be consistent with the obligation to tell the truth, any default rule would also be consistent with individual freedom, as long as the parties are allowed to change the rule by appropriate language. Consequently, some other principle must be invoked to decide which of the many possible default rules to adopt. The rule could be chosen by looking to sociological data to determine which rule most parties already expect in various circumstances; it could also be chosen by appealing to some substantive value such as economic efficiency, or Rawls' difference principle, or any other view about what makes some kinds of contractual relationships more valuable than others. Thus, even if those principles have been rejected as valid justifications for the binding force of promises, one or more of them must still be selected to provide the default rule for parties who have not unambiguously specified some other rule in their contract. . . .

[For example, Charles Fried] justifies the obligation to keep a promise primarily by viewing it as a necessary corollary of individual autonomy. . . . On the question of the appropriate remedy for breach, Fried supports the expectation measure of damages, which is designed to give the promisee the same benefits he would have received had the promise been kept. According to Fried, "[i]f I make a promise to you, I should do as I promise; and if I fail to keep my promise, it is fair that I should be made to hand over the equivalent of the promised performance."

A moment's consideration, however, will show that this conclusion cannot be derived solely from the value of individual freedom and autonomy. Fried may well be correct that, in order to give free rein to an individual's autonomy, "[i]t is necessary that I be able to make non-optional a course of conduct that would otherwise be optional for me."[See page 12 supra.] But almost any remedy—reliance damages, punitive damages, specific performance, etc.—makes the promised course of conduct non-optional to some degree, depending on the severity of the threatened penalty. There is surely nothing in the idea of individual autonomy that requires the exact degree of non-optionality provided by the expectation measure. The idea of individual autonomy does suggest that individuals should be allowed to make their conduct nonoptional to any extent they choose, by specifying one of these remedies in their contract. But the law must still select one of these remedies as the default rule, and nothing in the notion of individual autonomy gives any reason for favoring the expectation measure over any of the others.

Fried might, of course, have some other value in mind which explains why the expectation measure is to be preferred (unless the parties specify otherwise) over any of the other possible measures. For example, Fried might believe that the expectation measure promotes economic efficiency, or better satisfies Rawls' difference principle, or would better solve most coordination problems. However, no such argument is made anywhere in his book.

Alternatively, Fried might be appealing to data about people's existing beliefs to justify his preference for the expectation measure. He cannot be relying on existing nonlegal practices, for studies of those practices show that people often do not demand (or offer) expectation damages in cases of unexcused nonperformance. However, Fried might be taking existing *legal* practices as his normative benchmark, for Anglo-American law often does employ the expectation measure of damages. That is, Fried's argument might be that because the law adopts liability for expectation damages as one incident of the obligation of promising, anyone who promises thereby accepts that liability as one of the rules of the game. If this is Fried's argument, though, his theory cannot be what *justifies* the law's choice of the expectation measure. The same argument would work equally well to explain why an individual was obliged to respect any other damage rule the law happened to have adopted. . . .

In a nutshell, then, the difficulty with Fried's position on expectation damages is the difficulty identified earlier . . . . Any damage measure is consistent with the ideal of individual autonomy, as long as it is adopted solely as a default rule, since any default rule expands the promisor's options by making it easier for her to make a certain kind of promise. Fried must therefore invoke some other value in order to decide which of the many damage rules to select as a starting point. Moreover, that value will necessarily be one which Fried rejected as a possible justification for the binding force of promises, for the only value serving that role for Fried—the value of individual autonomy—is equally consistent with all default rules. To choose a default rule, then, Fried has to let one of the other values back into the analysis, and the only question is which one. . . .

Filling Gaps in Incomplete Contracts

IAN AYRES and ROBERT GERTNER

The legal rules of contracts and corporations can be divided into two distinct classes. The larger class consists of "default" rules that parties can contract around by prior agreement, while the smaller, but important, class consists of "immutable" rules that parties cannot change by contractual agreement. Default rules fill the gaps in incomplete contracts; they govern unless the parties contract around them. Immutable rules cannot be contracted around; they govern even if the parties attempt to contract around them. For example,

Ian Ayres & Robert Gertner, "Filling Gaps in Incomplete Contracts: An Economic Theory of Default Rules." Reprinted by permission of The Yale Law Journal Company and Fred B. Rothman & Company from The Yale Law Journal, vol. 94, pages 97–114. Copyright © 1989 by The Yale Law Journal Company.

under the Uniform Commercial Code (U.C.C.) the duty to act in good faith is an immutable part of any contract, while the warranty of merchantability is simply a default rule that parties can waive by agreement. . . .

This Article provides a theory of how courts and legislatures should set default rules. We suggest that efficient defaults would take a variety of forms that at times would diverge from the "what the parties would have contracted for" principle. To this end, we introduce the concept of "penalty defaults." Penalty defaults are designed to give at least one party to the contract an incentive to contract around the default rule and therefore to choose affirmatively the contract provision they prefer. In contrast to the received wisdom, penalty defaults are purposefully set at what the parties would not want—in order to encourage the parties to reveal information to each other or to third parties (especially the courts).

This Article also distinguishes between tailored and untailored defaults. A "tailored default" attempts to provide a contract's parties with precisely "what they would have contracted for." An "untailored default," true to its etymology, provides the parties to all contracts with a single, off-the-rack standard that in some sense represents what the majority of contracting parties would want. *The Restatement (Second) of Contracts'* approach to filling gaps, for example, provides tailored defaults that are "reasonable in the circumstances." "Reasonable" defaults usually entail a tailored determination of what the individual contracting parties would have wanted because courts evaluate reasonableness in relation to the "circumstances" of the individual contracting parties. In contrast, Charles Goetz and Robert Scott have proposed that courts should set untailored default rules by asking "what arrangements would *most* bargainers prefer? [see page 57 infra] . . .

An essential component of our theory of default rules is our explicit consideration of the sources of contractual incompleteness. We distinguish between two basic reasons for incompleteness. Scholars have primarily attributed incompleteness to the costs of contracting. Contracts may be incomplete because the transaction costs of explicitly contracting for a given contingency are greater than the benefits. These transaction costs may include legal fees, negotiation costs, drafting and printing costs, the costs of researching the effects and probability of a contingency, and the costs to the parties and the courts of verifying whether a contingency occurred. Rational parties will weigh these costs against the benefits of contractually addressing a particular contingency. If either the magnitude or the probability of a contingency is sufficiently low, a contract may be insensitive to that contingency even if transaction costs are quite low.

The "would have wanted" approach to gap filling is a natural outgrowth of the transaction cost explanation of contractual incompleteness. Lawmakers can minimize the costs of contracting by choosing the default that most parties would have wanted. If there are transaction costs of explicitly contracting on a contingency, the parties may prefer to leave the contract incomplete. Indeed, as transaction costs increase, so does the parties' willingness to accept a default that is not exactly what they would have contracted for. Scholars who attribute contractual incompleteness to transaction costs are naturally drawn toward choosing defaults that the majority of contracting parties "would have wanted" because these majoritarian defaults seem to minimize the costs of contracting.

We show, however, that this majoritarian "would have wanted" approach to default selection is, for several reasons, incomplete. First, the majoritarian approach fails to account for the possibly disparate costs of contracting and of failing to contract around different defaults. For example, if the majority is more likely to contract around the minority's preferred default rule (than the minority is to contract around the majority's rule), then choosing the minority's default may lead to a larger set of efficient contracts. Second, the received wisdom provides little guidance about how tailored or particularized the "would have wanted" analysis should be. Finally, the very costs of ex ante bargaining may encourage parties to inefficiently shift the process of gap filling to ex post court determination. If it is costly for the courts to determine what the parties would have wanted, it may be efficient to choose a default rule that induces the parties to contract explicitly. In other words, penalty defaults are appropriate when it is cheaper for the parties to negotiate a term ex ante than for the courts to estimate ex post what the parties would have wanted. Courts, which are publicly subsidized, should give parties incentives to negotiate ex ante by penalizing them for inefficient gaps.

This Article also proposes a second source of contractual incompleteness that is the focus of much of our analysis. We refer to this source of incompleteness as strategic. One party might strategically withhold information that would increase the total gains from contracting (the "size of the pie") in order to increase her private share of the gains from contracting (her "share of the pie"). By attempting to contract around a certain default, one party might reveal information to the other party that affects how the contractual pie is split. Thus, for example, the more informed party may prefer to have inefficient precaution rather than pay a higher price for the good. While analysts have previously explained incomplete contracting solely in terms of the costs of writing additional provisions, we argue that contractual gaps can also result from strategic behavior by relatively informed parties. By changing the default rules of the game, lawmakers can importantly reduce the opportunities for this rent-seeking, strategic behavior. In particular, the possibility of strategic incompleteness leads us to suggest that efficiency-minded lawmakers should sometimes choose penalty defaults that induce knowledgeable parties to reveal information by contracting around the default penalty. The strategic behavior of the parties in forming the contract can justify strategic contractual interpretations by courts. . . .

The Zero-Quantity Default

The diversity of default standards can even be seen in contrasting the law's treatment of the two most basic contractual terms: price and quantity. Although price and quantity are probably the two most essential issues on which to reach agreement, the U.C.C. establishes radically different defaults. If the parties leave out the price, the U.C.C. fills the gap with "a reasonable price." If the parties leave out the quantity, the U.C.C. refuses to enforce the contract. In essence, the U.C.C. mandates that the default quantity should be zero.

How can this be? The U.C.C.'s reasonable-price standard can be partly reconciled with the received wisdom that defaults should be set at what the parties would have contracted for. But why doesn't the U.C.C. treat a missing quantity term analogously by filling the gap with the reasonable quantity that the parties would have wanted? Obviously, the parties would not have gone to the expense of contracting with the intention that nothing be exchanged.

We suggest that the zero-quantity default cannot be explained by a "what the parties would have wanted" principle. Instead, a rationale for the rule can be found by comparing the cost of ex ante contracting to the cost of ex post litigation. The zero-quantity rule can be justified because it is cheaper for the parties to establish the quantity term beforehand than for the courts to determine after the fact what the parties would have wanted.

It is not systematically easier for parties to figure out the quantity than the price ex ante, but it is systematically harder for the courts to figure out the quantity than the price ex post. To estimate a reasonable price, courts can largely rely on market information of the type "How much were rutabagas selling for on July 3?" But to estimate a reasonable quantity, courts would need to undertake a more costly analysis of the individual litigants of the type "How much did the buyer and seller value the marginal rutabagas?"

The U.C.C.'s zero-quantity default is what we term a "penalty default." Because ex ante neither party would want a zero-quantity contract, such a rule penalizes the parties should they fail to affirmatively specify their desired quantity. Because the non-enforcement default potentially penalizes both parties, it encourages both of them to include a quantity term.

Toward a More General Theory of Penalty Defaults

Penalty defaults, by definition, give at least one party to the contract an incentive to contract around the default. From an efficiency perspective, penalty default rules can be justified as a way to encourage the production of information. The very process of "contracting around" can reveal information to parties inside or outside the contract. Penalty defaults may be justified as (1) giving both contracting parties incentives to reveal information to third parties, especially courts, or (2) giving a more informed contracting party incentives to reveal information to a less informed party.

The zero-quantity default, for instance, gives both contracting parties incentives to reveal their contractual intentions when it would be costly for a court to discover that information ex post. This justification—that ex ante contracting can be cheaper than ex post litigation—can also explain the common law's broader rule that "for a contract to be binding the terms of the contract must be reasonably certain and definite." Similarly, this rationale can explain corporate statutes that give incorporators an incentive to affirmatively declare the number of authorized shares, the address of the corporation for legal process and, indeed, the state of incorporation. Statutes that refuse to enforce corporate charters without these provisions create incentives similar to those created by the common law's refusal to enforce vague or indefinite contracts. In both cases, the parties can make these contractual choices more efficiently ex ante.

Lawmakers should select the rule that deters inefficient gaps at the least social cost. When the rationale is to provide information to the courts, the non-enforcement default is likely to be efficient. Non-enforcement defaults are likely to provide least-cost deterrence because they are inexpensive to enforce and give each party incentives to contract around the rule. It might seem that a penalty default set solely against one side of a contract would be sufficient to get both sides to reveal information. For example, a penalty default that makes the seller sell at one-tenth the market price would certainly encourage sellers to affirmatively fill any price gaps. But one side's penalty may be the other side's windfall. One-sided penalties can create incentives for opportunism. The non-penalized buyer in the above example would have incentives to induce sellers to enter indefinite contracts in order to extract the penalty rent. By taking each party back to her ex ante welfare, the non-enforcement default eliminates this potential for opportunism.

In contrast, when the rationale is to inform the relatively uninformed contracting party, the penalty default should be against the relatively informed party. This is especially true when the uninformed party is also uninformed about the default rule itself. If the uninformed party does not know that there is a penalty default, she will have no opportunistic incentives.

In some situations it is reasonable to expect one party to the contract to be systematically informed about the default rule and the probability of the relevant contingency arising. If one side is repeatedly in the relevant contractual setting while the other side rarely is, it is a sensible presumption that the former is better informed than the latter. Consider, for example, the treatment of real estate brokerage commissions when a buyer breaches a purchase contract. Such contracts typically include a clause which obligates the purchaser to forfeit some given amount of "earnest" money if she breaches the agreement. How should the earnest money be split between the seller and the broker if their agency contract does not address this contingency? Some courts have adopted a "what the parties would have wanted" approach and have awarded all the earnest money to the seller. We agree with this outcome, but for different reasons. The real estate broker will more likely be informed about the default rule than the seller. Indeed, the seller may not even consider the issue of how to split the earnest money in case of default. Therefore, if the efficient contract would allocate some of the earnest money to the seller, the default rule should be set against the broker to induce her to raise the issue. Otherwise, if the default rule is set to favor the broker, a seller may not raise the issue, and the broker will be happy to take advantage of the seller's ignorance. By setting the default rule in favor of the uninformed party, the courts induce the informed party to reveal information, and, consequently, the efficient contract results.

Although social welfare may be enhanced by forcing parties to reveal information to a subsidized judicial system, it is more problematic to understand why society would have an efficiency interest in inducing a relatively informed party to a transaction to reveal information to the relatively uninformed party. After all, if revealing information is efficient because it increases the value created by the contract, one might initially expect that the informed party will have a sufficient private incentive to reveal information—the incentive of splitting a bigger pie. This argument ignores the possibility, however, that revealing information might simultaneously increase the total size of the pie and decrease the share of the pie that the relatively informed party receives. If the "share-of-the-pie effect" dominates the "size-of-the-pie effect," informed parties might rationally choose to withhold relevant information.

Parties may behave strategically not only because they have superior information about the default, but also because they have superior information about other aspects of the contract. We suggest that a party who knows that a particular default rule is inefficient may choose not to negotiate to change it. The knowledgeable party may not wish to reveal her information in negotiations if the information would give a bargaining advantage to the other side.

How can it be that by increasing the total gains from contracting (the size-of-the-pie effect) the informed party can end up with a smaller share of the gains (the share-of-the-pie effect)? This Article demonstrates how relatively informed parties can sometimes benefit by strategically withholding information that, if revealed, would increase the size of the pie. A knowledgeable buyer, for example, may prefer to remain indistinguishable from what the seller wrongly perceives to be the class of similarly situated buyers. By blending in with the larger class of contractors, a buyer or a seller may receive a cross-subsidized price because the other side will bargain as if she is dealing with the average member of the class. A knowledgeable party may prefer to remain in this inefficient, but cross-subsidized, contractual pool rather than move to an efficient, but unsubsidized, pool. If contracting around the default sufficiently reduces this cross-subsidization, the share-of-the-pie effect can exceed the size-of-the-pie effect because the informed party's share of the default pie was in a sense being artificially cross-subsidized by other members of the contractual class. Under this scenario, withholding information appears as a kind of rent-seeking in which the informed party forgoes the additional value attending the revealed information to get a larger piece of the contractual pie. . . .

Notes on Selecting Default Rules

1. The readings in the subsequent chapters of this book analyze specific default rules, such as the remedies for breach (Chapter 2) or the conditions under which nonperformance will be excused (Chapter 3). Most of those readings apply one or more of the theories of default rules discussed in this section. Other recent discussions of default rules generally include Robert E.

Scott, A Relational Theory of Default Rules for Commercial Contracts, 19 Journal of Legal Studies 597 (1990); Jason Scott Johnston, Strategic Bargaining and the Economic Theory of Contract Default Rules, 100 Yale Law Journal 615 (1990); Ian Ayres and Robert Gertner, Strategic Contractual Inefficiency and the Optimal Choice of Legal Rules, 101 Yale Law Journal 729 (1992); and the symposium published in 3 Southern California Interdisciplinary Law Journal 1 (1993).

2. The excerpt by Ayres and Gertner discusses "majoritarian" default rules, or default rules that try to match what the parties would have agreed to if they had discussed the matter. These rules are often defended on utilitarian grounds: default rules chosen in this way can "minimize the costs of contracting" (page 23 supra) by saving parties from having to draft such a rule themselves.

It may also be possible to defend majoritarian default rules on nonutilitarian grounds. If the default rule matches what the parties would have agreed to, the rule can be said to reflect the parties' *hypothetical consent.* In the philosophical literature, moral arguments based on what parties would have agreed to in some more-or-less idealized setting are sometimes referred to as "contractualist" or "contractarian" moral arguments. See, for example, Thomas Scanlon, Contractualism and Utilitarianism, in Amartya Sen and Bernard Williams (eds.), Utilitarianism and Beyond (Cambridge: Cambridge University Press, 1982); David Gauthier, Morals by Agreement (Oxford: Clarendon Press, 1986).

An example of this contractarian approach can be found in the excerpt by Kim Lane Scheppele (infra at pages 166–70). Other discussions of the contractarian approach to the selection of default rules include Alan Schwartz, Proposals for Products Liability Reform: A Theoretical Synthesis, 97 Yale Law Journal 353, 357–60 (1988); Jules L. Coleman, Douglas D. Heckathorn and Steven Maser, A Bargaining Theory Approach to Default Provisions and Disclosure Rules in Contract Law, 12 Harvard Journal of Law and Public Policy 639, 640–49 (1989); David Charny, Hypothetical Bargains: The Normative Structure of Contract Interpretation, 89 Michigan Law Review 1815 (1991); Randy E. Barnett, The Sound of Silence: Default Rules and Contractual Consent, 78 Virginia Law Review 821 (1992); and Richard Craswell, Efficiency and Rational Bargaining in Contractual Settings, 15 Harvard Journal of Law and Public Policy 805 (1992).

3. Any contractarian analysis must devise some way of determining what rational parties would have agreed to. In economic analysis, this question is often answered by identifying the rule that maximizes the expected value of the transaction for the two parties. The following argument is illustrative:

> Since the object of most voluntary exchanges is to increase value or efficiency, contracting parties may be assumed to desire a set of contract terms that will maximize the value of the exchange. It is true that each party is interested only in the value of the contract to it. However, the more efficiently the exchange is structured, the larger is the potential profit of the contract for the parties to divide between them.

The use of economic efficiency as a criterion for legal decision-making is of course controversial. In the area of contract, however, the criterion is well-nigh inevitable once it is conceded that the parties to a contract have the right to vary the terms at will. If the rules of contract law are inefficient, the parties will (save as transaction costs may sometimes outweigh the gains from a more efficient rule) contract around them. A law of contract not based on efficiency considerations will therefore be largely futile.

Richard A. Posner and Andrew M. Rosenfield, Impossibility and Related Doctrines in Contract Law: An Economic Analysis, 6 Journal of Legal Studies 88, 89 (1977).

4. If the rule to which parties would have agreed is identified by this form of efficiency analysis, does a contractarian argument based on hypothetical consent add anything to a utilitarian argument based solely on maximizing the value of the exchange? Consider the following argument:

[T]he consent argument for the legal rule must be that ex ante the parties seeking to maximize overall wealth or utility *would have consented* to it, not that they actually *did* consent. . . . From the fact that a social state makes someone no worse off (that is, it is not irrational for him), we are to infer that the agent would have consented to it. Consent follows as a matter of *logic* from considerations of rationality.

Consequently, the concept of hypothetical consent expresses nothing that is not already captured in the idea of rational self-interest. The distinction between consent and rationality central to moral theory apparently evaporates. The claim that imposing obligations ex post is justified because the parties would have consented to them ex ante adds nothing to a defense of such a proposal that is not already expressed by the argument that imposing obligations ex post is justified whenever such obligations would have been *rational* for the parties ex ante. Thus, the reliance on ex ante rational bargaining provides a rationality or welfarist defense of the default rule, not a consensualist one.

Jules L. Coleman, Douglas D. Heckathorn and Steven M. Maser, A Bargaining Theory Approach to Default Provisions and Disclosure Rules in Contract Law, 12 Harvard Journal of Law and Public Policy 639, 645–46 (1989). See also David Charny, Hypothetical Bargains: The Normative Structure of Contract Interpretation, 89 Michigan Law Review 1815 (1991).

5. The Ayres and Gertner excerpt discusses the concept of "penalty" default rules, which are designed to force the parties to spell out their own wishes rather than requiring the courts to try to anticipate the parties' wishes for them. A similar concept was introduced (and labeled a legal "formality") in Lon L. Fuller, Consideration and Form, 41 Columbia Law Review 799 (1941). See also the excerpt by Duncan Kennedy (infra at pages 245–49). Fuller emphasized the role of penalty default rules, or formalities, in encouraging parties to signal their wishes more clearly *to the legal system,* thereby sparing courts the trouble of trying to figure out what the parties would have wanted. As Ayres and Gertner point out, penalty default rules can also encourage one party to signal its intentions more clearly *to the other*

party, especially where the other party might not know the applicable legal rule.

6. Default rules could also be selected by adopting the contract term believed to be fairest or most just, according to some theory of substantive fairness or justice, regardless of what most contracting parties wanted. For example, if one believes that expectation damages are inherently more consistent with individual autonomy than any other damage measure (see the excerpts by Charles Fried, supra at pages 9–14, and Richard Craswell, supra at pages 16–22), that might argue for making expectation damages the default remedy regardless of whether most parties wanted expectation damages. By contrast, the contractarian theory discussed supra in note 4 would support expectation damages as the default remedy *only* if that was the remedy most parties were thought to want.

Why should the law pick a default rule other than the one most parties would want if those who want something less fair or just can always contract around the default rule? It might be argued that those who want something less fair or just should have to bear the burden of contracting around the default rule. Another possible argument is that the law could have a "transformative" effect. That is, the law's endorsement of expectation damages (for example) might lead parties to view expectation damages in a more favorable light, with the ultimate result that most parties *would* prefer the expectation measure, even if that was not the measure most parties preferred before the law endorsed it. The strengths and weaknesses of these arguments are considered in David Charny, Hypothetical Bargains: The Normative Structure of Contract Interpretation, 89 Michigan Law Review 1815 (1991); and Alan Schwartz, The Default Rule Paradigm and the Limits of Contract Law, 3 Southern California Interdisciplinary Law Journal 389 (1993).

§ 1.3 Distributional Concerns

Passing on the Costs of Legal Rules

RICHARD CRASWELL

Many legal rules in many legal fields define the relations between sellers of goods or services and their customers. For example, contract law defines the respects in which a seller warrants the quality of its products, as well as the damages due if it breaks the contract. Products liability law determines a seller's responsibility for physical injury caused by its products. Landlord-tenant law defines the landlords' liability for unsafe or inadequate housing; it also limits the landlords' powers vis-à-vis tenants who damage the premises or are delinquent in paying the rent. Debtor-creditor law limits the analogous powers exercised by creditors outside the rental housing market.

The debate surrounding these rules often pits those who are concerned about a rule's efficiency against those who are concerned about its distributional effects. The law-and-economics literature, for example, usually aims only at economic efficiency: maximizing the joint welfare of buyers and sellers without distinguishing between benefits to sellers and benefits to buyers. . . .

To some, however, the identity of those who gain or lose is at least as important as the size of the total gains or losses. Especially where consumer transactions are concerned, people often attach more weight to the welfare of consumers than to the welfare of sellers. For expositional purposes, I will define the "pro-consumer" position (in somewhat extreme fashion) as the belief that a rule should be adopted if, and only if, it benefits consumers more than it costs them.

This pro-consumer position could of course be criticized directly, by asking why consumers ought to be favored over sellers. For example, some sellers are publicly-held corporations whose shareholders include insurance companies and pension funds, and there is no obvious reason why the beneficiaries of these funds deserve less concern than consumers. Some of the sellers' costs may also be borne by their employees if the increased costs lead to reduced sales and, hence, to reduced employment in the industry.

My concern here, however, is a very different criticism which is based on the argument that the pro-consumer analysis will usually lead to the same result as an efficiency analysis once the effect of a rule on the seller's price is taken into account. According to this argument, a rule that benefits consumers by $10 while increasing sellers' costs by $50 will not benefit consumers for very long, because sellers will eventually pas on their higher costs to consumers in the form of higher prices. Thus, consumers will end up as net winners (according to this argument) only if their direct benefits exceed the direct costs of the rule to sellers, for only in such a case will they still be better off even after they have paid the higher prices. . . .

The Standard Analysis of Cost Pass-Ons

The principal objection to this passing-on argument is that it assumes that all of the sellers' costs or benefits will be passed on in the form of higher or lower prices. If all $50 of the sellers' costs are passed on as higher prices, then buyers will indeed by worse off unless they receive direct benefits of $50 or more from the rule, so the distributional and efficiency goals will indeed coincide. But if only $40 of sellers' costs can be passed on, buyers might end up better off even if they only received $45 worth of direct benefits. A pro-consumer advocate would approve of such a rule, in spite of the fact that it fails a Kaldor-Hicks efficiency test because the $50 cost to sellers exceeds the $45 gain to consumers.

Thus, the asserted identity between the pro-consumer and overall efficiency positions seems to depend critically on the assumption that sellers can pass on all of their costs and benefits. However, the conventional economic wisdom is that sellers usually will *not* be able to pass on the entire amount of an exogenous cost increase. . . . Figure 1.1 illustrates the conventional economic analysis of the effect of an increase in the seller's costs of the sort that might be brought about by an increase in raw material prices or the imposition of a tax on sellers. Before the cost increase, the market price should have stabilized at the level at which the amount suppliers were willing to sell (shown by the line marked S1) just equalled the amount buyers demanded (shown by the line marked D). These will be equal only at the point where the two lines cross, which corresponds to a market-clearing price of P1.

If sellers' costs then increase by some amount c, sellers will no longer be willing to sell as much at any given price. This should cause the supply curve to shift upward by the amount of the cost increase, to become the new line marked S2. This, in turn, will cause the market price to rise until it reaches the new price P2, where supply and demand again become equal. Because demand is somewhat elastic, however, the increase in price from P1 to P2 is less than the increase in seller's costs from S1 to S2 (c). Thus, less than 100 percent of the cost increase has been passed on. . . .

Given this analysis, it seems easy to imagine a rule that produces total benefits equal to, say, only 80 percent of its total costs, but that nonetheless leaves buyers better off because buyers receive all of the rule's benefits while less than 80 percent of the costs are passed on. The problem, however, is that the graphs given [above] are incomplete. They show the seller's costs rising without any simultaneous increase in consumers' willingness to pay for the product. This is a perfectly reasonable assumption to make when analyzing an increase in raw material prices or the imposition of a tax on the seller, for such changes would increase the seller's costs without making the product any more attractive to consumers. An implied warranty, however, would normally make the product more attractive to consumers, thereby causing consumer demand (as well as sellers' costs) to increase. To determine the effects of rules such as implied warranties, we need a more complex analysis. . . .

A Simple Model with Homogeneous Consumers

In this section of the article, I make . . . [some] important assumptions to prevent the model from becoming too complex. First, I assume that each consumer would pay exactly the same amount for the addition of a warranty. In other words, consumers may differ in their tastes regarding the product itself, but they have identical preferences concerning the presence or absence

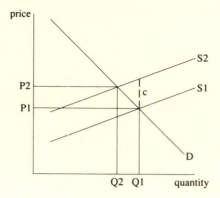

Figure 1.1. Effects of a cost increase

of a warranty. While this assumption is something of an oversimplification, it aids the initial exposition considerably. . . .

Second, and more controversially, I further assume that the increase in the maximum consumers are willing to pay is an appropriate measure of the benefits consumers receive from the warranty. That is, I adopt the consumer sovereignty position that consumer welfare is to be judged solely by reference to consumers' own tastes and preferences. I also assume that those tastes and preferences can be meaningfully translated into a dollar amount and that the appropriate amount is whatever each consumer is willing to pay to satisfy those preferences. To make this assumption more plausible, I also assume consumers have perfect information about the presence or absence of the warranty and about its value to them. . . .

As noted earlier, a legal rule such as an implied warranty should raise both the demand and the supply curves, although not necessarily by the same amount. The demand curve will shift upward by an amount equal to the consumers' increased willingness to pay for the product/warranty combination, reflecting the benefits consumers derive from the warranty. The supply curve will rise by the expected costs of the warranty to sellers. Under the above assumptions about consumers' valuation of the warranty, the warranty's efficiency can be assessed simply by comparing the upward shift of the demand and supply curves. If the value of the warranty to consumers exceeds its cost to sellers, the demand curve will rise by more than the supply curve; if not, then demand will rise by less than supply.

Under these assumptions, consumers cannot gain if the warranty is inefficient. That is, if the value of the warranty to consumers is less than its cost to sellers, consumers will end up worse off as a result of the warranty, even if sellers cannot pass on all of their costs. Figure 1.2 illustrates this by depicting a shift in the demand curve that is less than the shift in the supply curve. The increase from P1 to P2 is clearly less than the total cost increase c, showing that only a portion of the cost increase is passed on to consumers. The price increase is small because significant numbers of consumers are no longer

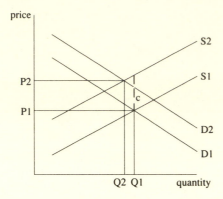

Figure 1.2. An inefficient warranty

willing to purchase the product/warranty package at the higher price, so the total quantity produced has declined from Q1 to Q2. The marginal consumers, who before the addition of the warranty were just willing to purchase the product alone, are now unwilling to pay a higher price for the product plus the warranty. These consumers must have valued the warranty at an amount less than the resulting price increase, so they have been made worse off by the introduction of the warranty.

Less obviously, consumers who continue to buy the product are also made worse off by the introduction of the warranty. I assumed earlier that all consumers placed the same value on the warranty; thus, if the marginal consumers value the warranty less than the price increase, so too must all the other consumers. The other consumers are still willing to purchase the product/warranty package because the value they place on the product alone is sufficiently greater than its price that the purchase is still worthwhile, even after the addition of a relatively unattractive warranty and price increase. But since they value the addition of the warranty by less than they value the price increase, even these consumers must end up with less satisfaction than if they were buying the product at the old price without the warranty. This can be verified by visual inspection of Figure 1.2: The amount by which consumers' valuation of the package has increased (shown by the increase from D1 to D2) is less than the amount by which the price has increased (shown by the increase from P1 to P2).

By contrast, Figure 1.3 illustrates a case in which consumers value the warranty by an amount that exceeds the cost c of providing the warranty. In this case, the total quantity sold has increased (from Q1 to Q2), as consumers who were unwilling to pay for the product alone are now willing to pay for the product plus the warranty. Sellers have had to increase their production of the product to meet this new demand, thereby raising their production costs (as indicated by the move upwards and to the right along the S2 line) over and above the extra costs added by the warranty itself. The result is that the price increase from P1 to P2 actually exceeds the cost of the warranty c; more than

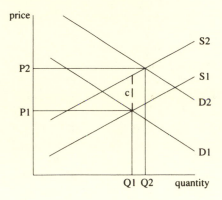

Figure 1.3. An efficient warranty

100 percent of the cost increase has been passed on. However, because the increase in consumers' valuation of the product (shown by the increase from D1 to D2) is greater than the price increase from P1 to P2, consumers still benefit from the introduction of this warranty. Indeed, the price increase exceeds the cost of the warranty in this case precisely because consumers find the product with the warranty so much more attractive, thereby requiring an increase in total production. . . .

To summarize, this section has shown that when consumers have identical preferences regarding a warranty or other legal rule, consumers will benefit from the rule if, and only if, the rule is efficient under a Kaldor-Hicks test. Moreover, the significance of sellers' ability to pass along their costs in such a market is exactly the opposite of what most people suppose. Under the model used in this section, if less than 100 percent of the costs are passed on, the rule has made the product/warranty package less attractive to consumers, which implies that the rule is not very good for consumers. Conversely, if more than 100 percent of the costs are passed on, the rule has made the product/warranty package more attractive to consumers, thereby increasing overall demand. Paradoxical as it may seem, the rules whose costs are most heavily passed on are also the rules that will benefit consumers the most.

Heterogeneous Consumers

. . . In this section . . . I assume that sellers are dealing with consumers who differ in their willingness to pay for a warranty. I also assume that sellers must charge all customers the same price for the product/warranty package—that is, I assume that it is impossible for the seller to engage in price discrimination. Under these assumptions, it is possible for some consumers to be made better off by the warranty while other consumers are made worse off. Moreover, it is also possible for consumers as a class to benefit from an inefficient warranty, or to lose from an efficient one—although the question of what it means for consumers "as a class" to benefit becomes more problematic when consumers differ in their willingness to pay. . . .

In markets where different consumers attach different values to a warranty, the size of the accompanying price increase will be determined largely by the valuations held by those consumers who are on the margin between buying and not buying the product/ warranty package. These consumers, the marginal consumers, will stop buying the package if its price rises; consequently, the willingness to pay of marginal consumers will determine how high the price of the combined package will rise.

If these consumers (on average) value the addition of a warranty by more than the warranty increases the sellers' costs, then overall demand for the combined package will increase, and the price will go up by more than 100 percent of the warranty cost. However, marginal consumers will still end up better off for the reasons discussed in [the preceding section]. While the price will increase by more than the sellers' costs, it will increase by less than the value the marginal consumers place on the warranty, so marginal consumers will still come out ahead.

Alternatively, if the marginal consumers value the warranty by less than the increase in sellers' costs, they will end up worse off as a result of the introduction of the warranty. The price increase will be less than the cost of the warranty to sellers because the reduction in demand will prevent sellers from passing along all of the costs of the warranty. However, the marginal consumers will still end up worse off since the reduction in demand means that some marginal consumers have dropped out of the market. In short, the marginal consumers will gain or lose depending on whether the value they place on the warranty is greater or less than the seller's costs—just as in the simple model presented in [the preceding section].

Determining whether the non-marginal (or "infra-marginal") consumers gain or lose from the introduction of a warranty is more difficult. The easiest case to analyze is one in which the marginal consumers value the warranty by an amount equal to the cost of the warranty to sellers. In that case, the price of the product/warranty package will rise by exactly the marginal consumers' valuation of the warranty. Consequently, any consumers of the product who value the warranty more highly than the marginal consumers will derive benefits which exceed the price increase and will be net winners from the introduction of the warranty. On the other hand, consumers who value the warranty less than the marginal consumers will derive benefits which are less than the price increase and will be net losers.

Some numerical examples may help illustrate this point. Suppose that the product without the warranty sold for $20, the warranty costs the seller $8 to administer, and the combined product/warranty package now sells for $28. Of the consumers who continue to buy the product after introduction of the warranty, those who value the warranty by more than $8 will have gained from its introduction. This clearly includes any consumer who valued the product at $18 and who also valued the warranty at $10 or more, for these consumers would not have been willing to buy the product at its old price of $20 without a warranty. It also includes other consumers who were previously willing to buy the product without the warranty and are now even happier to buy it with the warranty. For example, some consumers might have been willing to pay $24 for the product alone but are willing to pay $34 for the product/warranty package. These consumers derived a net surplus of $4 from the product without the warranty but a net surplus of $6 after the addition of the warranty. Thus, these consumers end up $2 better off as a result of the warranty. In effect, they are getting a warranty they value at $10 in exchange for a price increase of only $8.

At the same time, consumers of the product who value the warranty by less than $8 will be made worse off by the warranty's introduction. Some of these consumers will no longer buy the product—for example, those who valued the product alone at $22 but who are only willing to pay an extra $4 (or a total of $26) for the addition of the warranty. These consumers will not buy the product/warranty package at a combined price of $28 and, therefore, will lose the $2 surplus they received from buying the product alone. Other consumers may continue to buy the product, but they will nonetheless suffer a reduction in their net surplus. For example, consider a consumer who was willing to pay $27 for the product alone but will pay only $7 extra (or a total of $34) for the product/warranty combination. This consumer derived a net surplus of $7 ($27 minus $20) without the warranty but now receives a surplus of only $6 ($34 minus $28) after the introduction of the warranty. In general, consumers of the product who value the warranty by less than $8 will end up worse off as a result of the warranty, while those who value the warranty by more than $8 will end up better off. . . .

When consumers have identical preferences concerning the warranty, a warranty that benefits any one consumer must benefit all other consumers as well. When consumers differ in their willingness to pay for a warranty, however, this section has shown that the warranty can benefit some consumers while hurting others. In such a case, an unambiguous pro-consumer position is harder to define.

. . . Admittedly, in some cases, those who gain and those who lose from a warranty may be distributed randomly with respect to any social policy regarding redistribution. For example, consumers who are risk-neutral or risk-preferring place a relatively low value on protection against risks and are, therefore, likely to be among those who lose from the introduction of a warranty. If risk-aversion is distributed randomly throughout the population, a warranty that benefits the risk-averse while hurting risk-preferring or risk-neutral consumers would neither advance nor retard any current social policies concerning the proper direction of redistribution.

In other cases, however, the identity of the winners and losers may be correlated with wealth in a way that makes the resulting redistribution regressive. For example, critics of modern products liability law, ranging from Richard Abel on the left to George Priest on the right, have pointed out that wealthy people generally have more to lose from product-related accidents that result in lost earnings or in consequential damage to property. As a result, rich people will usually benefit much more than poor people from any form of seller liability that protects them from such losses. Indeed, even if the property at risk is the same for both rich and poor consumers, the rich may be willing to pay more for protection against that risk simply because they have more money with which to pay. In economic terms, the marginal value of money to the rich may be lower, even if the property at risk is the same. This difference does not guarantee that the rich will be willing to pay more for protection since rich consumers may also be less risk-averse than poor consumers. The higher risk-aversion of the poor could make them willing to pay more for a warranty, not less. However, if the difference runs the other way—as is almost certain to be true of extremely poor consumers, who could barely afford to pay *any* more money—the inter-consumer distributional effects might favor the rich at the expense of the poor. . . .

Notes on Distributional Concerns

1. The first and most influential of the legal articles analyzing the distributional effects of contract rules was Bruce Ackerman, Regulating Slum Housing Markets on Behalf of the Poor: Of Housing Codes, Housing Subsidies and Income Redistribution Policy, 80 Yale Law Journal 1093 (1971). This article prompted a flurry of responses, many of which dealt exclusively with the housing industry. Articles discussing distributional issues in contract law more generally include Richard S. Markovits, The Distributive Impact, Allocative Efficiency, and Overall Desirability of Ideal Housing Codes: Some Theoretical Clarifications, 89 Harvard Law Review 1815 (1976); and Duncan Kennedy, Distributive and Paternalist Motives in Contract and Tort Law, With Special Reference to Compulsory Terms and Unequal Bargaining Power, 41 Maryland Law Review 563 (1982). Portions of the Kennedy article are excerpted in Chapter 5 (infra at pages 308–12 and 315–19) in connection with the doctrine of unconscionability.

2. The Ackerman article demonstrated that, under certain extreme configurations of supply and demand, it is possible for *all* buyers to benefit from a warranty (or other contract term) even if the warranty is inefficient. This will happen if both (1) the marginal customers value the warranty or other term less highly than any other customers; and (2) the supply of the product is absolutely fixed, and cannot be reduced by sellers. Condition (2) guarantees that sellers will not raise their price more than the marginal buyers are willing to pay; otherwise, the marginal buyers would stop buying, and sellers would end up with unsold units. Condition (1) guarantees that if the price goes up by the amount that marginal customers value the warranty, the price increase will be less than the value placed on the warranty by all other customers, so the marginal customers will break even while all other customers will end up better off. Unfortunately, few if any real markets satisfy both conditions (1) and (2).

3. Economic analyses often measure the value of some contractual right by the maximum amount consumers would be willing to pay to receive that right. For example, the Craswell excerpt treats consumers as having been benefited by a warranty if and only if the higher price they have to pay is less than the maximum price they would have been willing to pay for that warranty. This method of measuring the benefits of a contractual right is, of course, controversial. Some alternative methods of measuring the value of contractual rights are discussed in a portion of the article not reprinted here, 43 Stanford Law Review at pages 385–98.

4. Both the Craswell excerpt and the earlier Ackerman article assumed that all sellers would face the same cost. If the cost in question is one that sellers can control, however—for example, if sellers can build better products that don't break down as often, thus reducing the number of claims on the warranty—then some sellers' costs may be lower than others. In a competitive market, only the minimum amount of cost (the cost of the marginal seller) will be passed on to customers. In cases where a careful seller can avoid the prohibited activity altogether, without incurring any additional administrative or monitoring costs, the most careful sellers will not see their costs rise at all (and competition will force other sellers to follow), so there will be no costs to be passed on.

More generally, one of the major sources of efficiency gains from legal rules is the effect those rules might have in giving sellers incentives to minimize certain costs. Some form of this argument can be found in most of the economically oriented readings in Chapters 2 and 3. In effect, the Craswell excerpt addresses the remaining distributional effect of legal rules *once these efficiency gains have been achieved,* and sellers have responded by reducing their costs to the lowest feasible level.

2

Remedies For Breach

Anglo-American law ordinarily awards expectation damages as the remedy for breach of contract. These damages are meant to make a promisee as well off as he or she would have been had the promise been performed. The readings in the first section of this chapter discuss some possible justifications for the expectation measure, including those based on the economic theory of "efficient breach." These excerpts also discuss possible criticisms of this economic theory.

The second and third sections then discuss the various alternatives to the expectation measure of damages. The second section discusses those alternatives that *limit* the expectation measure, thus deliberately leaving the promisee less well-off than if the promise had been performed. These include rules that prevent the promisee from recovering for damages he or she could have avoided by proper mitigation, legal limits on the recovery of consequential damages (including those that were "unforeseeable" at the time the contract was formed), and limits on the recovery of subjective or nonpecuniary losses.

The third section then turns to those alternative remedies that are *more favorable* to the promisee than the expectation measure. Some of these alternatives are deliberately designed to be more favorable to the promisee—for example, liquidated damage clauses set at a high amount, or punitive damages. These remedies generally exceed the expectation remedy as that remedy is actually applied under the limiting doctrines discussed in the preceding paragraph. These remedies could even exceed the theoretical goal of the expectation remedy, by leaving the promisee better off than if the contract had been performed.

Much the same can be said of the other alternative remedies discussed in the third section of this chapter: specific performance, rescission, and restitution. While these remedies will not always be more favorable than expectation damages, promisees are unlikely to request these remedies unless they are more favorable than the expectation remedy as it would actually be applied (with all its limits). In some cases, specific performance or rescission and restitution could even exceed the theoretical goal of expectation damages, by allowing the promisee to end up better off than if the contract had been performed. For this reason, questions about when promisees ought to be made better off than the expectation remedy would make them—and when it would be just or efficient to do so—are common to all of the readings in the third section of this chapter.

§ 2.1 **Expectation Damages**

Law and Economics

ROBERT COOTER AND THOMAS ULEN

Breach of contract can be regarded as something that happens as a consequence of the bargaining or transaction costs of forming contracts. If bargaining and drafting were costless, we could imagine that a contract would be created that explicitly provided for every contingency that could possibly arise. Included in the contract would be remedies for every type of non-performance under every possible circumstance. Since every type of non-performance in every possible circumstance would have a remedy explicitly attached to it in the contract, there would be no occasion for the law to prescribe a remedy.

Because bargaining and drafting are costly, an efficient contract will not explicitly cover every contingency. In fact, the majority of contracts do not specify *any* remedies for breach. The existence of such gaps in contracts creates a need for the law to supply remedies.

One of the most enlightening insights of law and economics is the recognition that there are circumstances where breach of contract is more efficient than performance. We define *efficient breach* as follows: *a breach of contract is more efficient than performance of the contract when the costs of performance exceed the benefits to all the parties.* We need to characterize the circumstances under which this will be true. The costs of performance exceed the benefits when a contingency arises such that resources necessary for performance are more valuable in an alternative use. These contingencies come in two types. First, a *fortunate contingency* or *windfall* might arise that makes non-performance even more profitable than performance. Second, an *unfortunate contingency* or *accident* might arise that imposes a larger loss for performance than for non-performance.

To illustrate a windfall, suppose that A promises to sell a house to B for $100,000. Let us assume that A values living in the house at $90,000, and B values living in the house at $110,000. Thus, at A's asking price, A realizes a seller's surplus of $10,000, B realizes a consumer's surplus of $10,000, and the total surplus from the exchange is $20,000. But suppose that before the sale is completed another buyer, C, appears on the scene and offers A $120,000 for the same house that he has contracted to sell to B for $100,000. C's appearance is a windfall that has increased the total available surplus from $20,000 to at least $30,000. . . .

Let us refine this notion of efficient breach by briefly discussing the question, "Which court-designed remedy—the payment of money damages or specific performance—will induce only efficient breach?" Analyzing this problem in terms of bargaining theory is useful. Recall that transferring the house from A to B creates a surplus of $20,000. Furthermore, transferring the house from B to C creates an additional surplus of at least $10,000. (Alternatively, the house could be transferred directly from A to C; in that event, the surplus would still be at least $30,000.) If bargaining is costless, the house will eventually end up being owned by C, regardless of whether the court-designated remedy is damages or specific performance. However, the distribution of the surplus from exchange is different under the two court-designed remedies.

To see why, assume that A has a binding contract to sell the house to B and the remedy for breach is damages. Further assume that the damages have been designed by the court so as to put B in the position he expected to be in if A had delivered the house, and B kept it. Because B anticipated a surplus of $10,000 from the performance of the contract, he is entitled to damages in that amount. Under this formulation of the damage remedy, A can breach the contract, pay $10,000 in damages to B, and sell the house to C for $120,000. As a result, A will enjoy $20,000 of the surplus and B will enjoy $10,000.

But suppose that the court-designed remedy is specific performance. Under that remedy, the court requires A to sell the house to B as promised. Thus, A will have to deliver the house to B for $100,000, who will then resell the house to C for $120,000. As a result, B will enjoy $20,000 in surplus and A will enjoy $10,000 in surplus, precisely the opposite shares of the surplus as occurred under court-designed damages. The important point of this example is that when bargaining costs are zero, efficient incentives for breach are created under either court-designed remedy, but the surplus from exchange is distributed differently.

An implication of this analysis is that there is a distinction between the efficiency of the court-designed remedies only when bargaining or transaction costs are not zero [as in the Coase Theorem] . . . .

The Efficient Breach Fallacy

DANIEL FRIEDMANN

> The only universal consequence of a legally binding promise is that the law makes the promisor pay damages if the promised event does not come to pass. In every case it leaves him free from interference until the time for fulfillment has gone by, and therefore free to break his contract if he chooses.

So wrote Oliver Wendell Holmes in his seminal discussion of contract remedies in *The Common Law*. That position, while widely discussed, is not acceptable as a normative (nor, as will be shown, as a positive) account of the question of contract remedies. Stated in a phrase, the weakness of Holmes's approach lies in its conclusion that the remedy provides a perfect substitute for the right, when in truth the purpose of the remedy is to vindicate that right, not to replace it. Holmes's analysis mistakenly converts the remedy into a kind of indulgence that the wrongdoer is unilaterally always entitled to purchase. As with any unifying ideal, Holmes's proposition is difficult to confine to the contract cases to which it was originally applied. Why not generalize the proposition so that every person has an "option" to transgress another's rights and to violate the law, so long as he is willing to suffer the consequences? The legal system could thus be viewed only as establishing a set of prices, some high and some low, which then act as the only constraints to induce lawful conduct.

The modern theory of "efficient breach" is a variation and systematic extension of Holmes's outlook on contractual remedy. It assumes that role because of the dominance that it gives to the expectation measure of damages in cases of contract breach: the promisor is allowed to breach at will so long as he leaves the promisee as well off after breach as he would have been had the promise been performed, while any additional gain is retained by the contract breaker. . . .

The essence of the theory is "efficiency." The "right" to break a contract is not predicated on the nature of the contractual right, its relative "weakness,"

Daniel Friedmann, The Efficient Breach Fallacy, 18 Journal of Legal Studies 1, 1–2, 4–6 (1989).
Copyright © 1989 by The University of Chicago. Reprinted by permission.

or its status as merely in personam, as opposed to the hardier rights in rem. Rather it is on the ground that the breach is supposed to lead to a better use of resources. The theory, therefore, is, in principle, equally applicable to property rights, where it leads to the adoption of a theory of "efficient theft" or "efficient conversion." To see the point, observe how this account of efficiency plays out in two cases. In the first, A promises to sell a machine to B for $10,000 but then turns around and sells it instead to C for $18,000. In the second, B owns a machine for which he has paid $10,000, which A takes and sells to C for $18,000.

To keep matters simple, assume that B values the machine at exactly $12,000 in both cases. If the willful contract breach is justified in the first case, then the willful conversion is justified in the second. In the first, B gets $2,000 in expectation damages and is released from paying the $10,000 purchase price. In the second, B obtains damages for conversion equal to $12,000 because he has already paid the $10,000 purchase price to his seller. The two cases thus look identical even though they derive from distinct substantive fields.

No doubt in the contract situation, A may negotiate with B a release from his contractual obligation. But this, in Posner's view, would lead to additional transaction costs. It is, therefore, preferable to permit the "efficient breach." But the property example is indistinguishable on this ground, for in the second, A, when he takes the machine from B, avoids the transaction cost of having to purchase it from him. The similarity between the two situations (breach of contract and conversion) becomes more striking if the converter did not wrongfully deprive the owner of his possession. Thus, suppose that A is a bailee who keeps B's goods. C offers A for the goods an amount that exceeds their value to B. A can negotiate with B for the purchase of the goods and, if he is successful, sell them to C at a profit. The cost of this transaction could be saved, just as in the contract example, if A were allowed to sell the goods to C, while limiting his liability to B's expectation-like damages. Nevertheless, the better rule, which has been universally adopted by Anglo-American law, is that the plaintiff is entitled to recover in restitution the proceeds of the sale from the defendant who converted the plaintiff's property and sold it to a third party. Efficient breach theory does not provide an explanation why the promisee in a contract of sale should not be accorded similar rights.

There are, of course, refinements. Where the promisor is a merchant engaged in selling these types of goods, he may be in a better position to find a buyer willing to pay a higher price for them, so that his transaction costs may be somewhat lower. This, however, is not necessarily the case, and in any event it does not justify the breach. Again, the situation can be compared to conversion. The fact that A, for example, is a car dealer who is likely to know that C is an excellent buyer for B's car does not justify him to take B's car from his driveway in order to sell it to C. Nor if B's car has been left with A for repairs can A sell it to C.

The real issue in both the conversion and the breach situation is who should benefit from C's willingness to pay a high price for the goods owned by B (the conversion example) or promised to him (the breach example). In principle, there should be in both situations only one transaction; in my view, it should be between C and B (the owner or the promisee). If A promised to sell a piece of property to B for $10,000 and C is willing to have it for $18,000, he should negotiate its purchase from B. A is simply not entitled to sell to C something he has promised to return or transfer to B, and A is therefore not the right party to negotiate with. Consequently, if C negotiates such a purchase, he may be exposed to liability toward B, the promisee. Similarly, with a bailment, C must negotiate with B (the owner) and not with the bailee. Hence, the question of additional transaction costs does not arise.

It is, of course, conceivable that a person (in the above example, A) would like to take advantage of a potential transaction between two other parties (B and C). In some instances this can legitimately be done. A may know that C is the best buyer for B's property (or for the property promised to B), while B and C are unaware of each other. A may buy the property from B and sell it to C, or he may reach an agreement (with B or C or with both) for the payment of a commission. If this is done, the inevitable result would be that the transfer from B to C would involve two or more transactions (and, arguably, additional transaction costs). This course of dealing is not objectionable. What is, however, objectionable is an attempt by A to obtain through the commission of a wrong (breach of contract or a tort such as conversion) the benefit of a transaction that should have been concluded between B and C. . . .

Contract Remedies, Renegotiation, and the Theory of Efficient Breach

RICHARD CRASWELL

Redistribution and Risk-Aversion

Even if noncompensatory contract remedies do not distort a seller's decision to perform or breach—say because costless ex post renegotiation is available to prevent inefficient breaches and enable efficient ones—the remedies can still produce distributional effects by affecting the parties' bargaining status

Richard Craswell, Efficiency, Renegotiation, and the Theory of Efficient Breach, 61 Southern California Law Review 629, 633–36, 640–47, 650–53, 665–67 (1988). Copyright © 1988 by the University of Southern California. Reprinted with the permission of the Southern California Law Review.

in post-breach negotiations. Consider the situation modeled in the standard literature, where a seller is deciding whether or not to break a contract and sell to another buyer at a higher price. If the remedy is exactly compensatory, a seller who decides to break the contract will only have to pay the first buyer a compensatory amount. If the remedy is overcompensatory, however, a seller who decides to break the contract may have to pay the first buyer a larger amount to be released from the contract. If the remedy is under-compensatory, a breaching seller will be able to get away with paying the first buyer less.

For example, suppose that a good is only worth $100 to the first buyer (who has contracted with the seller), but the seller now has a chance to earn $300 by selling the good to a second buyer who values the good more highly than does the first buyer. If the law lets the first buyer insist on a remedy that would cost the seller $500, this will not prevent the seller from selling the good to the second buyer (if renegotiation costs are low), for the seller could offer to pay the first buyer to be released from the contract. However, the amount the seller will have to pay under such a damage rule will be somewhere between $100 and $300. The buyer will not accept anything less than $100 to give up the right to performance, as that is what performance is worth to the buyer; the seller will not agree to pay anything more than $300, as that is the maximum the seller could get by breaching. The exact amount the parties agree on will be somewhere between these figures, depending on their respective bargaining abilities, but the amount will clearly be at least as large as the amount the seller would have to pay under a regime of compensatory damages ($100). Thus, even if ex post renegotiation is costless so that only efficient breaches take place, the legal remedy will still affect the distribution of wealth that the ex post negotiations produce.

From the standpoint of efficiency, this redistribution is often viewed as neither desirable nor undesirable, which may be why most economic analyses have not concerned themselves with it. Instead, the redistributional effects have been of more concern to noneconomists, who have sometimes seen them as a reason for favoring overcompensatory or punitive remedies. Since these remedies put the breacher in the weakest bargaining position, they redistribute wealth away from the "guilty" breacher and toward the "innocent" non-breacher, a result which is often seen as normatively desirable.

However, the actual redistributive effects of contract remedies are not this simple, for the legal remedy will also affect the price that the first buyer will have to pay. If the legal rule allows the seller to keep most of the gains in those cases where a better offer is later received, the seller will be able to quote a lower price than that which could be quoted under a rule allowing buyers to capture most of those gains. In the former case, the seller's operations can be subsidized with the profits from the occasional profitable breach; in the latter, the amount of that subsidy will be reduced and the price will have to be raised to compensate. Thus, the availability of punitive remedies does not really "give" buyers the right to more of the profits. Instead, it "sells" that right to them.

This effect is easiest to see if all parties are risk-neutral, in which case the amount of the price increase should equal the actuarial value of the chance at a share of the profits from breach. It would be like participating in a lottery: every buyer will pay extra to create a pool that is just large enough to compensate sellers for the loss of those profits that will eventually have to be paid back to the "winners" (that is, those buyers who are lucky enough to have a second buyer appear and offer a higher price for the goods specified in their contracts). For example, if there is a one in ten chance that another buyer will come along and offer $200 more for the product, and if the first buyer would get none of that $200 under a compensatory remedy but will be able to negotiate for half of it ($100) if some more favorable remedy is available, then the introduction of the more favorable remedy will be exactly equivalent to making the seller give the buyer a lottery ticket along with each purchase: a ticket that pays the buyer an extra $100 in one out of every ten cases. Needless to say, the seller will not include such a ticket without demanding a higher price for it. In a market with risk-neutral buyers and sellers, the price increase would usually equal the ticket's actuarial value of $\frac{1}{10}$ times $100, or $10.

If buyers are really risk-neutral, then they are indifferent between paying for such a lottery ticket and going without one, so the introduction of such a remedy would leave them no better or worse off than before. This may be another reason why many economic analyses ignore this effect, since it is often plausible to assume that real actors (especially businesses) are risk-neutral. However, since risk-neutral buyers will be indifferent concerning the availability of such remedies, their preferences should be ignored in deciding which remedies will be most efficient. The most efficient remedy as far as the allocation of risk is concerned is that which best satisfies the risk-preferences of those buyers and sellers who are not risk-neutral, as these are the only ones whose welfare will be affected by the redistribution effect.

There are only a few combinations of attitudes toward risk that might justify punitive remedies and a transfer of more of the gains from the breaching to the nonbreaching buyers. First, if buyers are risk-preferring and sellers are risk-neutral (or less risk-preferring than buyers are), then buyers might be willing to "gamble" by paying a higher price in exchange for a chance to receive extra compensation in the event of breach. As noted earlier, this redistribution is equivalent to including a lottery ticket with each purchase, and risk-preferring buyers (by definition) will want such a lottery. However, this is a very weak argument for overcompensatory remedies, for most risk-preferring buyers can find other ways to satisfy their taste for gambling—by going to the horse races, for instance. Unless the buyers have some particular reason for wanting to gamble on the chance that the seller will find a profitable opportunity for breach, rather than on the outcome of a horse race, overcompensatory remedies are not needed to satisfy these buyers' tastes.

Similarly, overcompensatory remedies might be justified if sellers are risk-averse and buyers are risk-neutral (or again, less risk-averse than sellers are). A risk-averse seller would prefer to give up any chance of an extra gain in the event of breach (by agreeing to pay all of those gains over to the buyer) in exchange for a higher price up front, since the higher price is guaranteed and the chance for extra gains from breach may never material-ize. But a seller with these tastes has other ways to exchange those risky potential profits for their equivalent in certain forms of compensation—for example, by selling on a commission basis rather than as a profit taking entrepreneur, or by selling stock to outsiders who are willing to take more of the business's risk. The only reason an overcompensatory remedy succeeds in transferring some of this risk is that it requires *buyers* to invest in the seller's business, by making them pay a higher price up front in exchange for extra gains if the seller finds a more profitable resale opportunity. And while it is possible for all of a seller's customers to want to invest in the profitabil-ity of the business (and to want to invest in exact proportion to their pur-chases), there is no reason to expect this to be the case. A better solution is to unbundle the investment from the sales contract, and let those who want to invest do so by other means. . . .

Thus, to the extent that the redistribution effect provides an argument either way, it probably argues against making buyers pay a higher price for the chance to impose overcompensatory or punitive remedies. My main point, however, is more fundamental. The level of contract damages will have an effect on the distribution of risks between the parties, and this effect will be independent of the ease of ex post renegotiation when the seller is considering whether to breach. Even when ex post renegotiation is costless, the choice of contract remedies can still make a difference.

The Precaution Decision

As noted above, many traditional analyses modeled the seller's decision to breach upon the option of selling to a second buyer. In such a situation, the only decision affecting breach is the seller's final decision about to whom to sell. A different decision, however, is implicated by breaches that take place when, for example, a product turns out to be defective, or a building collapses because the foundation was improperly laid. The decision that has to be optimized in these cases is not a decision about to whom to sell, but rather a decision about how much to spend on precautions to prevent such a problem.

The efficient level of precautions can be defined by principles similar to the Learned Hand negligence formula. For example, if nonperformance would inflict a $10,000 loss on a risk-neutral buyer, and a given precaution would reduce the chance of non-performance from three percent to two per-cent, that precaution is efficient if and only if it costs less than $100 ($10,000 × (.03−.02)). If the precaution costs more than $100, it would be better on balance to accept the incrementally increased risk and spend the $100 on something else. This is the decision rule that maximizes the expected surplus created by the contract; it is therefore also the rule to which the parties would negotiate if they could. If the legal system is attempting to replicate what most parties would want, it should provide a remedy that will induce this efficient level of precaution.

Many economists have pointed out that perfectly compensatory remedies will create incentives for the seller to take an efficient level of precautions, as they make the seller liable for all of the buyer's costs. In the numerical example given above, the seller will realize that a reduction in the chance of nonperformance from three percent to two percent will reduce by one percent the chance of having to pay compensatory damages of $10,000. The seller will thus take that precaution if and only if it costs less than $100, just as efficiency would dictate. The argument is essentially the same as the argument for the efficiency (with respect to the defendant's precaution decision) of strict liability in torts. Indeed, this similarity should be expected, as liability for breach of contract is itself a form of strict liability.

If the law makes punitive remedies available, though, the seller may not choose an efficient level of precaution. For example, suppose that in the event of a breach the seller will be liable for a remedy that costs $15,000—or, equivalently, a remedy that would cost some larger amount, but from which the seller expects to be released by paying the buyer $15,000 in ex post negotiations. The desire to avoid being subjected to so costly a remedy will lead the seller to take the precaution even if it costs as much as $150 ($15,000 × (.03−.02)). In other words, any remedy that costs the seller more than the subjective value of performance to the buyer will induce excessive precautions against breach. The opposite would occur if the law limits the buyer to undercompensatory remedies, for in that case the seller would take too few precautions. For example, if the seller is only liable for $5,000 in the event of breach, it will not be worth the seller's while to spend any more than $50 ($5,000 × (.03−.02)) on the precaution. . . .

The Selection Decision

Noncompensatory remedies can also distort another decision: the choice of a contracting partner. Suppose, for example, that certain parcels of land are especially prone to earthquakes, erosion, or other natural disasters; and that there is nothing either party can do to prevent the disaster, so that the precaution decision discussed in the preceding subsection is not an issue. To continue the numerical example used before, suppose that the risky parcels of land have a .03 chance of disaster while other, safer parcels have only a .02 chance—and suppose that the sellers of each parcel of land know the risk associated with their parcel, while potential buyers do not. Suppose further that if the disaster does occur it will damage the buyer (who plans to build on the land) to the extent of $10,000.

These assumptions do not necessarily imply that it is a bad idea for buyers to build on the riskier parcels of land. They only imply that this difference in risk should be taken into account, and that buyers should not build there unless the riskier parcels have some advantage that more than offsets the extra risk. Compensatory remedies will induce just this kind of consideration. If the sellers know they will be liable if the disaster occurs, each seller will have to charge a price that is high enough to cover the risk of liability. The price charged by the seller of the risky land will reflect a .03 × $10,000 = $300 risk; the price charged by the seller of the safer land need only reflect a .02 × $10,000 = $200 risk, so the seller of the risky land will be at a $100 disadvantage. This will make it unprofitable for the seller of the risky land to sell that land, unless it has some offsetting advantage that makes buyers willing to pay a $100 higher price. This is exactly the result that efficiency requires.

However, a punitive remedy will distort this process. If the seller must pay (say) $15,000 in the event of a $10,000 disaster, the price for the risky land will have to reflect a .03 × $15,000 = $450 risk, while the price for the safer land need only reflect a .02 × $15,000 = $300 risk. This gives the seller of the risky land a $150 disadvantage, thus deterring buyers from purchasing that land unless its offsetting advantages are enough to overcome the $150 difference. If the risky land has offsetting advantages worth less than $150 but more than $100, the risky land is actually more desirable when all the advantages and disadvantages are taken into account, but the $15,000 remedy would discourage the purchase of that land. In effect, the punitive remedy exaggerates the actual risk posed by each seller, thus distorting buyers' purchase decisions.

The opposite is true of an undercompensatory remedy. If the seller is liable for only $5,000 if the disaster occurs, the seller of the safer land will have expected liability costs of .02 × $5,000 = $100, while the seller of the risky land will have expected liability costs of .03 × $5,000 = $150, leaving a difference between the two of only $50. Thus, just as punitive remedies would overstate the difference between the two sellers' risks, undercompensatory remedies would understate the differences. Moreover, both will have this effect regardless of the ease or difficulty of ex post renegotiation, since ex post renegotiation (after the disaster takes place) will clearly be too late to change the decision about which parcel to build on. The ease or difficulty of ex post renegotiation is therefore irrelevant to this effect as well.

This analysis of the selection decision should also be familiar from products liability law. The notion of "enterprise liability" rests in part on the idea that the accident costs associated with risky products should be internalized by their manufacturers and thereby reflected in their prices, so that the use of risky products will be appropriately deterred. Steven Shavell's more rigorous analysis of the effect of strict liability in inducing consumers to adapt the level of their consumption of different products to each product's riskiness rests on a similar analysis, as does much work in the economics of product warranties. Indeed, the analysis here is similar in many respects to the analysis of the

incentive to take precautions . . . . The only difference is that the "precaution" decision takes the other party to the contract as given, and attempts
to reduce the risks involved in dealing with that party. The "selection" decision is an attempt to reduce the risk by finding some other, less-risky party to
deal with. . . .

The distinction between the precaution and selection decisions also illuminates another problem case: sellers who have cost-effective precautions available, but who are unlikely to take those precautions regardless of liability
because, for example, the question of what precautions to take is never rationally considered, or because they do not know that the law will make them
bear any resulting loss. If there is really no way the law can correct the sellers'
behavior, then the risks resulting from their lack of precautions might as well
be classed with other "inherent" or "unavoidable" risks, in the sense that
there is nothing the law can do about them. But even in this situation, there
can still be efficiency gains from inducing others not to deal with the risky
party (assuming that no offsetting advantages are offered), and in eventually
driving that party from the market. This, too, is a familiar conclusion in the
torts literature, where analysts have often questioned the actual effect of legal
incentives on changing peoples' precaution decisions.

In short, even in cases where the precaution decision seems to be irrelevant
to an efficiency analysis, the selection decision may still be very important. . . .

Punishing Inefficient Breachers

The argument to this point has implicitly assumed that any legal remedy must
apply to all breachers, whether or not they have behaved efficiently. This is, of
course, the normal rule for contract remedies, for most contract liability is
based on something more akin to strict liability than to negligence. For example, a seller will be liable for failing to deliver a product even if the excuse is
that the product was sold to another buyer who valued it more highly. The
seller will also be liable for selling a product that fails to live up to its warranty,
even if it can be shown that the optimal amount was spent on quality control
to try to prevent the defect.

This "strict liability" aspect of contract damages is crucial to the economic
analysis of the preceding parts. Under strict liability, if even those sellers who
behave perfectly efficiently will occasionally find it desirable to breach (or will
take a level of precautions that occasionally results in a breach), then even the
best of sellers will have to build the risk of liability into their price. This then
gives rise to the "lottery" objection . . . and to the distortion of the selection
decision discussed [earlier in the excerpt] . . . . The risk of having to pay that
extra liability will also lead sellers to take too many precautions to avoid being
put in that situation, thus distorting the precaution decision . . . .

These distortions would disappear, however, if liability for the punitive
remedy could be conditioned on the seller's having behaved inefficiently. For
example, one could imagine a rule that imposed liability for overcompen-

satory damages only if it were shown that the seller had sold to another buyer who valued the good less highly than did the plaintiff, or if it were shown that the seller had taken a suboptimal level of precautions. This would make the seller's liability for the punitive remedy turn on something more like negligence than strict liability—and under a perfectly operating negligence system, even damages far above the compensatory level should not distort the defendant's level of precautions. As long as the seller can entirely avoid that liability by taking the optimal level of care, the threatened liability can (in theory) be increased to infinity without inducing the seller to take more than the optimal level; this is so because once the optimal level of care is taken, the seller (by assumption) will no longer be exposed to the punitive damages. This ability to reduce the risk of punitive liability to zero also means that an efficient seller would have no need to reflect that risk in the price, so that there would be no distortion of the selection decision and no bundling of a lottery with the underlying contract.

However, this argument is only valid as long as there is no chance that a defendant who behaves efficiently will be subjected to the punitive remedy. If this condition is not met, then either (1) the punitive remedy will deter some defendants from behaving efficiently; or (2) defendants will still behave efficiently but the risk of extra liability will have to be reflected in their price, thus giving rise to all of the problems discussed above. Consequently, this case for punitive remedies must depend on the remedies' being applied only to truly inefficient conduct. For example, a rule making punitive remedies available only for deliberate or "willful" breaches would not suffice, for even efficient breaches can be deliberate—unless perhaps "willful" is interpreted to mean something like "willful and unjustified."

Indeed, even a test like "willful and unjustified would only work if defendants could be confident that the application of that test by judges and juries would never result in its application to efficient conduct. Otherwise, efficient defendants would still face a risk of incorrectly imposed punitive remedies, thus requiring them to take this into account in their prices and their precaution decisions, and thereby giving rise to all of the distortions discussed above. Of course, most "negligence" type tests (including the Learned Hand cost-benefit analysis) do not yield such certainty in their application. Consequently, making liability contingent on the defendant's having been found to violate a cost-benefit standard would still create some risk of punishing efficient behavior.

Notes on Expectation Damages

1. The economic analysis of expectation damages has been presented more rigorously in economics articles employing mathematical models. Important articles in the development of this technical literature include Steven Shavell, Damage Measures for Breach of Contract, 11 Bell Journal of Eco-

nomics 466 (1980); A. Mitchell Polinsky, Risk Sharing Through Breach of Contract Remedies, 12 Journal of Legal Studies 427 (1983); and Steven Shavell, The Design of Contracts and Remedies for Breach, 99 Quarterly Journal of Economics 121 (1984). The most accessible nontechnical summaries are A. Mitchell Polinsky, An Introduction to Law and Economics, chapters 5 and 8 (Boston: Little, Brown & Company, 2d ed. 1989); and Lewis A. Kornhauser, An Introduction to the Economic Analysis of Contract Remedies, 57 University of Colorado Law Review 683 (1986).

2. The "Coase Theorem" referred to in the excerpt by Cooter and Ulen (supra at page 44) is usually attributed to Ronald A. Coase, The Problem of Social Cost, 3 Journal of Law & Economics 1 (1960), although the theorem was not stated explicitly in that article. The Coase Theorem holds that if bargaining costs are sufficiently low, resources that can be bargained over will end up in the hands of whichever party values them most, regardless of which party is initially assigned the right to those resources. This is why Cooter and Ulen assert that if transaction costs are low, it won't matter (as far as the decision to perform or breach is concerned) whether the law gives the performing party a right to breach and pay damages, or whether it gives the nonperforming party the right to block the other party's breach through a remedy of specific performance. If breach is efficient (in the sense defined by Cooter and Ulen) and the law does not give the performing party the right to breach, that party will simply purchase that right from the nonperforming party.

This aspect of the Coase Theorem will be relevant to many of the other remedies discussed in this chapter. See, for example, the subsequent excerpts by Charles J. Goetz and Robert E. Scott (infra at pages 56–58), Timothy J. Muris (infra at pages 81–86), and Alan Schwartz (infra at pages 93–96). Whenever a remedial rule requires conduct that seems inefficient, you should consider whether the party who would lose the most from that inefficiency will be able to pay the other party to abandon his or her insistence on the inefficient remedy.

3. As the Craswell excerpt points out, remedial rules can affect many other aspects of the parties' relationship, not just their conduct after an opportunity for breach has arisen. These issues did not arise in the excerpts by Friedmann and by Cooter and Ulen because those excerpts began their analysis at a point *after* the contract was signed, when an opportunity for a profitable breach had arisen. As a consequence, those excerpts did not consider what effect different remedies might have on the price set in the initial contract (for example), or on the parties' incentives to take precautions against accidents that might lead to breach.

4. The effect of contract remedies on the price set in the initial contract suggests a possible distinction between cases of breach of contract and cases of "efficient theft" of the sort discussed in the Friedmann excerpt. If theft is deterred by subjecting thieves to harsher penalties, the thieves cannot respond by charging their victims a higher price to make up for their lost

opportunity to steal. By contrast, if breaches of contract are deterred by subjecting breachers to harsher penalties, potential breachers may refuse to enter into the contract unless they are paid a higher price to compensate them for either (1) having to forego profitable breach opportunities, or (2) having to pay harsh penalties. Can this distinction justify adopting smaller penalties for breach of contract than for outright theft?

5. A related question is whether the law should care, in selecting damage remedies, about how high a price contracting parties would have to pay under various damage rules? Under an opposing view, the degree of protection to be given to a promisee's contractual rights should be decided deontologically— that is, without regard to the consequences to the promisee (such as the effect on contract prices) of higher or lower levels of protection. The Friedmann excerpt can be read as making a deontological argument: theft is theft, and breaking a contract violates the promisee's rights. On this view, neither theft nor breaches of contract should be permitted, regardless of whether the effect of prohibiting them is to raise the price that promisees have to pay. Is this argument persuasive?

In response, it might be argued that allowing A to break the contract and pay B damages does not violate B's rights at all *unless* we first decide that the contract gave B an absolute right to performance, rather than merely the right to performance of the contract or (at A's election) the payment of an equivalent amount of damages. In other words, on this view the question of what rights B has can only be answered *after* deciding how the contract between A and B ought to be interpreted. If the contract had explicitly said that B had the right to absolute performance, or if it had explicitly given A the right to substitute damages for performance, then the language of the contract might be invoked to settle this interpretation question. But if the contract did not explicitly settle between these two options, the law would have to select a "default rule" to decide what rights B acquires in the absence of a specification to the contrary. If it is more efficient to have a default rule allowing B to breach and pay expectation damages, the law should adopt expectation damages as its default rule. If the law does adopt expectation damages as its default rule, B can no longer claim that A's breach of contract violates B's rights as long as A pays B's expectation damages.

Viewed from this standpoint, questions about the proper remedy for breach are really questions about whether contract law's default rules should be selected on deontological or utilitarian grounds. As a consequence, the earlier readings concerning the selection of default rules (supra at pages 16–30) will be relevant to many of the readings in this chapter as well.

6. The Craswell excerpt's analysis of extra-large damage remedies for inefficient breaches (supra at pages 52–53) raises issues that will be discussed at more length in the readings on punitive damages (infra at pages 127–37). For a more detailed discussion of these issues, see Robert Cooter, Prices and Sanctions, 84 Columbia Law Review 1523 (1984).

§ 2.2 Limits on the Expectation Measure

§ 2.2.1 Mitigation and Reliance by the Promisee

The Mitigation Principle

CHARLES J. GOETZ AND ROBERT E. SCOTT

Most contract rules are permissive, applying only if the parties do not otherwise agree. By providing standardized and widely suitable risk allocations in advance, the law enables most parties to select a preformulated legal norm "off-the-rack," thus eliminating the cost of negotiating every detail of the proposed arrangement. Atypical parties remain free to bargain for customized provisions, much as a person with an unusual physique may purchase custom-tailored garments for a premium rather than accept a standard size and cut available at a lower price.

Ideally, the preformulated rules supplied by the state should mimic the agreements contracting parties would reach were they costlessly to bargain out each detail of the transaction. Using this benchmark raises two separable issues: First, what arrangements would most bargainers prefer? And, second, what atypical arrangements should be supported as benign alternatives?

The model developed in this article will show that the contractual obligee and obligor would agree in advance to minimize the joint costs of adjusting to prospective contingencies, assigning the responsibility of mitigating to whoever is better able to adjust to the changed conditions. The occurrence of contingencies requiring adjustment, however, may encourage strategic behavior by both parties: the obligor may attempt to evade his performance responsibilities while the obligee may bargain opportunistically whenever his cooperation is requested. Any effort legally to regulate one manifestation of this strategic behavior almost inevitably exacerbates the other. But where a developed market for substitute performances exists, the potential for opportunism is negligible; parties can therefore focus on eliminating evasion of contractual obligations without losing the benefits of cooperation. The tension between performance and mitigation responsibilities is most keen in situations lacking a good substitute market; parties in such environments must balance the costs of evasion and opportunism, knowing that no single solution will eliminate the tension.

Charles J. Goetz & Robert E. Scott, The Mitigation Principle: Toward a General Theory of Contractual Obligation, 69 Virginia Law Review 967, 971–75, 982–83 (1983). Copyright © 1983 by the Virginia Law Review Association. Reprinted by permission.

The Principle of Joint-Cost Minimization

Formulating the ideal mitigation principle requires one first to identify the kinds of costs contracting parties might want to reduce. The parties recognize many of the costs of promissory activity at the time of contracting and allocate these within the scope of the defined contractual rights and obligations. For instance, they may condition alternative modes of performance or excuse from performance upon the occurrence of certain contingencies. It is one thing, however, to perceive a risk in a manner sufficient to allocate its consequences to one party or the other; it is quite another to work out definitively the optimal responses to all future contingencies. As time passes and information increases, parties reassess the risk associated with certain future contingencies. Such reassessments may follow a change in the probability of an event, the magnitude of its consequences, or both. Inevitably, the party who perceives an increase in prospective cost will regret the initial assignment of risks.

A regretting promisor will react to such a "readjustment contingency" by selecting the least costly of the following alternatives: (1) he may continue to pursue the original performance obligation and absorb whatever loss results from his higher performance costs; (2) he may breach and pay compensatory damages; or (3) he may attempt, by renegotiation or otherwise, to modify the original contract. Although a regretting promisor will naturally seek the least costly alternative, interparty cooperation is frequently essential to minimize adjustment costs. In other words, both parties may have to adjust in order to exploit fully the net benefits of contracting.

Once a contract has been made, an obligee may seem to have little interest in the obligor's excess costs. But a party who anticipates bearing excess costs will presumably negotiate for a more costly return promise to compensate for those inflated costs. Because the terms acceptable to a risk-bearing obligor will reflect the expected magnitude of his potential regret costs, both parties gain if they agree in advance to provisions that will reduce expected future costs. One can therefore derive a broad principle of mitigation by predicting how contractors would agree to cooperate if charged explicitly with designing a policy to cope with readjustment contingencies. The resulting mitigation principle would require each contractor to extend whatever efforts in sharing information and undertaking subsequent adaptations that are necessary to minimize the joint costs of all readjustment contingencies.

The doctrine of avoidable consequences confirms this cost-minimizing conception of the mitigation principle, requiring a mitigator to bear the risk of his failure to minimize losses. It denies a mitigator recovery for losses he unreasonably failed to avoid, but allows him full recovery for costs incurred through any reasonable affirmative efforts to minimize losses. The courts seem implicitly to have adopted a joint-cost minimization construction of "reasonable." In one illustrative case, breaching subcontractors argued that the plaintiff prime contractor should have mitigated damages by withdrawing from a building contract and forfeiting the one percent bid bond. The court rejected their claim, reasoning that "[t]he duty of the plaintiff to keep the damages as low as reasonably possible does not require of it that it disregard its own interests [in maintaining good will] or exalt above them those of the defaulting defendants." In other words, although the doctrine of avoidable consequences requires a mitigator to minimize the joint costs of breach, it does not require minimizing the defendant's loss in a way that imposes a still greater loss on the mitigator himself. . . .

The Problem of Opportunism in Renegotiation

Parties are not limited to autonomous readjustments. If mutual cooperation is necessary to minimize costs, such cooperation can be achieved consensually through the third option of *renegotiation*. By renegotiation, the parties can reallocate the rights and duties which have become inefficient because of intervening events. For example, Buyer could agree to delay his occupancy until the strike is settled. Seller would thus solicit Buyer's cooperation in making adjustments Seller could not achieve alone. The maximum payment Seller will offer Buyer is the difference between Seller's position with and without Buyer's cooperation.

Renegotiation, however, creates a moral hazard in addition to the obligee's indifference: the obligee may actually threaten to exacerbate damages unless the obligor purchases his cooperation at a premium. For instance, Buyer might engage in opportunistic behavior to extract the full "value" of his cooperation in adjusting to the strike. He could accomplish this goal by foot-dragging, by inflating the estimates of mitigation costs, or by manifesting any other sign of reluctance to cooperate. Of course, Seller has analogous motives to induce Buyer's cooperation at minimum cost, perhaps by exploiting the potential for evasion as an implicit or explicit threat.

Would strategic behavior affect renegotiations more than original negotiations? Although both situations involve carving up gains from trade, renegotiations will provoke more costly strategies if parties have become "contractually specialized" and face substantially restricted alternate arrangements. At best, renegotiations impose significant transaction costs on the parties. Especially when opportunism magnifies them, renegotiation costs tend to impede readjustments that offer potential benefits for both parties. Parties will hesitate to trade information necessary for readjustments if bargaining over such transfers may itself alert the potential buyer to all or part of the very information that one might wish to "sell." Moreover, even when the parties ultimately achieve cooperative readjustment, the associated renegotiation costs remain a deadweight loss reducing the potential benefits of the contractual relationship.

Unity in Tort, Contract, and Property

ROBERT COOTER

Even when necessary or unavoidable, an accident, breach of contract, taking, or nuisance causes harm. The affected parties, however, can usually take steps to reduce the probability or magnitude of the harm. The parties to a tortious accident can take precautions to reduce the frequency or destructiveness of accidents. In contract, the promisor can take steps to avoid breach, and the promisee, by placing less reliance on the promise, can reduce the harm caused by the promisor's breach. Similarly, for governmental takings of private property, the condemnor can conserve on its need for private property, while property owners can reduce the harm they suffer by avoiding improvements whose value would be destroyed by the taking. Finally, the party responsible for a nuisance can abate; furthermore, the victim can reduce his exposure to harm by avoiding the nuisance.

Generalizing these behaviors, I extend the ordinary meaning of the word "precaution" and use it as a term of art in this Article to refer to any action that reduces harm. Thus the term "precaution" includes, for example, prevention of breach and reduced reliance on promises, conservation of the public need for private property and limited improvement of private property exposed to the risk of a taking, and abatement and avoidance of nuisances. These examples are, of course, illustrative, not exhaustive.

When each individual bears the full benefits and costs of his precaution, economists say that social value is internalized. When an individual bears part of the benefits or part of the costs of his precaution, economists say that some social value is externalized. The advantage of internalization is that the individual sweeps all of the values affected by his actions into his calculus of self-interest, so that self-interest compels him to balance all the costs and benefits of his actions. According to the marginal principle, social efficiency is achieved by balancing all costs and benefits. Thus, the incentives of private individuals are socially efficient when costs and benefits are fully internalized, whereas incentives are inefficient when some costs and benefits are externalized.

In situations when both the injurer and the victim can take precaution against the harm, the internalization of costs requires both parties to bear the full cost of the harm. To illustrate, suppose that smoke from a factory soils the wash at a commercial laundry, and the parties fail to solve the problem by private negotiation. One solution is to impose a pollution tax equal to the harm caused by the smoke. The factory will bear the tax and the laundry will

Robert Cooter, Unity in Tort, Contract, and Property: The Model of Precaution. Copyright © 1985 by the California Law Review Inc. Reprinted from the California Law Review, volume 73, number 1 (January, 1985), pages 3–4, 11–16, by permission.

bear the smoke, so pollution costs will be internalized by both of them, as required for social efficiency. In general, when precaution is bilateral, the marginal principle requires both parties to be fully responsible for the harm. The efficiency condition is called double responsibility at the margin.

One problem . . . however, is that compensation in its simplest form is inconsistent with double responsibility at the margin. In the preceding example, justice may require the factory not only to pay for harm caused by the smoke, but also to compensate the laundry for that harm. Compensation, however, permits the laundry to externalize costs, thereby compromising efficiency. Thus, a paradox results: If the factory can pollute with impunity, harm is externalized by the factory; if the factory must pay full compensation, harm is externalized by the laundry; if compensation is partial, harm is partly externalized by the factory and partly externalized by the laundry. Assigning full responsibility for the injury to one party or parceling it out between the parties cannot fully internalize costs for both of them. Thus, there is no level of compensation that achieves double responsibility at the margin. In technical terms, when efficiency requires bilateral precaution, strict liability for any fraction of the harm, from zero percent to 100 percent, is inefficient.

Rules that combine compensation for harm with incentives for efficient precaution are therefore patently difficult to formulate. The problem confronted in this Part of the Article is to explain how the law combines compensation with double responsibility at the margin. The law has evolved three distinct mechanisms for achieving this end, which I will sketch by reference to the law of torts, contracts, and property. . . .

Breach of Contract

Yvonne and Xavier enter into a contract in which Yvonne pays for Xavier's promise to deliver a product in the future. There are certain obstacles to Xavier's performance that might arise, and if severe obstacles materialize, Xavier will not be able to deliver the product as promised. The probability of timely performance depends in part on Xavier's efforts to prevent such obstacles from arising. These efforts are costly.

One purpose of contracting is to give Yvonne confidence that Xavier's promise will be performed, so that she can rely upon his promise. Reliance on the contract increases the value to Yvonne of Xavier's performance. However, reliance also increases the loss suffered in the event of breach. The more the promisee relies, therefore, the greater the benefit from performance and the greater the harm caused by breach.

To make this description concrete, suppose that Xavier is a builder who signs a contract to construct a store for Yvonne by the first of September. Many events could jeopardize timely completion of the building; for example, the plumbers union may strike, the city's inspectors may be recalcitrant, or the weather may be inclement. Xavier can increase the probability of timely completion by taking costly measures, such as having the plumbers work overtime before their union contract expires, badgering the inspectors to finish on time, or rescheduling work to complete the roof before the rainy season arrives. Yvonne, on the other hand, must order merchandise for her new store in advance if she is to open with a full line on the first of September. If she orders many items for September delivery and the store is not ready for occupancy, she will have to place the goods in storage, which is costly. The more merchandise she orders, the larger her profit will be in the event of performance, and the larger her loss in the event of nonperformance.

As thus described, the structure of the contractual model is similar to the model developed for tortious accidents. The precaution taken by the potential tortfeasor against accidents parallels the steps taken by the promisor to avoid obstacles to performance. The parallel between the tort victim and the promisee, however, is more subtle. *More* precaution by the tort victim is like *less* reliance by the contract promisee, because each action reduces the harm caused by an accident or a breach. Therefore, the tort victim's precaution against accidents and the contract promisee's reliance upon the contract are inversely symmetrical.

If Xavier does not perform, then a court must decide whether a breach has occurred or whether nonperformance is excused by circumstances. Among the excuses that the law recognizes are: that the quality of assent to the contract was too low due to mistake, incapacity, duress, or fraud; that the terms of the contract were unconscionable; or that performance was impossible or commercially impractical. If the court narrowly construes excuses, usually finding nonperformance to be a breach, then Xavier will usually be liable. If the court construes excuses broadly, usually finding nonperformance to be justified, then Xavier will seldom be liable.

The incentive effects of a broader or narrower construction of excuses are similar to the effects of strict liability and no-liability rules in tort. If defenses are narrowly construed and perfect expectation damages are awarded for breach, the promisee will rely as if performance were certain. Specifically, Yvonne will order a full line of merchandise as if the store were certain to open on the first of September. A promisee's reliance to the same extent as if performance were certain corresponds to a tort victim's failure to take precaution against harm.

A broad construction of excuses has the symmetrically opposite effect: the promisor expects to escape liability for harm caused by his breach, so he will not undertake costly precautions to avoid nonperformance. Specifically, if Xavier is unconcerned about his reputation or the possibility of future business with Yvonne, and if nonperformance due to a plumber's strike, recalcitrant inspectors, or inclement weather will be excused, say, on grounds of impossibility, then Xavier will not take costly precautions against these events. The promisor's lack of precaution against possible obstacles to performance corresponds to the injurer's lack of precaution against tortious accidents.

As explained, the narrow and broad constructions of excuses for breach of contract affect behavior in ways that parallel no liability and strict liability in tort. Furthermore, the effects of these constructions on cost internalization and efficiency are also parallel. Specifically, if excuses are broadly construed, allowing the promisor to avoid responsibility for breach regardless of his precaution level, the promisor will externalize some of the costs of breach. As a result, his incentives to take precaution against the events that cause him to breach are insufficient relative to the efficient level. If, on the other hand, excuses are narrowly construed and full compensation is available for breach, the promisee can externalize some of the costs of reliance. Insofar as the promisee can transfer the risk of reliance to the promisor, her incentives are insufficient to provide efficient reliance and, therefore, reliance will be excessive.

To illustrate, social efficiency requires Xavier to hire the plumbers to work overtime if the additional cost is less than the increase in Yvonne's expected profits caused by the higher probability of timely completion. Suppose, however, that there are circumstances in which tardiness will be excused regardless of whether or not Xavier hired the plumbers to work overtime. Suppose for example that inclement weather excuses tardiness on grounds of impossibility. In the event inclement weather provides Xavier with an excuse, the extra cost of hiring the plumbers to work overtime, which is valuable to Yvonne, has no value to Xavier. Anticipating this eventuality, Xavier may not hire the plumbers to work overtime, even though social efficiency may require him to do so.

Social efficiency also requires Yvonne to restrain her reliance in light of the objective probability of breach. To be more precise, social efficiency requires her to order additional merchandise until the resulting increase in profit from anticipated sales in the new store, discounted by the probability that Xavier will finish the store on time, equals the cost of storing the goods, discounted by the probability that Xavier will finish the new store late. Suppose, however, that Xavier must compensate Yvonne for her storage costs in the event that the goods must be stored. From a self-interested perspective, Yvonne has no incentive to restrain her reliance in these circumstances. Anticipating this possibility, instead of weighting the cost of storage by the objective probability of breach, Yvonne will weight it by the probability of breach without compensation. Since in this example the probability of breach is greater than the probability of breach without compensation, the weight Yvonne gives to the possibility of storage cost is too small. Therefore, her reliance will be excessive and thus inefficient.

In general, the possibility of successful excuses may externalize the costs of not taking precaution, so that the promisor takes too little precaution and the probability of breach is excessive. Similarly, the possibility of compensation may externalize the costs of reliance, so the promisee relies too heavily and the harm that materializes in the event of breach is excessive. This is an aspect of the paradox of compensation that arises in tort with respect to no liability and strict liability. As with tort law, contract law has a solution to the paradox, but the contract solution is different from the tort solution. To illustrate the characteristic remedy in contracts, consider the liquidation of damages. If the contract stipulates damages for breach requiring Xavier to remit, say, $200 per day for late completion, then the promisor will have a material incentive to prevent breach. Specifically, Xavier may find that paying the plumbers to work overtime is cheaper than running the risk of late completion. If the promisee receives the stipulated damages as compensation, then the level of her compensation is independent of her level of reliance, so she has a material incentive to restrain her reliance. Specifically, if Yvonne receives $200 per day in damages for late completion whether or not she orders the bulky merchandise, she may avoid the risk of bearing storage costs by not ordering it.

Like a negligence rule in tort, liquidation of damages in a contract imposes double responsibility at the margin: the promisor is responsible for the stipulated damages and the victim is responsible for the actual harm. By adjusting the level of stipulated damages, efficient incentives can be achieved for both parties. Stipulated damages are efficient when they equal the loss that the victim would suffer from breach if her reliance were efficient. To illustrate, assume that efficient reliance requires Yvonne to order the compact merchandise and not the bulky merchandise. Furthermore, assume that if Yvonne orders the compact merchandise she will lose $200 in profits for each day that Xavier is late in completing the new store. Under these assumptions, liquidating damages at $200 per day for late completion provides efficient incentives for both Xavier and Yvonne.

Under the stated assumptions, stipulating damages at $200 per day will cause Yvonne to order the compact merchandise and not the bulky merchandise. Consequently, the actual harm that Yvonne will suffer in the event of breach is $200 per day. Thus the stipulation of damages at the efficient level is a self-fulfilling prophecy: the stipulation of *efficient* damages causes the actual damages to equal the stipulation. Since Xavier internalizes the actual harm caused by breach, and Yvonne bears the risk of marginal reliance, there is double responsibility at the margin as required for efficiency.

Since liquidation of damages provides an immediate solution to the problem of overreliance, it would seem that liquidation clauses should be found in contracts where efficiency requires restraints on reliance. In fact, rather than liquidating damages, most contracts leave the computation of damages until after the breach has occurred. When damages are not liquidated in the contract and restraint of reliance is required by efficiency, various legal doctrines are available that can accomplish the same end as liquidation of damages. Liquidated damages restrain reliance by making damages invariant with respect to reliance. Courts restrain reliance by applying other legal doctrines that make damages similarly invariant.

To illustrate, the goods supplied by different firms in a perfectly competitive market are, by the definition of perfect competition, perfect substitutes. When the promisor fails to perform in a competitive market, damages are ordinarily set equal to the cost of replacing the promised performance with a close substitute (the replacement-price formula). Specifically, if the seller breaches his promise to supply a good at a specified price, the damages paid to the buyer may include the additional cost of purchasing the good from someone else. In technical terms, damages in such a case will equal the difference between the spot price and the contract price for that particular good. In a competitive market, no single buyer or seller can influence these prices. Consequently, damages computed by the replacement-price formula are invariant with respect to the level of the promisee's reliance. Thus, replacement price damages in a competitive market have the same efficiency characteristics as liquidated damages.

For noncompetitive markets, doctrinal alternatives are available to reduce or eliminate the effects of variations in damages due to reliance. To illustrate, recovery may be limited to damages that were foreseeable at the time the promise was made. It is but a short step to argue that reliance that is excessive in efficiency terms is also unforeseeable. Thus, the foreseeability doctrine can be used to avoid compensation for excessive reliance.

There are other doctrinal approaches to damages that have similar effects. For example, suppose that Xavier fails to complete the building on the first of September as promised, and Yvonne has to rent temporary space elsewhere. The court might award damages based in part on the additional rent, if it finds Yvonne's calculation of lost profits too speculative. If damages are based on the additional rent, and if the additional rent varies less than Yvonne's profits with respect to her reliance, then her incentive to overrely is reduced. As another example, failure to perform on a franchise agreement may result in an award of damages equal to the profit of similar franchise establishments, but not the "speculative profits" lost by the particular plaintiff. The general point of these two examples is that if compensation is restricted to nonspeculative damages, and if nonspeculative damages vary less with respect to reliance than the actual harm, then restricting compensation to nonspeculative damages reduces the incentive to overrely.

Notes on Mitigation and Reliance by the Promisee

1. In the final section of their excerpt, Goetz and Scott note that efficient mitigation might be induced even without any legal duty to mitigate because the promisor could always go to the promisee and bribe him or her to take any cost-effective steps that would reduce the losses caused by the breach. If the promisor had to make such an offer to get the promisee to mitigate damages—that is, if the promisee had a legal right to refuse to mitigate unless the promisor came forth with an acceptable bribe—this would allow the promisee, rather than the promisor, to capture more of the benefits created by mitigation activity.

This question of who should be allowed to capture the gains created by efficient conduct is very similar to one of the issues discussed in the readings on expectation damages and efficient breach (supra at pages 42–55). In the mitigation context, Goetz and Scott argue that, even when renegotiations over efficient mitigation are successful, the costs of the renegotiations add unnecessary transaction costs. On the other hand, the transaction costs involved in implementing a legal duty to mitigate may also be significant, so it is not clear which regime will have the lowest transaction costs overall. This was one of the points made in the excerpt by Cooter and Ulen on efficient breach (supra at pages 42–44).

2. Even if renegotiation were always costless—that is, even if the Coase Theorem applied (see page 54 supra)—a rule allowing the promisee to demand an extra share of the gains from mitigation could have other effects. For example, under such a rule promisees might have to agree to a higher contract price to get promisors to enter into the initial contract, to make up for the risk that promisors might occasionally end up paying large sums to induce efficient mitigation. The threat of having to pay large sums to induce mitigation might also induce promisors to take a higher level of precautions against accidents that could leave them in a position in which, as a result of being unable to perform, they would end up having to pay the promisee a larger sum to induce efficient mitigation. This was one of the points of the Craswell excerpt on efficient breach (supra at pages 46–53).

3. If renegotiation is easy, one of the conclusions of the Cooter excerpt may also have to be modified. Cooter argued (supra at pages 63–64) that any stipulated damage measure in which the amount the promisee collects does not depend on the extent of the promisee's reliance will give the promisee an incentive to choose the optimal level of reliance. (A similar assertion is made in the excerpt by Richard Epstein on consequential damages, infra at page 69).

This conclusion is correct if ex post renegotiation is too costly to be feasible. It is also correct if the only possible cause of breach does not leave any room for renegotiation—for example, when a plumbing strike or some other catastrophe makes timely performance impossible, no matter how much the promisee is willing to pay. In other contexts, however, the promisor may claim to have suffered a mild increase in costs and may therefore refuse to perform unless the promisee agrees to pay a higher price. If legal remedies were always fully compensatory, the promisee might not have to give in to this demand, but there are many circumstances where the promisee's remedies will not be fully compensatory. (See the discussion in the excerpt by Aivazian, Trebilcock, and Penny, infra at pages 212–13.) Moreover, if the promisee has already relied heavily on the contract—say, by investing in assets that would be wasted if the promisor pulls out of the deal—that reliance would make the promisee willing to pay quite a high price to keep the deal alive, thus strengthening the promisor's bargaining position. Moreover, the threat of being exposed to this ex post bargaining strength could give promisors an incentive to limit their exposure by engaging in too little reliance, rather than too much reliance (as the Cooter excerpt suggests).

The possibility of ex post renegotiation will be discussed again in connection with specific performance (infra at page 94) and in the excerpt by Aivazian, Trebilcock, and Penny (infra at pages 211–19). For formal economic analyses of the effect of renegotiation on a promisee's incentives to rely, see William P. Rogerson, Efficient Reliance and Damage Measures for Breach of Contract, 15 Rand Journal of Economics 39 (1984); and Aaron S.

Edlin, Specific Investments, Holdups and Efficient Contract Remedies, Stanford Department of Economics Working Paper (May 1992).

4. Both the Cooter and the Goetz and Scott excerpts analyze these issues from the standpoint of efficiency. However, it might also be possible to make noneconomic arguments analogous to those made in the earlier excerpt by Friedmann (supra at pages 44–46). For example, why not give a promisee such as Yvonne an absolute right to rely to the fullest on Xavier's promised performance? That is, why should Yvonne have to moderate *her* behavior solely because of the risk that the other party, Xavier, might fail to perform? Similarly, in the mitigation context analyzed by Goetz and Scott, why should a promisor be allowed to impose on the promisee the duty of taking cost-effective steps to mitigate the losses following the promisor's breach. If the promisor wants the promisee to do something that would reduce the promisor's liability, why not make the promisor get the promisee's permission, as discussed in note 1?

On the other hand, note 2 pointed out that such a rule might raise the price that promisees such as Yvonne would have to pay to get promisors to contract in the first place. Should this be a concern in deciding what rights promisees ought to have? (The analogous issue concerning expectation damages was discussed in the notes supra at page 55.)

5. As these excerpts illustrate, there are many similarities between a promisee's failure to mitigate damages after a breach, on the one hand, and a promisee's excessive reliance before a breach has occurred, on the other. Both actions increase the losses that will be suffered if a breach occurs; thus, both must somehow be limited in order to maximize the expected value (or minimize the expected losses) of the contract. Moreover, if the promisee is guaranteed full compensation in the event of a breach, the promisee will have no incentive to limit his or her conduct in either of these respects. This is the "moral hazard" discussed by Goetz and Scott (supra at page 58).

In theory, a "reasonableness" standard, of the sort discussed by Goetz and Scott in the mitigation context, might be applied to a promisee's prebreach reliance. That is, a court could try to determine whether a promisee such as Yvonne had relied more than was efficient and refuse to compensate Yvonne for any part of her reliance that was not efficient. In another article, Goetz and Scott suggested just such an interpretation of legal doctrines that employ the notion of *reasonable* reliance. Charles J. Goetz and Robert E. Scott, Enforcing Promises: An Examination of the Basis of Contract, 89 Yale Law Journal 1261, 1280 n. 42 (1980). See also Cooter's statement, supra at page 161, that "[i]t is but a short step to argue that reliance that is excessive in efficiency terms is also unforeseeable." (The rules limiting recovery of *unforeseeable* damages, and those rules' effects on promisees' incentives, are discussed in the following section of this chapter.)

Are standards such as this administratively workable? That is, do you think courts would be very good at determining whether the level of reliance chosen by a promisee was above or below the efficient level of reliance?

6. Issues can also arise concerning the proper *timing* of mitigation efforts—for example, if the promisor repudiates the contract before the time for performance has arrived, or if the time for performance is past but the promisor wants to continue trying to perform rather than having the promisee cut his or her losses and start to mitigate. Some of these timing issues are discussed in portions of the Goetz and Scott article not reprinted here. See also Thomas H. Jackson, "Anticipatory Repudiation" and the Temporal Element of Contract Law: An Economic Inquiry into Contract Damages in Cases of Prospective Nonperformance, 31 Stanford Law Review 69 (1978); and Richard Craswell, Insecurity, Repudiation, and Cure, 19 Journal of Legal Studies 399 (1990).

§ 2.2.2 Consequential Damages

Beyond Foreseeability: Consequential Damages in the Law of Contract

RICHARD A. EPSTEIN

[T]he common law (of both England and the United States) has exhibited something of an uneasy dualism. With regard to the primary obligation of the parties—to buy, to ship, to work—the tendency has been to allow the parties themselves to specify the subject matter of the agreement, including any price term. This inquiry is case specific. After breach, the opposite tendency has emerged for the selection of remedy. While the judicial practice is far from uniform, courts and commentators often treat the required damage rules as though they were generated by some normative theory external to the contract itself. From this premise, the implicit understanding has grown up that damages are resolved more by "rules of law" and less by default rules of construction.

This attitude is most evident in typical statements about the function and purpose of damage awards for breach of contract. Here discussions of the damage rules often begin from a social norm that the innocent party should be made whole after breach. The question whether it is possible to contract out of the rules is suppressed, so that it is often unclear how the issue of contracting out would be resolved if raised expressly. Samuel Williston articulated the

Richard A. Epstein, Beyond Foreseeability: Consequential Damages in the Law of Contract, 18 Journal of Legal Studies 105, 106–8, 114–18, 120–21 (1989). Copyright © 1989 by The University of Chicago. Reprinted by permission.

classical position as follows. "In fixing the amount of these damages, the general purpose of the law is, and should be, to give compensation: that is to put the plaintiff in as good a position as he would have been in had the defendant kept his contract." The statement takes a stance external to the will of the parties, for it is the purpose the law "should" have that seems dispositive. Countless cases take much the same view. Thus it has been rendered axiomatic that "the function of the award of damages for a breach of contract is to put the plaintiff in the same position he would have been in had there been no breach." Charles Fried has argued for the normative power of this proposition on fairness grounds: "If I make a promise to you, I should do as I promise; and if I fail to keep my promise, it is fair that I should be made to hand over the equivalent of the promised performance. In contract doctrine this proposition appears as the expectation measure of damages for breach."

The dominance of the expectation interest is sometimes challenged, but usually on the wrong ground. Lon Fuller, in his classic study, "The Reliance Interest in Contract Damages," takes it as "obvious" that the expectation measure of damage is entitled to less protection than the reliance interest—that is, compensation that places the plaintiff in the same position he would have enjoyed had there been no contract. In addition, he further claims that the reliance interest is entitled to less protection than the restitution interest, by which the defendant is made to surrender the gains obtained under the contract. "It is as a matter of fact no easy thing to explain why the normal rule of contract recovery should be that which measures damages by the value of the promised performance." Fuller's criticism rests on the supposed superiority of one legal norm over another. His position does not challenge the (implicit) assumption that the choice of proper remedial rules is largely a judicial function. Yet as a matter of basic contract theory, there is, in principle, no abstract way to resolve this battle of intuitions. Damage rules are no different from any other terms of a contract. They should be understood solely as default provisions subject to variation by contract. The operative rules should be chosen by the parties for their own purposes, not by the law for its purposes.

Consequential Damages

With consequential damages, the superiority of the expectation measure of damages is far from self-evident. Whatever the abstract law of contract, many ordinary sales contracts contain warranty provisions that specify both the obligation of the supplier and the remedy in the event of breach. The provisions widely adopted across different industries stipulate for liquidated damages or call for the repair or replacement of damaged goods at the option of the seller. While the full expectation measure of damage may require the seller to bear the cost of delivery to the firm, contracts often leave that loss on the buyer, who likewise may receive no compensation for any interim loss of use. Often the warranty is voided where the product was damaged by excessive or impermissible use by the buyer. Most notably, all liability for personal injury or property damages is excluded. Without question, these warranties call for a risk-sharing arrangement, which leaves the buyer worse off when the seller honors the warranty in full than he, the buyer, would have been if the product had worked perfectly in the first place. Warranty restrictions similar to these led the New Jersey Court in *Henningsen v. Bloomfield Motors, Inc.* to invalidate these warranties as limitations on recovery for personal injury and to usher in the modern age of product liability law by imposing full tort liability for consequential damages.

Against this backdrop of express contractual provisions, there is ample reason to doubt that the expectation measure of damage of the classical common law maximizes the joint gains of the parties ex ante. If it did, we should expect to observe it frequently in practice, which is decidedly not the case. The failure to observe this standard in practice cannot easily be attributed to the systematic ignorance of buyers and sellers in all product markets, for someone must have the incentive to break the logjam if making the plaintiff whole on breach is the ideal contract measure of damages. The better approach, therefore, is to ask why it is that informed parties might not choose to use this damage measure. A closer inspection of the expectation measure of damages reveals some costs of its application.

First, under the orthodox view, the basic rule provides for full consequential damages, but (like the tort rules of contributory negligence it parallels) it then allows the defendant an affirmative defense, where the plaintiff is in breach of some condition precedent. The two halves of this rule thus raise two high-stakes issues (plaintiff's and defendant's breach) on which enormous liabilities can turn. As a general matter, the parties' investment in litigation increases with both the uncertainty of the outcome and the size of the stakes. Ex ante, the parties wish to avoid this cost, as it represents a deadweight loss to both sides. In addition, full consequential damages raise the real risk that the plaintiff, while in the better position to avoid the loss, will, in fact not take the right steps to do so. Any money that is spent on further loss reduction is his own, while the money that is saved is the defendant's. The temptation to maximize private gain results in the systematic externalization of losses: why should I spend my money to reduce his damages? The law recognizes this and imposes a duty of mitigation of damages to counter this all too-tenacious human tendency. But that duty is a very imprecise tool to use against so persistent a business practice, for defendant's monitoring of plaintiff conduct, whether under a misuse or a mitigation doctrine, is both costly and error prone. The expectation rules with affirmative defenses may not offer the best prospect of minimizing the total costs of contractual failure.

The weaknesses of the expectation damage system are thrown into high relief by comparing it with an alternative regime that gives fixed damages without making any provision for separate affirmative defenses. This system of damages could be superior for both parties because of the way in which it reduces the joint costs of litigation while preserving the incentives on both sides to perform as agreed. The plaintiff whose level of recovery is fixed in advance has a powerful incentive to mitigate his loss. Any rule of fixed damages, independent of subsequent events, makes him face the identical incentives of a single owner of all relevant inputs. Every dollar he spends in mitigation now results in a dollar's saving for himself from the reduction in consequential damages. He will therefore behave exactly as he would if he had been the sole and original author of his own harm. While one should not expect perfection in the delicate task of mitigation—the innocent party may be a complex firm with agency-cost problems of its own—clearly there is nothing that the legal rule could do to improve performance once it has eliminated this potential conflict of interest posed by any ad hoc mitigation rule. There is no need to build mitigation doctrines explicitly into legal rules in order to create the correct incentives to mitigate. Any fixed lump-sum damage award has just that effect, regardless of the level at which it is set.

The amount of the fixed damages is critical, however, in influencing the frequency of breach by the plaintiff. If damages were exceedingly high, the level of precautions taken would be low, given the fixed rate of return that the plaintiff could expect to receive. Similarly, the risk of fraudulent or dubious claims of seller's breach would increase as well. Where the seller is suing for the buyer's failure to accept goods or to pay for goods accepted, the question of seller's breach is relatively unimportant, so that all incentives can be directed toward the conduct of the single wrongdoer. Matters are far more complicated with lost profits, personal injury, and property damages, for here the possibility that both sides will be in breach is far more likely. Setting damages below actual loss is far more important in this context precisely because there is frequent need to restrain abuses by defendants and plaintiffs simultaneously.

Any effort to constrain misbehavior by the buyer invites misbehavior by the seller. But the critical question is the rate of substitution between buyer and seller incentives for breach. Here there are at least two reasons to think that the control of buyer's misconduct will often be the more important issue, pointing to a lower level of fixed damages. First, reputational constraints tend to operate far more powerfully upon institutional defendants than they do upon individual buyers. Major failures are perceived by the market, resulting in a loss of future sales that can best be avoided by maintaining product quality.

Second, it is important to note that even low damage awards can exert a considerable incentive on a defendant to perform his contracts. Consider the standard repair and replacement warranties set out above. Here defendant must lose from any individual transaction if required, say, to refund the consideration received or to make repairs while still having to bear other costs under the contract. In principle, the defendant will have no incentive to supply defective products so long as his breach costs him more than he gains. Even a low level of nonperformance is sufficient to remove all the profits that the defendant derives from other contracts that are performed successfully. The combination of reputational and financial losses helps keep the defendants in check.

The last point concerns administrative costs. Fixing these damages reduces

the associated uncertainty and, therefore, the costs of administering the remedial provisions. Nonetheless, if there is any serious question whether the
defendant was in breach, the costs of litigation will increase as the size of the
damage award increases, even if damages are fixed. These administrative
costs will become far larger if the level of damages is both large and uncertain
and subject to reduction for plaintiff's misconduct and failure to mitigate. All
in all, the optimal contracting strategy does not appear to call for the high
consequential damages, subject to defense rules, that courts have tended to
adopt. Less clearly, within the class of fixed damage awards, there is reason to
expect these damages to be kept relatively limited, which is what the express
contracts have typically provided. . . .

Common Carriers

[This] argument helps explain the contractual provisions governing the liability of common carriers for lost profits attributable to the delayed shipment of
goods. The expectation measure of damage calls forth the question of extensive litigation over both plaintiff's and defendant's conduct. In contrast, the
use of a fixed tariff regardless of circumstances helps advance the joint interests of the parties. It gives the plaintiff the incentive to mitigate losses, while
imposing on the defendant some financial incentives to perform as promised,
which are doubtless augmented by the fear of reputational loss. The small
level of damages awarded reduces the costs of both litigation and settlement.

   In principle, it is possible that expectation-measure damages could dominate over the fixed tariff, but there is evidence from the structure of real
contracts that suggests the opposite. The modern Federal Express standard
form, for example, calls for a return of the contract price when the package is
not delivered—a restitution form of damages. But it also contains a liquidated
damage provision for loss or damage to goods that limits recovery to $100 or
actual losses, whichever is lower—even when caused by the company's
negligence—with opportunity to purchase additional coverage for stated
rates. The company then completes its list of limitations with boldface disclaimers of all liability for incidental, consequential, or special damages,
"whether or not we knew that damages might be incurred."

   The form contains all the explicit limitations on damages that high-minded
judges and academics often deplore, but it is wholly admirable in both its
conception and execution. The difference in damages between delay and loss of
packages is about eight- or tenfold (if the $100 maximum is awarded), which
bears at least rough correlation to the differential level of losses. Moreover,
while that $100 damage figure for loss or damage may look puny, in fact it
conveys very powerful information to the consumer about the firm's reliability
in making deliveries. If the net profit per standard transaction is, say, 5 percent,
then for a $10 transaction, the gain is about fifty cents. A damage award of $100
thus translates into a situation in which a failure rate of half of 1 percent strips
the firm of all its profits, even if there were (a) no reputational losses to the firm

and (b) no additional costs to trace lost packages or to process complaints and the like. A rate of failure of less than one in a thousand is probably above the break-even point for the firm and far higher than its actual nondelivery rate. With its sophisticated client base and the competitors snapping at its heels, the limitations on damages found in the Federal Express agreement cannot be attributed to some mysterious contract of adhesion; after all, why stop at disclaiming expectation damages if, as dictator, you could just knock out all liability by dictating contract terms? Instead, the terms limiting liability are best understood as a way to maximize joint benefits to the parties ex ante: they minimize the costs of administration and mitigation that otherwise arise with breach while insuring a certain high level of contract performance by the firm.

Contract Damages and Cross-Subsidization

GWYN D. QUILLEN

It has been argued that the purpose of the rule in *Hadley v. Baxendale* is to deny recovery to a plaintiff who could have avoided the loss at a lower cost than could the defendant. The argument is that in general, with respect to extraordinary or *unforeseeable* losses, the buyer is in the best position to avoid the losses, either by taking precautions against them (if the buyer is the cheapest-cost-avoider) or by informing the seller of the unusual risks so that the seller can take the appropriate precautions. Only the buyer knows the particular use to which the product or service will be put, and therefore only the buyer knows what steps should be taken to avoid the loss.

Even if the breaching party is the cheapest-cost-avoider, it may not always be desirable to allow the non-breaching party to recover full expectation damages. An increasingly common transaction . . . involves a form contract that is offered by a mass seller of products or services. The prices and terms are the same for every buyer, and it would be prohibitively expensive to negotiate an individual contract for each transaction. Even though the buyers have varying uses for the product or service, the seller cannot distinguish among their particular uses.

Suppose that sellers are liable for "expectation" damages upon breach, including all consequential damages flowing from the breach whether foreseeable or not. Since the sellers cannot distinguish among buyers with varying uses, they will charge all the buyers the same price. Assuming the market is competitive, the price sellers charge will be the marginal cost of providing the product or service, plus the average cost of damages. The same price will be paid by buyers who will suffer minimal damages if the seller breaches as by purchasers who will suffer large damages in the event of breach.

For example, suppose that the Community Car Wash expects one car to be damaged for every thousand cars that go through the car wash. The managers' fancy cars and the workers' cheap cars go through the wash in equal numbers and are the only cars washed. The average total cost is $200 when a cheap car is damaged and $2,000 when a fancy car is damaged. Thus, the expected loss is $0.20 for each cheap car driven through the car wash and $2.00 for each fancy car. Assuming the marginal cost to Community Car Wash of washing one car is $3.90, the workers should pay $4.10 and the managers should pay $5.90. However, the car wash is unable to charge different prices depending on the value of the car, so it must base its price on the average expected loss of $1.10 and charge $5.00 to all customers. Thus, for each car wash, the workers are cross-subsidizing the managers by $0.90.

This blanket price method raises both fairness and efficiency concerns. In the Community Car Wash example and in general, it is likely that the high risk buyers (by definition those with the most to lose) are the wealthiest, and the low risk buyers the poorest. The poorer would therefore be cross-subsidizing the wealthier.

The inefficiency caused by cross-subsidization stems from the seller's inability to indicate the true cost to each individual buyer, resulting in inefficient consumption and production decisions by the buyer. Under an efficient pricing scheme, each buyer should pay the marginal cost of the product or service plus the expected marginal cost of damages to the particular buyer if the seller breaches. Thus, the high risk buyer will have to pay a higher total price than will the low risk buyer.

Holding the car wash liable causes it to raise its price to cover not only this loss but expected future liabilities to the few customers who would suffer large losses due to the car wash's equipment malfunction. Fancy car owners would tend to prefer using the Community Car Wash over what might otherwise be a lower cost system; this is because fancy car owners would not pay the full expected cost of their damages in addition to the marginal cost of the Car Wash. On the other hand, owners of cheap cars might use a different method of cleaning their cars even though the car wash would otherwise be cheaper, since these owners must pay a price in excess of the full expected cost of their damages in addition to the marginal cost of delivery.

Suppose Hand Car Wash, up the street from Community, cleans cars without machines and therefore only damages one out of a million cars. Hand Car Wash has a total cost of $5.50 for washing a fancy car. Since the total expected cost of having a car washed at Community is $4.10 for cheap cars and $5.90 for fancy cars, efficiency would require the fancy car owners to go to Hand Car Wash which can wash a fancy car for $0.40 less than Community can. However, since the price Community charges ($5.00) is actually $0.90 less than its total cost of washing a fancy car, the fancy car owners will continue to go to Community. An efficiency loss of $0.40 per car wash thus occurs.

The fact that owners of cheap cars must pay more than their expected damages plus the marginal cost of washing a car may also lead to an efficiency loss. Suppose the workers value their time at $4.00 per hour. The workers can wash their own cars with one hour of time and $0.50 worth of material for a total cost of $4.50, which is $0.40 more than the total cost if Community washed their cars. Under an efficient pricing scheme, Community would charge $4.10, and the workers would use Community. However, if Community is liable for damages to fancy cars and cannot charge different prices to reflect differential risks, it must charge $5.00 for a car wash, and the workers would wash their own cars for an efficiency loss of $0.40.

When a seller is liable for expectation damages and cannot individually price the contract even though customers have varying risks, some low risk buyers (here, the owners of cheap cars) will no longer buy the seller's product or service because the value of the product or service (including complete compensation for losses caused by breach) is less than the price they must pay when they are required to cross-subsidize high risk buyers. High risk buyers (fancy car owners) for whom the value is greater than the cross-subsidized price will buy more of the seller's product or service. This "adverse selection" will cause an increase in the expected cost of damages. All buyers will continue to buy if the value of the product or service, including compensation for losses caused by a breach, exceeds the price charged by sellers. Among these buyers, some will be low risk buyers and some will be high risk buyers, so that the cross-subsidization problem remains.

The foregoing discussion illustrates that imposing a rule of full expectation damages upon a breaching seller who cannot make distinctions among buyers of varying risks is inefficient and unfair. The question then is how damages should be limited to avoid the cross-subsidization and adverse selection problems, and what effect limiting damages would have on incentives to avoid losses.

Cross-Subsidization Versus Efficient Precautions

To completely eliminate the cross-subsidization problem in situations where the seller cannot distinguish among buyers according to risk and change the contract price to reflect that risk, the maximum damages that should be awarded are damages that are common to all buyers, such as replacement of the defective product or return of the price paid. For example, there will be no cross-subsidization problem if the maximum amount that a car owner can recover from the Community Car Wash is $200. Since the Car Wash has a 0.10 percent chance of causing damage, the expected loss per transaction is $0.20, and the price of the service, including insurance against losses up to $200, is $4.10. In that case, all of the customers are receiving an equal amount of "insurance" against losses caused by the car wash. The fancy car owner, of course, either will have to bear the cost or insure against the remaining $1,800 of loss. The owners of cheap cars will now prefer to use the Community Car

Wash rather than wash their cars themselves. Fancy car owners, however, will now have a total cost of $5.90, and will therefore prefer Hand Car Wash which charges $5.50. Thus, by limiting damages to those that are common to all buyers, the economically correct result is achieved. . . .

While this severe limitation on damages can eliminate the cross-subsidization problem, it may create undesirable incentives for sellers and thereby violate the cheapest-cost-avoider principle. In the Community Car Wash example, assuming equal numbers of high and low risk buyers, the average loss caused by a malfunction is $1,100. This is the amount of loss which the seller, if the cheapest-cost-avoider, should consider in deciding on the level of precautions. If the seller's liability is limited to $200, however, the seller may not have the incentive to take the same level of precautions that would be justified based on average losses of $1,100. Where the *seller* is the cheapest-cost-avoider, relieving the seller of liability may cause a welfare loss; if the seller does not internalize all of the costs, full precautions will not be taken. Thus, the total cost of losses and precautions against losses will be higher whenever the cheapest-cost-avoider fails to take the appropriate measures to prevent loss. If the *buyer* is the cheapest-cost-avoider, limiting damages is efficient both from the cross-subsidization standpoint and from the cheapest-cost-avoider standpoint.

The relative importance of using liability to place appropriate incentives on sellers to take precautions depends on several factors. When sellers can distinguish among various types of buyers without incurring any transaction costs, and when markets are perfectly competitive, the level of precautions the seller will take is unaffected by who bears the liability. Under these conditions, there will be neither cross-subsidization nor incentive problems and thus no reason to limit liability. As a result, the Community Car Wash will charge fancy car owners $5.90 and cheap car owners $4.10, and can take the proper precautions in each case so as to minimize losses. On the other hand, if sellers cannot distinguish among buyers, making sellers liable would lead to cross-subsidization problems. This suggests that sellers' liability should be limited, unless to do so would cause problems with sellers' incentives to take precautions—a question which depends in part on the level of information available in the market about the likelihood that the seller will breach.

If sellers are not liable for damages caused by breach, and buyers have insufficient information about the likelihood of the sellers' breach, a "market for lemons" can develop, giving sellers incentives to take too few precautions. Sellers will have no incentive to avoid breaching because they are not legally responsible for damages caused by their breach; they will not suffer in the market since buyers do not know which sellers are more likely to breach. The overall cost of sellers' taking too few precautions will be the increase in the sum of the costs of breaches and the costs of precautions taken by the buyers over the sum of those costs had the sellers taken the precautions.

If perfect competition exists and buyers have perfect information about the likelihood that a seller will breach, market forces will cause the seller to take optimal precautions, thus minimizing the total cost of the seller's product or service (including risks of buying from this seller). Under these conditions, arguments for making the seller liable are less compelling, and damages can be limited to avoid cross-subsidization. . . .

[In summary,] where sellers can distinguish among buyers and charge different prices (that is, when sellers have perfect information), making sellers liable will provide appropriate incentives and will not result in cross-subsidization. Where sellers cannot charge different prices but buyers know the likelihood that a particular seller will breach (that is, when buyers have perfect information), limiting sellers' liability will eliminate cross-subsidization without leading to an inefficiently low level of precautions. If both buyers and sellers have perfect information, the liability rule does not matter. Finally, if neither buyers nor sellers have perfect information, there is no clear rule.

To decide whether it will be economically desirable to allow the incentive effects rather than the cross-subsidization effects to govern, it is necessary to compare the efficiency gain resulting from placing liability on the proper party with that resulting from the elimination of cross-subsidization and adverse selection problems. Since the factors will balance differently in every case, no strict rule should be applied.

Information and the Scope of Liability for Breach of Contract

LUCIAN AYRE BEBCHUK AND STEVEN SHAVELL

We attempt in the present article to analyze systematically the effects and the social desirability of [the rule of *Hadley v. Baxendale*], as opposed to the rule of unlimited liability for breach. To this end, we study a stylized model with buyers and sellers. A seller can reduce the likelihood of breach by taking precautions, but these will involve additional expense or effort. There are two types of buyers: a minority who place a high value on performance, and the majority who place a low value on it. Whether a buyer places a high or low valuation on performance is not observable to sellers. Buyers, however, may choose to identify themselves—that is, make representations to the seller about the value they place on performance. (As will be discussed, for a buyer to identify himself as having low valuation will imply that the seller's liability would be limited to such valuation.) The addition of such buyer representations to the contracting process involves transaction costs, which we will call "communication costs."

Lucien Ayre Bebchuk & Steven Shavell, Information and the Scope of Liability for Breach of Contract: The Rule of *Hadley v. Baxendale,* 7 Journal of Law, Economics, & Organization 284, 285–86 (1991). Copyright © 1991 by Oxford University Press. Reprinted by permission.

Accordingly, two types of decision are made in the model: buyers' decisions about communication of their valuations; and sellers' decisions about the level of precautions to reduce the likelihood of nonperformance. We identify the decisions that are socially optimal for buyers and sellers to make, and then compare such decisions to those that the parties in fact make under the limited and unlimited liability rules for breach.

The gist of our conclusions from the model is as follows. First, if it is socially desirable that sellers possess information enabling them to distinguish between buyers' types, then high valuation buyers alone should communicate their valuation to sellers; buyers who do not communicate will then be known by sellers to be of the low valuation type. This way of transferring information minimizes transactions costs. The two other possible ways—for all buyers to communicate their valuations, or for low valuation buyers alone to do so—are wasteful, for they involve greater communication costs. Furthermore, it is socially desirable that sellers obtain information enabling them to distinguish between buyers' types if and only if the resulting social benefits—which inhere in sellers' taking different precautions for low and high valuation buyers—exceed the communication costs incurred.

Second, if transfer of information about buyers' types is socially desirable, then the limited liability rule of *Hadley* will result in socially optimal behavior. Under the rule, high valuation buyers will find it beneficial to identify themselves to secure full protection against breach even though they will have to pay a high contract price. And, informed of a buyer's high valuation, sellers will take proper measures to increase the likelihood of performance.

Third, if transfer of information about buyers' types is socially desirable, then the unlimited liability rule will produce behavior that differs from the socially optimal. Under the unlimited liability rule, high valuation buyers will have no reason to identify themselves (indeed, doing so would be costly, as it would result in sellers' raising the contract price). Thus, the rule may lead to a situation in which sellers are unable to determine buyers' types and consequently do not take added precautions for high valuation buyers. Alternatively, the rule may lead *low* valuation buyers to identify themselves in order to enjoy a reduction in the price. In the latter case, sellers will have the information necessary to distinguish between buyers' types, but the costs of transferring the information will not be minimized.

Fourth, if transfer of information about buyers' types is not desirable—because communication costs are higher than the benefits from differential precautions—then neither the limited liability rule nor the unlimited liability rule will necessarily lead to socially optimal behavior. . . .

Notes on Consequential Damages

1. The rule of *Hadley v. Baxendale* is usually said to release the breacher from liability for "unforeseeable" damages. One difficulty in applying this test

is that "unforeseeable" is a vague term. It seems to have something to do with the ex ante probability of damages being that high, but there is no bright-line test about just how low that probability must be to make the damages legally unforeseeable.

A more fundamental difficulty is that the probability of any given event depends critically on how that event is described. For example, there might have been a 15 percent probability that a lost package would cause damages of $500 or more—but it could equally well be said that there was a 7 percent probability that a lost package would cause damages of $800 or more, a 4 percent probability that a lost package would cause damages in the $800–$850 range, and a microscopic probability that a lost package would cause damages of exactly $837.29. Thus, even if the threshold for legal foreseeability were defined precisely at (say) 5 percent, this still would not tell us whether damages of $837.29 represented a loss that was legally unforeseeable.

This problem is familiar in tort and criminal law because of the analogous role played by "foreseeability" in theories of legal and moral responsibility. See generally H. L. A. Hart and A. M. Honoré, Causation in the Law (Oxford: Oxford University Press, 1959). For a discussion of this issue in a contractual setting, see Alan Schwartz, Products Liability, Corporate Structure, and Bankruptcy: Toxic Substances and the Remote Risk Relationship, 14 Journal of Legal Studies 689, 693 (1985).

2. Perhaps for the reasons given above, the law has not always looked only to the ordinary sense of "foreseeability" (i.e., to the ex ante likelihood of the loss) in applying the rule of *Hadley v. Baxendale*. For an early recognition of this, see L. L. Fuller and William R. Perdue, Jr., The Reliance Interest in Contract Damages, 46 Yale Law Journal 52, 85 (1936):

> [I]t is clear that the test of foreseeability is less a definite test itself than a cover for a developing set of tests. As in the case of all "reasonable man" standards, there is an element of circularity about the test of foreseeability. "For what items of damage should the court hold the defaulting promisor? Those which he should as a reasonable man have foreseen. But what should he have foreseen as a reasonable man? Those items of damages for which the court feels he ought to pay." The test of foreseeability is therefore subject to manipulation by the simple device of defining the characteristics of the hypothetical man who is doing the foreseeing.

For a more recent discussion of this phenomenon, see Melvin Aron Eisenberg, The Principle of *Hadley v. Baxendale,* 80 California Law Review 563 (1992).

3. The "information forcing" argument referred to at the outset of the Quillen excerpt (supra at page 72) has been stated by Richard Posner as follows:

> A commercial photographer purchases a roll of film to take pictures of the Himalayas for a magazine. The cost of development of the film by the manufacturer is included in the purchase price. The photographer incurs heavy ex-

penses (including the hire of an airplane) to complete the assignment. He mails the film to the manufacturer but it is mislaid in the developing room and never found.

Compare the incentive effects of allowing the photographer to recover damages for the full consequences of the breach and limiting him to recovery of the price of the film. The first alternative creates few—maybe no—incentives to avoid similar losses in the future. The photographer will take no precautions; he will be indifferent between the successful completion of his assignment and the receipt of full compensation for its failure. The manufacturer of the film will probably take no additional precautions either, because he cannot identify the films whose loss would be extremely costly, and unless there are many of them it may not pay to take additional precautions on *all* the film he develops. The second alternative, in contrast, should induce the photographer to take precautions that turn out to be at once inexpensive and effective: using two rolls of film or requesting special handling when he sends the roll in to be developed.

The general principle is that if a risk of loss is known to only one party to the contract, the other party is not liable for the loss if it occurs. This principle induces the party with knowledge of the risk either to take appropriate precautions himself or, if he believes that the other party might be the more efficient preventer or spreader (insurer) of the loss, to reveal the risk to that party and pay him to assume it.

Richard A. Posner, Economic Analysis of Law 126–27 (4th ed. 1992).

Is it always worthwhile to communicate information about the size of the loss at stake in any particular transaction? The excerpt by Bebchuk and Shavell discusses one situation where this communication may not be worthwhile: when the costs of communicating the information exceed the benefits that would be obtained (in terms of fine-tuning the defendant's level of precautions). Another possible counterexample was discussed by Richard Danzig:

At least in mass-transaction situations, the modern enterprise manager is not concerned with his corporation's liability as it arises from a particular transaction, but rather with liability when averaged over the full run of transactions of a given type. In the mass-produced situation the run of these transactions will average his consequential-damages pay-out in a way far more predictable than a jury's guesses about the pay-out. In other words, for this type of entrepeneur—a type already emerging at the time of *Hadley v. Baxendale,* and far more prevalent today—there is no need for the law to provide protection from the aberrational customer; his own market and self-insurance capacities are great enough to do the job.

Richard Danzig, *Hadley v. Baxendale:* A Study in the Industrialization of the Law, 4 Journal of Legal Studies 249, 281 (1975). Mass transaction situations, where the seller is unable to adjust his or her price on a customized basis, were also the focus of the Quillen excerpt.

4. If the breaching seller can charge different buyers different prices *and* the breaching seller is a monopolist, the "information forcing" argument encounters further difficulties. In monopolized markets, buyers may resist dis-

closing the amount they have riding on the contract, for fear that the seller
will respond by charging them a higher price once he or she learns just how
high a value the buyer attaches to performance. For analyses of how the
Hadley rule might affect markets with a monopoly defendant, see Ian Ayres
and Robert Gertner, Strategic Contractual Inefficiency and the Optimal
Choice of Legal Rules, 101 Yale Law Journal 729 (1992); Jason Scott Johnston, Strategic Bargaining and the Economic Theory of Contract Default
Rules, 100 Yale Law Journal 615 (1990); and Louis E. Wolcher, Price Discrimination and Inefficient Risk Allocation Under the Rule of *Hadley v. Baxendale,* 9 Research in Law & Economics 9 (1989).

5. The Epstein excerpt noted that limiting the breacher's liability could
have the effect of increasing the nonbreacher's incentives to mitigate his or
her potential damages. For example, a limited liability rule might give mill-
owners an incentive to keep a spare shaft on hand if a single slipup by a
transportation company would cause them to lose huge profits. This point is
analytically the same as the suggestion in the earlier Cooter excerpt (supra at
page 63) that limits on a breacher's liability could increase the buyer's incentives to avoid relying too heavily on the prospect of performance, thus check-
ing what might otherwise be an incentive toward overreliance.

6. The excerpts in this section analyzed the breaching party's liability for
consequential damages from the standpoint of economic efficiency. However,
some of these arguments have obvious noneconomic parallels. For example, if
full liability for a car wash results in owners of inexpensive cars paying higher
rates to subsidize owners of expensive cars, this seems undesirable from the
standpoint of virtually any moral theory, not merely from the standpoint of
efficiency.

It is hard to find analyses of consequential damages from a uniquely
noneconomic perspective. Many judicial opinions do reflect the intuition
that it is in some way unfair to hold defendants responsible for losses they
were not warned of in advance. Can a theory of fairness be developed that
explains *why* this is unfair—and why it is any fairer to leave that loss to be
borne by the plaintiff, for whom the loss may have been equally unforeseeable? (This issue, too, will arise again in Chapter 3's consideration of impracticability and mistake. See, in particular, the excerpt by Charles Fried, infra
at pages 145–49).

In *Globe Refining Co. v. Landa Cotton Oil Co.,* 190 U.S. 540 (1903),
Justice Holmes suggested a standard of hypothetical consent, based on "what
the parties probably would have said if they had spoken about the matter." Is
this approach significantly different from the economic theory of holding
contracting parties who have not specified otherwise to whatever rule would
be most efficient? (Cf. the notes on selecting default rules, supra at page 29,
discussing the relationship between efficiency and hypothetical consent).

7. In markets where full liability for sellers would be undesirable, either
on economic or noneconomic grounds, that result can be avoided in several
ways. First, the law could reduce the extent of the seller's liability—say, under

the rule of *Hadley v. Baxendale*—as discussed in the excerpts in this section. Second, the law could limit the seller's liability under some other limit on damages—say, by ruling that the buyer had not proved his or her damages with reasonable certainty, or by finding that the buyer had not properly mitigated. Third, and perhaps most significantly, the law could refuse to hold the seller liable at all—for example, by holding that the seller had not warranted the product's performance, or by excusing the seller on the ground of mistake or impracticability. Many of these same issues will resurface in this connection in Chapter 3 in the readings dealing with mistake, impracticability, and implied warranty.

§ 2.2.3 **Subjective Losses**

Cost of Completion or Diminution in Market Value
TIMOTHY J. MURIS

Any decision to ignore subjective value cannot rest on the ground that such damages are either unreal or frivolous—this argument is patently false. Instead, the decision must rest on more prudential considerations, namely, that the costs of determining subjective value exceed any allocative benefits that the determination might yield. . . .

Subjective value necessarily exists if the original price of the performance exceeds the value in the market of that performance to other potential purchasers (hereinafter, the fair market value). The Restatement of the Law of Contracts contains a well-known illustration:

> P [the performing party] contracts to construct a monumental fountain in N's [the nonbreaching party] yard for $5000, but abandons the work after the foundation has been laid and $2800 has been paid by N. The contemplated fountain is so ugly that it would decrease the number of possible buyers of the place. The cost of completing the fountain would be $4000. N can get judgment for $1800, the cost of completion less the part of price unpaid.

In this example, the original purchase price ($5,000) exceeds the fair market value (less than $0). The purchase price is a market price in that supply and demand determine its amount. Fair market value is also market deter-

Timothy J. Muris, Cost of Completion or Diminution in Market Value: The Relevance of Subjective Value, 12 Journal of Legal Studies 379, 382, 384–92 (1983). Copyright © 1983 by The University of Chicago. Reprinted by permission.

mined, reflecting the gain or loss to buyers other than N from the expenditure of the $5,000. . . .

Even if the purchase price does not exceed the fair market value, the buyer may still value the product more than others. For example, the completed fountain in the Restatement illustration may increase the market value of the house by $5,000 yet increase N's value of the house by even more. Even if the performing party (P) knew that N would be willing to pay even more, the threat of N's buying from P's competitors constrains P from raising the price above the cost of production, including a reasonable accounting profit.

Two other points about the relation of the original purchase price and the fair market value are relevant. First, if courts protect subjective value and if the performing party (the builder in the construction case and the lessee in the mining case) knows of N's subjective value, the purchase price to N will rise. If there is some possibility that P will breach, the expected costs to P of the contract increase because court protection of subjective value increases N's recoverable damages. Second, it will often be difficult in practice to determine whether the purchase price exceeded fair market value. One problem is that a court would need to assess these amounts at some time after the contract was signed, when changed conditions may make evidence of current prices and costs unreliable guides for determining past values. More important, the performance will often be part of a package of tasks for which N paid one appropriate price and for which P does not normally make separate cost estimates. In such a case, determining the appropriate value for the original purchase price may be difficult. . . .

Costs and Benefits of the Two Awards

Case 1: Original purchase price exceeds fair market value, and cost of completion equals original price minus cost of work already performed. The Restatement illustration discussed above involving the monumental fountain provides an example. N's payment in the original contract indicates that subjective value exists. Because the cost award equals the cost of the balance of the work as originally planned, unless N's valuation of performance dropped after the contract was formed, cost is the proper award, one clearly superior to diminution. Subjective value is protected because the cost figure is the cost of specific performance in the market. Awarding only diminution verges on making promises not legally binding, although—except when diminution equals $0— the effect is probably not as drastic. To award diminution, thereby undercompensating N, would cause at least some parties in N's position to rely more on devices such as stipulated damages, credit bureaus, bonding, experience gained from past dealings, and other devices that reduce uncertainty. As has been detailed elsewhere, these devices appear to increase the cost of contracts more than an adequate judicial award of damages by (at the least) increasing the number of agreements necessary for a party such as N to achieve a given objective.

Case 2: Original purchase price equals fair market value. Consider the following facts, based upon an actual case in which I was involved: N contracts with P to purchase one of P's townhouses for use as N's family dwelling. P builds from a standard floor plan but agrees to N's request for changes when N protests that a few minor features of the plan are "unappealing." Other than the minor inconvenience of changing the plan and any additional risk that these changes may cause, no additional cost to P of making the changes results because the house has yet to be built. Assume that these costs, in effect the purchase price of the changes, are negligible. Because the changes in the floor plans do not affect the market value, the purchase price and fair market value are identical. Partway through construction, however, N discovers that P is not making the agreed-upon changes, perhaps because his employees misread the plans or because the employees deliberately diverted resources elsewhere. To make the changes at this stage, P would have to undo some of the work already done at an expense of $3,000. If the house is completed per the original specifications, the market value will not drop. Thus, at this stage in the contract, cost ($3,000) exceeds diminution ($0).

The question arises whether cost or diminution best protects N's subjective value, if any. The answer is complex, depending upon how the parties will act under the alternative awards. The cost award protects subjective value, because cost is equivalent to awarding specific performance. Awarding cost, however, may cause expenses that make protection of subjective value not worth the effort. To understand this point, one must first realize that, unlike case 1, here the cost of completion does not necessarily measure subjective value. Indeed, in case 2, cost could greatly exceed the amount necessary to protect subjective value even if the diminution award is inadequate, as cost may vary with factors unrelated to subjective value, such as how much work must be redone.

Similarly, if N has no subjective value or if cost would overcompensate N's subjective value, the cost award will cause costs of its own. Assume for the sake of illustration that $500 would make N indifferent between receiving the revised performance or the money. N and P then have an incentive to negotiate a settlement of N's claim against P for some amount between $3,000 and $500. N has no incentive to reveal the true amount for which he will settle, and, to force negotiations, P may have to threaten to complete performance. However the process occurs, settlement negotiations could be quite costly, particularly given that no legal principles determine the expected judicial damages award. Indeed, it is now an elementary principle in economics that such forced bargaining can lead to significant costs.

Another cost of an overcompensatory award is that P will take more care to perform according to the specifications. Because the cost award will increase liability (relative to the diminution award) by the negotiation costs plus the amount the settlement exceeds diminution, P will increase expenditures to prevent breach up to the expected savings, which equal these extra costs discounted by the probability of breach. For example, P may exercise more supervision over his employees or he may negotiate a stipulated damage clause to limit his damages upon breach. Whether these expenditures are significant will vary with the factual circumstances. The full amount of these expenditures by P are not merely added to the forced bargaining costs, however, because their existence reduces the need for forced bargains. If P negotiates a limit on damages, the risk of being subjected to forced bargaining would be eliminated.

The cost-of-completion award thus has costs as well as benefits. The relevant issue for protecting subjective value, therefore, involves comparing these costs and benefits with those that would occur with the diminution award. Awarding the diminution in market value avoids forced bargaining, but it sometimes underprotects subjective value. With diminution, innocent parties who value specific performance more than the market have an incentive to protect themselves. For example, they can increase expenditures on investigating those with whom they contract, or they can negotiate stipulated damage clauses for the amount of subjective value. If stipulated damages are used, N has the incentive to set the damages at the proper level, for he has no desire to pay for protection that he does not in fact want. Yet if diminution is the judicial standard, there is some risk that the clause itself—if far above the diminution level—will be challenged (at least under current doctrine) as an illegal penalty clause, thus inviting a costly and uncertain legal proceeding.

In contrast, cost as a presumptive benchmark has one clear advantage given the current judicial attitudes to penalty clauses. Even with cost as the presumptive judicial measure, the parties should wish to calculate subjective value properly. With cost, then, any negotiated clause will be normally for a lesser amount than the law itself allows. Legal challenges to the validity of the clause will therefore be more difficult to mount, given that these limitations upon damages are routinely enforced, while estimated damage clauses are subject to far greater scrutiny. A second advantage of cost over diminution is that the cost of completion may often be easier to calculate than the drop in market value. Although resources spent on determining diminution will be limited by the spread between the reasonable high and the reasonable low estimate for that measure, parties can simply venture into the market to calculate cost. To calculate diminution, they must use costly expert testimony.

At least one other argument favors the cost award. If the law awards diminution, individuals likely to have subjective value might systematically underprotect themselves because they did not understand the law enough to negotiate a stipulated damages clause. They may seek legal advice once breach occurs, but by then it is too late. If the law awards diminution and there are too few damages clauses, parties in the place of P will systematically breach too much, because they do not face the true cost of their actions in cases when N's subjective value exceeds market value. If the law awarded cost, it might be objected that P could simply stipulate damages at diminution, causing the same problem. In this case, however, the existence of the damages clause will reduce the asymmetry by informing consumers of the consequences of breach. Thus informed, consumers will have an incentive to shop for contracts more protective of their subjective value. The extent, indeed even the existence, of such asymmetry of information is unknown. Nevertheless, consumers are more likely to have subjective value than are businesses, and consumers are more likely to know less law.

In those cases, therefore, in which an explicit negotiation of a damage award is not undertaken there are several clear advantages to using cost as the uniform damage measure. First, the uniform rule avoids uncertainty as to the legal status of individual cases. Second, the rule tends to strengthen the likelihood that stipulated damage clauses will be respected, and, third, it tends to encourage the efficient transmission of information from producers to consumers in those cases in which subjective value is important, doubtless an appreciable number.

Nonetheless, the matter is further clouded because there are other arguments that point against use of cost as the uniform award. In those transactions in which there is no element of subjective value, the cost award tends to overcompensate individual purchasers. These payments have only distributional consequences. They may, however, generate substantial costs in settlement, especially if the difference between market value (now equal to subjective value) and cost of completion is great. Diminution therefore avoids these joint costs of exchange, making both parties better off at the time of contract formation.

The same point often arises when the difference between subjective value and diminution is small, while that between subjective value and the cost of completion is great. This situation underlies the so-called economic waste exception to awarding cost, at least as properly interpreted. Many courts and some commentators argue that, in construction cases, the nonbreaching party should receive the cost of completion unless that award will result in excessive "economic waste." The concern is that awarding cost would mean undoing, and therefore wasting, work already completed. As has been noted elsewhere, in this form the argument is erroneous. The analysis of case 2 reveals that undoing of work would occur only if it is beneficial. If N values exact performance at less than cost, he will offer P a settlement between cost and his own subjective value if P insists upon performing to specification. Nor will N spend the money received on completing performance under this assumption. Only if N values exact performance at or above the cost of completion will the work be redone. Under this assumption, rework is not wasteful in the sense of producing value not worth the expense to the buyer.

Nonetheless waste can occur, at least if that term is not confined to the meaning given it in the decided cases. As before, expensive negotiations are the culprit, as the parties try to settle the case at some figure between cost and subjective value. The greater the gap between these two numbers, the greater the expenses incurred upon breach, and so too the economic waste. At least two factors are relevant in determining that the cost award is so excessive as to indicate that settlement costs from awarding cost is likely to exceed any benefits from protecting subjective value. First is the reason that the cost of completion exceeds the drop in market value. If market value has dropped and the cost has not changed over time (meaning that the original purchase price still reflects the cost of the entire job), cost is probably not an excessive award, because the original price reflects N's subjective value. The second factor is the amount of undoing and redoing work that comprises the cost figure. Because these factors are not correlated with N's subjective value, the higher the percentage they comprise, the more likely it will be that the cost of completion greatly exceeds N's subjective value, all else equal.

Proposals for Products Liability Reform

ALAN SCHWARTZ

A firm that compensates consumers for the harms its product causes will reflect the expected compensation cost in the purchase price. An element of the price thus is an insurance premium, whose size ideally varies with the amount of "coverage" against loss that consumers demand. The provisions of the optimal contract respecting defect risks therefore will reflect the amount of insurance against these risks that consumers prefer. It is customary to identify this amount on the basis of three principal assumptions. First, a consumer will choose an insurance contract that maximizes his expected utility. Second, consumers' utility functions are "state dependent": They depend on the state of affairs arising after purchase of the product. The consumer's utility is lower in the state of the world in which the product is defective than in the state of the world in which it works perfectly. Third, firms offer insurance at actuarially fair prices; the amount of their premium equals the expected value of the risk against which the person insures.

Given these assumptions, a consumer's goal is to equalize the marginal utility of money to him in both states of the world he may face. The marginal utility of money is the rate at which the consumer's satisfaction from wealth changes with changes in the amount of wealth he holds. The amount of satisfaction a dollar yields is a function of the importance of the needs it satisfies: A dollar that helps buy a meal for a poor and hungry person yields greater satisfaction than a dollar that would help the same person buy a yacht if he were rich. Since the satisfaction dollars bring changes with the significance of the needs dollars meet, the marginal utility of money varies with income.

To see why consumers who maximize expected utility will attempt to equalize the marginal utility of money in all possible states of the world, let the expected marginal utility of money be higher in possible state of the world A

Alan Schwartz, Proposals for Products Liability Reform: A Theoretical Synthesis. Copyright © 1988 by The Yale Law Journal Company. Reprinted by permission of The Yale Law Journal Company and Fred B. Rothman & Company from The Yale Law Journal, volume 97, pages 362–67.

than in possible state of the world B because the consumer has a lower income or greater demands on his income in the former state. Then the consumer would want to shift marginal dollars from state B to state A, in which the addition of new dollars will yield greater satisfaction than the satisfaction lost in state B. Utility is maximized when no further shifts of wealth between possible social states would increase utility; this outcome is reached when marginal utilities are equal in all possible states.

Consumers equalize expected marginal utilities by purchasing insurance. For example, an accident may increase a person's marginal utility of money by creating a need for medical care. A seriously injured person is likely to use marginal dollars to satisfy more important needs, such as for medical services, than he would have satisfied with such dollars if no accident had occurred. If an accident would increase the marginal utility of money in this way, the competent, informed consumer would insure against it by buying a contract requiring a firm to provide him with extra dollars if he is hurt. Insurance thus shifts wealth from the state of the world in which the marginal utility for money is relatively low—the state in which no injury occurs—to the state in which it is relatively high—the state in which an injury happens. This analysis predicts, therefore, that consumers will insure against those risks whose materialization would increase their marginal utility for money. Identifying these risks can be a difficult empirical inquiry, but some aspects of the question seem obvious.

Certain forms of loss have two significant properties: They increase the marginal utility of money, and they are replaceable, dollar for dollar, by insurance. An accident that causes a consumer to lose wages creates such a loss. The consumer's marginal utility for wealth is higher in the state in which such an accident occurs than in the state in which it does not because the consumer has less wealth in the former state than in the latter and so will use marginal dollars to satisfy more urgent needs, such as for shelter or medicine. Further, since the accident causes only a monetary loss, it is fully replaceable by insurance. In consequence, the consumer could equalize his marginal utility for wealth across states of the world by purchasing full insurance against losses—for example, of wages—that both increase the consumer's marginal utility of money and can be completely erased by monetary payments. Losses with these two properties constitute what lawyers call "pecuniary" loss or harm; insurance theory thus predicts the existence of substantial private insurance against pecuniary loss. The wide use of major medical and disability insurance is consistent with this prediction. Therefore, the default rule should require firms to compensate consumers fully for pecuniary loss.

A more difficult question is whether, and to what extent, consumers would insure against other forms of loss. It sometimes is difficult to know whether accidents will increase or reduce a person's marginal utility for money. Consider a business executive who runs recreationally and who loses a foot in an accident. Suppose that she insured fully against her "replaceable" losses, such as medical expenses and temporary lost wages. Apart from these losses, the injury could increase the marginal utility of money for this consumer if it caused her to substitute travel or the symphony for running because these activities are more expensive. Her marginal utility could fall, however, if she substitutes reading for running. In the latter case, the consumer not only would want no insurance, but would prefer to shift dollars from the injury to the noninjury state—the reverse of the cases above—by betting against the accident happening. It is therefore difficult to say, as a general proposition, that people will insure against events that would only induce them to substitute other activities for those activities that accidents preclude.

This discussion implies that consumers will not insure against harms that reduce, or do not affect, the marginal utility of money. This implication has considerable normative significance because some theorists claim that pain and suffering and emotional distress exemplify such harms. Consider an accident that causes no financial loss but is very painful for two weeks, or an accident that kills a person's relative. Theory holds that a consumer will insure against such losses only if they would increase the consumer's marginal utility for money. This could occur, for example, if persons who expect to be in pain would want to put themselves in the position where they could eat caviar or wear mink as compensations. But whether people actually want the ability to console themselves in these ways, and so will buy insurance to permit such consolations, or whether people would choose just to suffer is an empirical issue. In the wrongful death context, the appropriate resolution of this issue is obvious. The motive for insuring is to be able to purchase substitutes in the state of the world in which one is injured, and this motive vanishes when it would be impossible to make such purchases. A person obviously could not buy substitutes were her suffering to culminate in death. Hence, the current practice of awarding pain and suffering damages in survival actions, as recompense for the suffering the victim experienced before death, is incorrect if the legal award is meant to provide the insurance coverage people would have been willing to buy before the accident. When the injured person survives, the pain and suffering issue is more difficult to resolve.

Commentators sometimes claim that people prefer not to insure against "mental" losses because they almost never do insure against them. This evidence is inconclusive, however, because supply-side difficulties may prevent firms from offering "mental loss" coverage. In particular, adverse selection problems could be very significant in the sale of insurance against mental harms. The intensity of pain or distress at another's harm varies considerably across persons. Insurance companies cannot distinguish among potential insureds by the insureds' capacities to suffer. Hence, the companies cannot charge higher premiums to persons who are more likely to experience mental pain and thus make claims. The penchant of consumers to "select adversely" against insurance companies—to need and buy more insurance if they are likely to suffer more—can prevent the creation of insurance markets. Insurance companies respond to their inability to distinguish among insureds on claim-related factors by charging such high rates to everyone that low-risk persons often refuse the coverage. It is difficult to sell insurance profitably when the pool of buyers is exclusively constituted of those who are high risks.

Therefore, that consumers now do not buy pain and suffering insurance does not imply that they voluntarily eschew it.

Some evidence suggests that people may want coverage against pure suffering. Consumers, for example, sometimes purchase accidental death and dismemberment insurance that protects against particular dramatic events whose occurrence is easy to verify, such as the loss of a leg. A partial motive for this insurance probably is to receive dollars that in some sense will ease the mental pain of these traumatic losses. On the other hand, the premium volume for this insurance is so small that the insurance cannot reflect a large pain and suffering component. And people do not routinely insure against the loss of children, even though such losses cause great emotional pain. It thus is difficult to infer whether people want insurance against pain and suffering losses from observing what they actually buy.

In addition to these empirical uncertainties, other factors also argue against including insurance for pain and suffering in a compensation-based rule. Consumers would prefer less than full insurance against accidents that cause only mental pain, even when these accidents would increase the marginal utility of money, because of "income effects." Recall that such accidents increase the marginal utility of money only when victims will purchase expensive substitute activities to assuage the utility losses from suffering. Such substitutes are sought in the states of the world in which accidents happen. The demand for most goods and services has positive income elasticity; people increase their consumption as their incomes rise. Because accidents make people poorer in a utility sense, people will purchase lesser amounts of substitute activities in "accident states" than they would have purchased if they had not been injured but instead had to give up goods that they then valued as much as they valued not suffering. Informed consumers will anticipate wanting lesser amounts of substitute activities in accident states than they would otherwise want, and so will make provision to buy less. In other words, consumers will not purchase full insurance. Therefore, the ideal legal rule regulating accidents causing mental losses that increase people's marginal utility for money would award victims partial damages. These damages would reflect the partial insurance consumers would want ex ante. The level of partial insurance consumers would want varies among people, however, so the law's manageable choices are full insurance—overcompensation—or no insurance—undercompensation. The issue is on which side to err. The remaining factors that should influence courts in deciding what consumers want imply erring on the side of undercompensation.

Intuition suggests that people would want to buy slight or no coverage against purely mental harms. As the runner illustration above showed, there is no good reason to suppose that, apart from causing pecuniary harm, accidents commonly increase persons' marginal utility for money. In addition, to buy mental loss coverage is, in effect, to sacrifice considerable present wealth in the form of insurance premiums to consume expensive vacations that will assuage whatever emotional distress accidents may cause. In the absence of evidence that spending money is a typical, or even common, response to grief and suffering, this motive for insurance seems unlikely. Finally, pain and suffering losses are difficult for firms to anticipate and verify. The likely response of firms to these problems is to charge high prices for the coverage. These prices make pain and suffering insurance a bad buy for most people, whether it is sold by manufacturers or by insurance companies. These three factors together imply that the more purely mental the loss, the less likely a consumer will want to insure against it.

To summarize, the optimal contract concerning product-related risks would pay firms to provide insurance against the core pecuniary losses that defective products could cause, such as lost wages, medical expenses, or property damage. It is unlikely, though not certain, that this contract would require any insurance against what now are sizable and common elements of the standard products liability judgment: pain and suffering and emotional distress. The appropriate default rule, then, probably should allocate the risk of incurring pecuniary harm to firms, and the risk of incurring nonpecuniary harm to consumers.

Notes on Subjective Losses

1. The Muris excerpt begins with an assertion that there is no reason in principle for contract law not to respect parties' subjective losses. Judges (and laypeople) sometimes discuss subjective losses as though they were less tangible or less important than other kinds of losses, such as a decline in the market value of a piece of property. However, the market value of a piece of property is merely a reflection of the subjective value that *other* people—specifically, potential purchasers of the property—assign to the property in question. As a consequence, most economists regard the property owner's subjective values as just as real, and just as worthy of protection, as the property's market value. As Richard Posner has put it:

> If I refuse to sell for less than $250,000 a house that no one else would pay more than $100,000 for, it does not follow that I am irrational, even if no "objective" factors such as moving expenses justify my insisting on such a premium. It follows only that I value the house more than other people. This extra value has the same status in economic analysis as any other value.

Richard A. Posner, Economic Analysis of Law 56 (Boston: Little, Brown & Co., 4th ed. 1992). See also Donald Harris, Anthony Ogus, and Jennifer Phillips, Contract Remedies and the Consumer Surplus, 95 Law Quarterly Review 581 (1979).

2. Is the Schwartz excerpt correct when it suggests (albeit tentatively) that most people probably would not be willing to pay for insurance against nonpecuniary losses? Think of some thing, or some person, whose loss would grieve you greatly. Would you be willing to pay, say, $100 for the right to receive a large sum of money if you ever lost that person or thing—knowing,

of course, that the large sum of money you would receive could not bring that person or thing back again?

The conditions under which utility-maximizing individuals would be willing to pay for insurance against nonpecuniary losses have been analyzed in mathematical terms in a number of economic articles. See Philip J. Cook and Daniel A. Graham, The Demand for Insurance and Protection: The Case of Irreplaceable Commodities, 91 Quarterly Journal of Economics 143 (1977); Daniel A. Graham and Ellen R. Peirce, Contingent Damages for Products Liability, 13 Journal of Legal Studies 441 (1984); Samuel A. Rea, Jr., Nonpecuniary Loss and Breach of Contract, 11 Journal of Legal Studies 35 (1982).

3. Economists have also pointed out that, in cases where most people would not be willing to pay for insurance against nonpecuniary losses, the legal rule that provides the optimal level of insurance may not be the same as the rule that would induce the optimal level of precautions by the potential breacher. The problem is that the efficient level of precautions depends on the size of the total losses (including nonpecuniary losses) induced by a breach. Thus, in any industry where market incentives alone are insufficient to induce an efficient level of precautions, efficient precautions can usually be induced only by holding the breaching party fully responsible for *all* resulting losses, as discussed in the earlier excerpts by Cooter (supra at pages 59–64) and Craswell (supra at pages 46–53). But this full compensation remedy may provide more than the optimal level of insurance, for the reasons Schwartz discusses in this excerpt.

In theory, this tension could be resolved by holding breaching parties fully liable for all resulting losses (including nonpecuniary ones), with that portion of the judgment representing the nonpecuniary losses being paid to the state rather than to the injured plaintiff. Potential breachers would still charge a price reflecting their total anticipated liability (to both the victim and to the state) with the result that, if nothing more were done, customers would find themselves paying the costs of unwanted insurance without even being able to collect the reward. However, the state could then put its share of the damage awards into subsidies to bring down the price of the product or service in question. This would free buyers from having to pay for unwanted insurance, while still making sellers pay the full costs caused by any breach, thereby giving sellers an incentive to take the optimal level of precautions. For theoretical analyses of such a system, see Steven Shavell, Economic Analysis of Accident Law 233–35 (Cambridge, Massachusetts: Harvard University Press 1987); Michael Spence, Consumer Misperceptions, Product Failure, and Product Liability, 64 Review of Economics Studies 561 (1977); see also the Schwartz article, 97 Yale Law Journal at pages 408–11. The obvious practical difficulties involved in a system of fines and subsidies make it difficult to imagine how such a system could ever be implemented.

4. In one of Muris's examples, the builder fails to make the agreed-upon changes "perhaps because his employees misread the plans or because the employees deliberately diverted resources elsewhere" (supra at page 83).

Should it matter whether the reason for the breach involves a deliberate choice or a mere accident? For an argument that the deliberateness of the breach should indeed matter, see Patricia H. Marschall, Willfulness: A Crucial Factor in Choosing Remedies for Breach of Contract, 24 Arizona Law Review 733 (1982); see also the discussion of punitive damages later in this chapter by Bruce Chapman and Michael Trebilcock (infra at pages 127–35). Keep in mind, though, that it is not an easy matter to decide which breaches ought to be counted as deliberate. For example, how would you classify a breach resulting from a builder's deliberate decision to take very few precautions to make sure that the employees read the plans properly? What about a breach resulting from a builder's deliberate decision to take exactly the cost-effective level of precautions—no more, and no less—against such an accident?

5. Other arguments made by Muris parallel those considered earlier in this chapter in the readings dealing with the expectation remedy and efficient breach. For example, at page 83 Muris argues that the parties can always avoid inefficient construction or inefficient breaches by agreeing to a settlement, but that the costs of negotiating to that settlement must be taken into account. (Cf. Cooter and Ulen's analysis of renegotiation, supra at pages 42–44.) Muris also notes that the threat of having to pay more than the buyer's true losses as part of such a settlement might give the builder an incentive to take too many precautions to avoid committing a breach. (Cf. the discussions of precautions in the excerpts by Cooter, supra at pages 59–64, and Craswell, supra at pages 46–53).

6. The Muris excerpt also suggests a possible case for a "penalty default rule" of the sort discussed by Ayres and Gertner in Chapter 1 (supra at pages 25–27). At page 84, Muris notes that if diminution in market value were made the default rule, sophisticated commercial promisors would have no incentive to negotiate around the rule, as they benefit from the lower measure of damages. While promisees might benefit from stipulating some other measure of damages in their contract, unsophisticated promisees might not know enough law to realize that any such stipulation was needed. This might justify adopting the cost of completion measure of damages as the default rule, *even if we thought that most parties would prefer to end up with some other measure of damages,* in order to induce such a stipulation by the only party with enough legal knowledge to do so.

7. Depending on the facts of the case, subjective losses could also raise most of the other issues discussed earlier in this chapter. For example, some nonpecuniary losses may be more easily prevented by the nonbreacher, bringing into play the analysis of Goetz and Scott (supra at pages 56–58), Cooter (supra at pages 59–64), and Epstein (supra at pages 67–72). If potential breachers are unable to adjust their prices on an individual basis, slightly damaged buyers may be forced to subsidize heavily damaged buyers if the breaching party is held fully liable for the entire range of nonpecuniary losses, as discussed by Quillen (supra at pages 72–76).

2.3 Alternatives to Expectation Damages

§ 2.3.1 Specific Performance

The Case for Specific Performance

ALAN SCHWARTZ

[T]here are three reasons why [specific performance] should be routinely available. The first reason is that in many cases damages actually are undercompensatory. Although promisees are entitled to incidental damages, such damages are difficult to monetize. They consist primarily of the costs of finding and making a second deal, which generally involve the expenditure of time rather than cash; attaching a dollar value to such opportunity costs is quite difficult. Breach can also cause frustration and anger, especially in a consumer context, but these costs also are not recoverable. . . .

Second, promisees have economic incentives to sue for damages when damages are likely to be fully compensatory. A breaching promisor is reluctant to perform and may be hostile. This makes specific performance an unattractive remedy in cases in which the promisor's performance is complex, because the promisor is more likely to render a defective performance when that performance is coerced, and the defectiveness of complex performances is sometimes difficult to establish in court. Further, when the promisor's performance must be rendered over time, as in construction or requirements contracts, it is costly for the promisee to monitor a reluctant promisor's conduct. If the damage remedy is compensatory, the promisee would prefer it to incurring these monitoring costs. Finally, given the time necessary to resolve lawsuits, promisees would commonly prefer to make substitute transactions promptly and sue later for damages rather than hold their affairs in suspension while awaiting equitable relief. The very fact that a promisee requests specific performance thus implies that damages are an inadequate remedy.

The third reason why courts should permit promisees to elect routinely the remedy of specific performance is that promisees possess better information than courts as to both the adequacy of damages and the difficulties of coercing performance. Promisees know better than courts whether the damages a court is likely to award would be adequate because promisees are more familiar with the costs that breach imposes on them. In addition, promisees generally know more about their promisors than do courts; thus they are in a better position to predict whether specific performance decrees would induce their promisors to render satisfactory performances.

In sum, restrictions on the availability of specific performance cannot be justified on the basis that damage awards are usually compensatory. On the contrary, the compensation goal implies that specific performance should be routinely available. This is because damage awards actually are undercompensatory in more cases than is commonly supposed; the fact of a specific performance request is itself good evidence that damages would be inadequate; and courts should delegate to promisees the decision of which remedy best satisfies the compensation goal. . . .

Post-Breach Negotiations

. . . [One] efficiency argument for restricting the availability of specific performance is that making specific performance freely available would generate higher post-breach negotiation costs than the damage remedy now generates. For example, suppose that a buyer (B1) contracts with a seller (S) to buy a widget for $100. Prior to delivery, demand unexpectedly increases. The widget market is temporarily in disequilibrium as buyers make offers at different prices. While the market is in disequilibrium, a second buyer (B2) makes a contract with S to purchase the same widget for $130. Subsequently, the new equilibrium price for widgets is $115. If specific performance is available in this case, B1 is likely to demand it, in order to compel S to pay him some of the profit that S will make from breaching. B1 could, for example, insist on specific performance unless S pays him $20 ($15 in substitution damages plus a $5 premium). If S agrees, B1 can cover at $115, and be better off by $5 than he would have been under the damage remedy, which would have given him only the difference between the cover price and the contract price ($15). Whenever S's better offer is higher than the new market price, the seller has an incentive to breach, and the first buyer has an incentive to threaten specific performance in order to capture some of the seller's gains from breach.

The post-breach negotiations between S and B1 represent a "deadweight" efficiency loss; the negotiations serve only to redistribute wealth between S and B1, without generating additional social wealth. If society is indifferent as to whether sellers or buyers as a group profit from an increase in demand, the law should seek to eliminate this efficiency loss. Limiting buyers to the damage remedy apparently does so by foreclosing post-breach negotiations.

This analysis is incomplete, however. Negotiation costs are also generated when B1 attempts to collect damages. If the negotiations by which first buyers (B1 here) capture a portion of their sellers' profits from breach are less costly than the negotiations (or lawsuits) by which first buyers recover the market contract differential, then specific performance would generate lower post-breach negotiation costs than damages. This seems unlikely, however. The difference between the contract and market prices is often easily determined, and breaching sellers have an incentive to pay it promptly so as not to have their extra profit consumed by lawyers' fees. By contrast, if buyers can threaten specific performance and thereby seek to capture some of the sellers' profits from breach, sellers will bargain hard to keep as much of the profits as they can. Therefore, the damage remedy would probably result in quick payments by breaching sellers while the specific performance remedy would probably give rise to difficult negotiations. Thus the post-breach negotiation costs associated with the specific performance remedy would seem to be greater than those associated with the damage remedy.

This analysis makes the crucial assumption, however, that the first buyer, B1 has access to the market at a significantly lower cost than the seller; though both pay the same market price for the substitute, B1 is assumed to have much lower cover costs. If this assumption is false, specific performance would not give rise to post-breach negotiations. Consider the illustration again. Suppose that B1 can obtain specific performance, but that S can cover as conveniently as B1. If B1 insists on a conveyance, S would buy another widget in the market for $115 and deliver on his contracts with both B1 and B2. A total of three transactions would result: S-B1; S-B2; S2-S (S's purchase of a second widget). None of these transactions involves post-breach negotiations. Thus if sellers can cover conveniently, the specific performance remedy does not generate post-breach negotiation costs.

The issue, then, is whether sellers and buyers generally have similar cover costs. Analysis suggests that they do. Sellers as well as buyers have incentives to learn market conditions. Because sellers have to "check the competition," they will have a good knowledge of market prices and quality ranges. Also, when a buyer needs goods or services tailored to his own needs, he will be able to find such goods or services more cheaply than sellers in general could, for they would first have to ascertain the buyer's needs before going into the market. However, in situations in which the seller and the first buyer have already negotiated a contract, the seller is likely to have as much information about the buyer's needs as the buyer has. Moreover, in some markets, such as those for complex machines and services, sellers are likely to have a comparative advantage over buyers in evaluating the probable quality of performance and thus would have lower cover costs. Therefore, no basis exists for assuming that buyers generally have significantly lower cover costs than sellers. It follows that expanding the availability of specific performance would not generate higher post-breach negotiation costs than the damage remedy. . . .

[A possible objection to this analysis] assumes that sellers breach partly because their cover costs are higher than those of their buyers; it then argues that when cover costs do diverge, allowing specific performance seemingly is less efficient than having damages be the sole remedy. Returning to the widget hypothetical, let C_b = the first buyer's (B1's) cover costs; C_s = the seller's cover costs. Assume that S has higher cover costs than B1, that is,

Cs > Cb. If specific performance were available, B1 could threaten to obtain it, so as to force S to pay him part of the cover cost differential, Cs − Cb. If B1 made a credible threat, S would be better off negotiating than covering. Because only the availability of specific performance enables B1 to force this negotiation, one could argue that it is less efficient than having damages as the sole remedy.

This objection is incorrect, even if differential cover costs influence seller decisions to breach. A credible threat by B1 to seek specific performance would usually require preparing or initiating a lawsuit. This would entail costs of lost business time, lost goodwill and lawyer's fees, and these costs usually exceed any cover cost differential (Cs − Cb) that may exist. This is because the magnitude of cover costs—and hence of the differential—are low in relation to legal costs. Locating and arranging for substitute transactions are routine, relatively inexpensive business activities. Since the legal and related costs necessary for a credible threat commonly exceed the cover cost differential, it would rarely pay buyers to threaten specific performance to capture part of this differential. Thus no post-breach negotiations would be engendered by any differences in the parties' cover costs.

The second objection to the conclusion that post-breach negotiation costs are no higher under specific performance than under damages follows from the fact that in some cases sellers cannot cover at all. In these cases, buyers can always compel post-breach negotiations by threatening specific performance. There are two situations in which a seller cannot cover: if he is a monopolist or if the goods are unique. In either event, the first buyer would also be unable to cover. If neither the seller nor the first buyer can cover, no reason exists to believe that there would be higher post-breach negotiation costs with specific performance than with damages. If specific performance were available, B1 and S would negotiate over B1's share of the profit that S's deal with B2 would generate, or B1 would insist on a conveyance from S and then sell to B2. If only the damages remedy is available, B1 would negotiate with S respecting his expected net gain from performance rather than over the contract market difference, because he could not purchase a substitute. This expected gain is often difficult to calculate, and easy for the buyer to exaggerate. There is no reason to believe that negotiations or litigation over this gain would be less costly than the negotiations over division of the profit that B2's offer creates, or the costs of a second conveyance between B1 and B2. Thus even when the seller cannot cover, specific performance has not been shown to generate higher post-breach negotiation costs than damages. Moreover, when neither party can cover—the case under discussion—buyers have a right to specific performance under current law.

To summarize, if the initial buyer has access to the market at a significantly lower cost than the seller, a damages rule generates lower post-breach negotiation costs than a rule that makes specific performance routinely available. It seems likely, however, that both parties will be able to cover at similar, relatively low cost, or that neither will be able to cover at all. In either event, post-breach negotiation costs are similar under the two rules.

Notes on Specific Performance

1. This excerpt raises many of the same issues discussed earlier in connection with expectation damages and efficient breach. For example, the discussion of transaction costs should remind you of the earlier excerpt by Cooter and Ulen (supra at pages 42–44) discussing the relationship between efficient breach and relative transaction costs. Cooter and Ulen suggested that efficient breaches could be achieved even under remedies other than the expectation measure if the parties were able to renegotiate (without significant transaction costs) after an opportunity for breach arose. The Schwartz excerpt makes a similar point about specific performance. It concludes that when renegotiation costs are low, specific performance should be no less efficient than expectation damages (at least insofar as its effect on efficient breach is concerned) and might even be superior. Other articles discussing this aspect of specific performance include William Bishop, The Choice of Remedy for Breach of Contract, 14 Journal of Legal Studies 299 (1985); and Thomas S. Ulen, The Efficiency of Specific Performance: Toward a Unified Theory of Contract Remedies, 83 Michigan Law Review 341 (1984).

2. When renegotiation costs are low, specific performance may also be superior to expectation damages in its effect on nonbreachers' reliance decisions. As discussed earlier in the excerpt by Cooter (supra at page 62), the efficient level of reliance depends in part on the probability that the contract will be performed: buyers should rely less heavily on risky ventures than on more certain ones. Unfortunately, the expectation measure creates incentives for too much reliance because it guarantees that buyers will be fully compensated for their reliance whether or not the seller performs. This gives buyers an incentive to ignore the risk of nonperformance in deciding how heavily to rely.

If buyers are entitled to specific performance, however, they may not have quite as much of an incentive to overrely. This is because specific performance allows buyers to capture more of the gains from efficient breach. As discussed in the Friedmann excerpt on expectation damages (supra at pages 44–46), the expectation measure guarantees the buyer his or her expectation interest and not a penny more, thus allowing the breaching seller to capture all the additional surplus from a profitable breach. A rule of specific performance lets the buyer veto such a breach unless the seller agrees to share at least some of the gains with the buyer. The amount of the gains the buyer can capture will depend on his or her bargaining ability, but it should be somewhere between the seller's total gains from the breach and the buyer's total losses from nonperformance.

This is why specific performance gives buyers an incentive to temper their overreliance. If buyers rely too heavily, their losses from nonperformance will be quite large (because they will have sunk too much in the prospect of performance). This, in turn, will reduce the amount buyers can gain by bargaining with a breaching seller because the buyer's only gains from bargaining will be whatever the buyer receives over and above his or her losses from nonperformance. As a result, when buyers are deciding how heavily to rely, specific performance gives buyers an incentive to temper their reliance because of the possibility that the seller might want to breach (in which case they could bargain for a large net payoff). This incentive counteracts, to some extent, the buyer's temptation to overrely because of the possibility that the seller might perform (in which case the reliance would be valuable). These are exactly the two possibilities that buyers ought to balance in order to choose the optimal level of reliance.

For economic analyses of the specific performance remedy in this regard—though most of this literature is highly mathematical—see William P. Rogerson, Efficient Reliance and Damage Measures for Breach of Contract, 15 Rand Journal of Economics 39 (1984); Tracy R. Lewis, Martin K. Perry, and David E. M. Sappington, Renegotiation and Specific Performance, 52 Law and Contemporary Problems 33 (1989); Benjamin E. Hermalin and Michael L. Katz, Moral Hazard and Verifiability: The Effects of Renegotiation in Agency, 59 Econometrica 1735 (1991); Aaron S. Edlin, Specific Investments, Holdups and Efficient Contract Remedies, Stanford Department of Economics Working Paper (May 1992).

3. While specific performance may have desirable effects on the non-breacher's reliance decisions (when renegotiation is possible), the effects on the *breacher's* investment decisions will not always be so favorable. Specifically, when a breach could be caused by factors within the breacher's control, the availability of specific performance may give the breacher an incentive to invest too much in precautions to reduce the likelihood of a breach. This is because of the same factor discussed in the preceding note. If specific performance allows the nonbreacher to negotiate for a payoff that exceeds the nonbreacher's expectation interest, the effect (from the breacher's point of view) is much the same as if the breacher were subject to punitive damages, or to any other supercompensatory remedy.

For example, suppose that a builder who used the wrong brand of pipe in a house was legally obliged to tear down the house and rebuild it using the proper brand of pipe, even if the cost of the rebuilding greatly exceeded the value to the buyer of the proper brand of pipe. While the builder might be able to negotiate out of rebuilding the house by offering the buyer a large payment, the threat of having to make that large payment could give the builder an incentive to take too many precautions against using the wrong brand of pipe. See the discussion of the breacher's precautions in the excerpts by Cooter (supra at pages 59–64) and Craswell (supra at pages 46–53). See also the Muris excerpt on the cost-of-completion remedy (supra at pages 81–86); and its application to specific performance in Timothy J. Muris, The Costs of Freely Granting Specific Performance, 1982 Duke Law Journal 1053. The upshot of all this is that when both parties have investment decisions to make—that is, when the buyer must choose a level of reliance and the seller must choose a level of precautions—it is difficult to tell which remedy creates the best incentives overall.

4. Independently of its effect on the reliance and precaution decisions, specific performance could also affect the division of risks. While specific performance gives nonbreachers the ability to demand large side-payments in

the event of a breach (as discussed above), potential breachers may realize this in advance and may demand a more favorable price in order to be compensated for accepting that risk. Breachers will discount that risk by the probability that they will have to make the side-payment–for example, if there is a 10 percent chance of having to make a $100 payment to negotiate out of specific performance, breachers might raise their prices by $10. In effect, then, nonbreachers will be investing in a lottery ticket: they will be paying a $10 higher price in exchange for a 10 percent chance of being able to demand $100. If this $100 demand exceeds their true expectation interest, risk-preferring nonbreachers might be delighted by such an option, and risk-neutral nonbreachers would be indifferent—but risk-averse buyers would be left worse off by such a rule. (Recall the earlier discussion of risk-aversion in the Craswell excerpt on efficient breach, supra at pages 46–49. For a more extensive treatment, see A. Mitchell Polinsky, Risk Sharing Through Breach of Contract Remedies, 12 Journal of Legal Studies 427 [1983].)

5. Because specific performance moves the nonbreacher closer to full compensation—and perhaps even above full compensation if the nonbreacher can renegotiate to demand more of the surplus—specific performance could also raise many of the issues discussed in the second section of this chapter. For example, buyers may prefer not to be insured for full compensation for nonpecuniary losses (see the more recent article by Schwartz, supra at pages 86–90). Full liability might lead to lightly injured victims paying a higher price to subsidize the compensation received by heavily injured victims (see the Quillen excerpt, supra at pages 72–76). The earlier readings suggested that these factors might argue for an undercompensatory remedy rather than a fully compensatory (or overcompensatory) remedy. Are any of these factors likely to be significant in cases where specific performance has been requested?

6. Specific performance is often said to be available only when the promised goods are "unique" or only when the plaintiff's monetary damages would be "inadequate." For a discussion of the relationship between these requirements, see Anthony Kronman, Specific Performance, 45 University of Chicago Law Review 351 (1978). The continued vitality of these requirements was questioned in a recent article by Douglas Laycock, The Death of the Irreparable Injury Rule, 103 Harvard Law Review 687 (1990). As Laycock puts it (id. at page 691): "Courts have escaped the rule by defining adequacy in such a way that damages are never an adequate substitute for plaintiff's loss. Thus, our law embodies a preference for specific relief if plaintiff wants it." However, this preference is not an absolute one (according to Laycock), for specific relief may still be denied in certain limited situations. For example, if specific relief would impose a serious hardship on the breacher disproportionate to the benefit to the nonbreacher, courts may refuse to grant an injunction, especially if the degree of hardship that specific relief would impose on the breacher exceeds the degree of hardship that an inadequate damage measure would impose on the nonbreacher (id. at pages 749–52). For an example, would specific performance be awarded in the pipe example discussed supra in note 3? Should it be awarded in such a case?

7. What if the parties agree in their contract that the nonbreacher will be entitled to specific performance if the breacher fails to perform? Should such a provision be enforced by the courts? For a discussion, see Alan Schwartz, The Myth that Promisees Prefer Supercompensatory Remedies: An Analysis of Contracting for Damage Measures, 100 Yale Law Journal 369, 387–89 (1990). Many of the issues raised by such provisions are considered in the readings in the following section on liquidated damage clauses.

§ 2.3.2 Liquidated Damages

Liquidated Damages, Penalties, and the Just Compensation Principle

CHARLES J. GOETZ AND ROBERT E. SCOTT

Applying an efficiency analysis to contract damage rules suggest the following enforcement hypothesis:

In the absence of evidence of unfairness or other bargaining abnormalities, efficiency would be maximized by the enforcement of the agreed allocation of risks embodied in a liquidated damages clause.

This hypothesis is based on the assumption that liquidated damage provisions will (1) reduce transaction costs where the parties determine that the costs of negotiation are less than the expected costs of litigation upon breach; and (2) reduce the error costs produced upon breach when the promisee is denied recovery for his non-provable idiosyncratic value. It follows, unless enforcement produces other inefficiencies, that enforcing agreements negotiated *ex ante* will enhance efficiency by permitting the parties to minimize the cost of transacting. The current penalty rule seems to produce significant inefficient effects by limiting the possibilities of mutually beneficial exchange. In addition, negotiated damage agreements are now subject to post-breach attack as penal sanctions. This increases the direct costs of litigation in all cases–even where the agreement is upheld.

The situational model which will be used to test the hypothesis can be illustrated by the hypothetical *Case of the Anxious Alumnus*. Assume the following facts: Dean Smith, a 1957 graduate of the University of Virginia, is a loyal, some would say fanatical, fan of the University of Virginia Cavalier college basketball team. For the 1976 season, after years of second division performances in the highly competitive Atlantic Coast Conference, the Cava-

Charles J. Goetz & Robert E. Scott, Liquidated Damages, Penalties and the Just Compensation Principle: Some Notes on an Enforcement Model and a Theory of Efficient Breach. Copyright © 1977 by Directors of The Columbia Law Review Association. This article originally appeared at 77 Columbia Law Review, 554, 578–83, 588–93 (1977). Reprinted by permission.

liers finally produce a team that advances to the finals of the conference championship tournament at the end of the season. Through hard work and financial sacrifice, Smith acquires twenty-five tickets to the conference championship game in Landover, Maryland. Smith enters into contract negotiations with the Reliable Charter Service, Inc., to arrange for a bus to transport himself and twenty-four other Virginia fans to Landover on the day of the game. The standard price for this service is $500.

Smith considers his attendance at the game to be of supreme importance and does not relish the thought of anxious and sleepless hours worrying whether the bus will arrive and successfully accomplish the desired purpose. He is eager to quiet his fears by securing adequate protection in case Reliable fails to perform. However, under the current legal rule Smith cannot protect his unprovable reliance either by securing fully compensating post-breach damages or a bargained-for stipulation of the value of performance. Consequently, he is forced to consider other protective alternatives.

One option is to attempt to insure against the subjective consequences of breach with a third party (Lloyd's of London, for example). Assuming that a policy could be secured and enforced up to the assessed valuation of performance, adding the proceeds of the policy to the award of provable expectation damage which Smith could recover under existing law would provide him with full recovery for his idiosyncratic value upon breach by Reliable.

Alternatively, Smith could negotiate for direct insurance from Reliable or any of its competitors offering the same service. Dean might propose to pay Reliable $1,000 for the charter service if, in return for the additional premium, Reliable would agree to a penal sanction of $10,000 upon failure of performance. The stipulated sum of $10,000 would represent that amount at which Dean would be indifferent between performance and breach. Unfortunately, insurance purchased directly from the promisor, Reliable, is not a real alternative; the mere labelling of the idiosyncratic damages as pursuant to an "insurance" contract is unlikely to prevent a perceptive court from recognizing that such payments are de facto equivalent to a penalty. Hence, legally enforceable insurance for damages not recoverable as breach damages is in practice obtainable only from third parties.

This insurance model and the enforcement hypothesis pose the following issues for resolution:

1. As between the third-party insurance company and the promisor, which is the more efficient provider of insurance?
2. To what extent do the rationales supporting the "indemnity principle" suggest significant additional social costs due to enforcement of agreements at stipulated values?
3. Assuming changed conditions after an insurance agreement has been negotiated, such that the value of performance to the promisor is reduced, does enforcement of a penal sanction produce a high probability of inefficient effects?
4. What presumptions of unfairness can be developed to cope with those special classes of cases where bargaining abnormalities would produce inefficient effects if stipulated damage provisions were enforced?

The Efficient Insurer Model

Identifying the efficient insurer requires an analysis of the costs of providing insurance. At what price would a profit-making commercial enterprise be willing to offer Smith $10,000 worth of protection against the contingency of bus failure en route to the fabled final game of the championship?

Perhaps the overwhelming element in the cost of this insurance would be the expected value of the underwriting loss to the insurer. This expected value is defined as the product pR where p is the probability of nonperformance and R is the recovery payable to the insured. In addition, an insurer will also have other transactions costs, such as the costs of ascertaining the true probability p and the costs of negotiation and communication with the insured. These transaction costs will be subsumed in the portmanteau variable T, so that the total cost C of a policy paying R on the occurrence of an event with probability p can be summarized as

$$C = pR + T.$$

We assume that since the services in question are marketed competitively in the presence of alternative sellers, the cost of breach insurance to Smith will be $(1 + \alpha)$ C where α is the competitive rate of return or profit for the insurer. The question, then, is whether C would differ between the bus company and the third-party insurer.

An obvious focal point of interest is the transaction cost element T. Here, it is tempting to argue that the advantage lies with the bus company. In the first place, the bus company is in a superior position to know the breakdown probability p. Secondly, many of the other transaction costs normally incident to customer communication may be negligible when communication is already being undertaken relative to the carriage service itself. Hence, T may be lower for the bus company and thus so would C and the offering price of the insurance to Smith.

Actually, however, the transaction cost element is not the strongest argument in favor of the bus company as the most efficient insurer. The bus company's main advantage derives from its power to exercise some control over the breakdown probability p. . . .

Absent the $10,000 liquidated damage agreement, the bus company anticipates that D [its legal liability] will embody only the standard objective damage recovery which, let us assume, amounts to $1,000 for the Smith bus trip. [This anticipation leads the bus company to choose a relatively low level of care, which will result in a breakdown probability that can be labelled p_0.] In computing the expected underwriting loss, the third party insurance company will therefore arrive at a value $(p_0 \times \$10,000)$.

Suppose, however, that the bus company can offer the same insurance. The expected value of damages is now based, not on a D of $1,000, but on a D of $1,000 actual provable damages plus $10,000 insurance recovery. . . . The company will expend additional maintenance costs A, . . . and the breakdown probability will consequently decline to p_1.

What are the implications of these adjustments on cost? For the bus company, the insurance cost must now be modified to reflect the net benefits of possible risk-avoidance efforts. Hence, the appropriate cost function for the provision by the bus company of \$10,000 coverage is $C = (p_0 \times \$10,000) - [\$11,000 (p_0 - p_1) - A] + T$ where the terms in the square bracket are net gains from adjusting maintenance levels: $(p_0 - p_1)$ is the change in breakdown probability, its product with \$11,000 is the expected damage reduction, and A is the added cost of maintenance. We know that these net gains are positive from the nature of their computation and that they would not be achieved when the third-party insurance is purchased. Hence, even where the transaction cost component T is identical for the alternative insurers, the bus company has an efficiency advantage equal to the square-bracketed term in the equation above. . . .

The preceding argument has been that non-enforcement of liquidated damages provisions has the result of inducing individuals to protect against otherwise non-recoverable losses through special third-party insurance. This is likely to be an extremely inefficient alternative since there are strong economic arguments that suggest that the vendor is the lowest-cost insurer against non-performance. Although our argument has been framed in terms of the bus company example, a similar conclusion may be generalized to all cases in which the vendor has some control over the probability of externally caused non-performance.

In sum, many people may not want to make deals unless they can shift to others the risk that they will suffer idiosyncratic harm or otherwise uncompensated damages. To the extent that the law altogether prevents such shifts from being made or reduces their number by unnecessarily high costs, it creates efficiency losses; that is, it prevents some welfare-increasing deals from being achieved. . . .

Unfairness and Bargaining Abnormalities

Presumption of Fair Exchange

The underlying premise of the enforcement hypothesis is that, in the absence of bargaining unfairness, a stipulated damage clause reflects equivalent value. The possibility that a given provision does not reflect subjective compensation, but is penal in nature, is irrelevant to the question of enforcement unless this fact is caused by bargaining abnormalities. This premise is a derivative of what can be described as the flexibility principle of private exchange. Assuming no violation of process constraints, the subjective value of exchange is not amenable to judicial scrutiny. Except by controlling the subject matter, no neutral principle has been devised to evaluate the relative worth of a voluntary, freely-bargained exchange. Instead, contracts doctrine has developed fairness constraints which focus on the maintenance of process values—full access to information and competitive market opportunities. The enforcement hypothesis relies on this jurisprudential tradition by incorporating a presumption of fair exchange. Under this presumption, evidence that equivalent value was not exchanged for the liquidated damages provision would be relevant only to the extent that it permitted an inference as to the relative unfairness of the bargaining process. This analysis has been consistently applied by courts and legislatures to agreements for underliquidated or "limited" damages. These partial allocations of the risk of breach to the nonbreacher have been enforced absent specific evidence of unfairness. The enforcement hypothesis, by validating liquidated damage clauses which allocate similar risks to the breacher, does not raise any unique dangers of fraud or duress. Rather, the two situations present perfectly symmetrical fairness issues.

Unfairness and the Efficiency Criterion

The enforcement hypothesis identifies unfairness or other bargaining aberrations as the only limitations on the use of liquidated damages provisions. The normative notions of fairness implicit in the common law tradition are consistent with the analytical model of economic efficiency. Bargaining unfairness precludes the assumption of fair exchange and increases the risk of allocative inefficiencies. The inefficient effects of unfairness include an increase in the incidence of erroneously valued exchange as well as the increased social costs of fraud, misrepresentation, and duress. Asserting the inefficiencies of unfairness is not helpful analytically unless neutral principles can be identified within the fairness rubric. In the bargain context, two neutral principles may justify constraints on contracting flexibility.

Access to information at minimum cost is the first principle of bargain fairness. Where the bargain reflects processes which inhibit information exchange, the risk of allocative inefficiencies is enhanced. This constraint, identified in the unconscionability doctrine as "unfair surprise," would incorporate contracting behavior ranging from fraudulent exchange of false or misleading information to failures to reasonably disclose essential contract terms. This incentive to information exchange will maximize efficiency by reducing the transaction costs of acquiring information.

The second fairness principle supports the maximizing of competitive market opportunities. Bargaining aberrations which inhibit competitive exchange will tend to produce inefficient resource allocation. The identifiable bargaining abnormalities would encompass duress as well as the more traditional cases of monopoly. The fairness value of enhanced market opportunities has also traditionally been reflected in the unconscionability doctrine. Scrutinizing a bargain produced by "oppression" or "absence of meaningful choice" is a response to the perceived inefficiencies of reduced markets. The benefits of this response by the private law doctrine of unconscionability, however, remain indeterminate.

If this elaborated definition of unfairness is incorporated into the enforcement hypothesis, the following decision rule would be proposed:

> Liquidated damage provisions should be enforced in all cases unless evidence of information barriers or reduced competitive opportunities rebuts the presumption of fair exchange.

Party Sophistication and Presumptions of Unfairness

The jurisprudential anomaly of the penalty rule is the imposition of a second level fairness constraint. There is no reason to presume that liquidated damages provisions are more susceptible to duress or other bargaining aberrations than other contractual allocations of risk. Consequently, the extraordinary limitation seems to produce many more costly effects than are warranted by the perceived risk of unfairness. Nonetheless, it is clear that party sophistication will often be a relevant issue in determining the fairness of a stipulated damages provision. Many contracting parties may not be capable of calculating the risks necessary to bargain for the *in terrorem* clause at an equivalent price. It is clear that some parties are incompetent to act as direct insurers of idiosyncratic value.

The problem of status does not justify the current rule under which these agreements are conclusively unenforceable in all cases. Nonetheless, a presumption of unfairness (and unenforceability) might well be appropriate in those factual contexts where the expected unfairness costs exceed the expected gains from unlimited contracting flexibility. For instance, if there exists an identifiable class of cases where application of the enforcement hypothesis predictably produces a high incidence of unfairness, the social costs can be reduced by attaching the unfairness presumption to those cases alone. This less restrictive limitation on contracting flexibility could be rebutted by the promisee's demonstrating that the clause was a product of a fairly-bargained exchange. As part of his burden of proof, the promisee would be required to demonstrate that the parties had sufficient commercial sophistication and access to information to allocate fairly the identified risks.

Efficiency Implications of Penalties and Liquidated Damages

SAMUEL A. REA, JR.

Much of the confusion in applying the penalty doctrine occurs because of the failure to observe the distinction between ex ante and ex post evaluation of losses. If the predetermined damages equal the actual loss, the damages can be said to be "reasonable" ex post. In this situation, the penalty doctrine is irrelevant, except perhaps to shift the burden of proof concerning the actual

Samuel A. Rea, Jr., Efficiency Implications of Penalties and Liquidated Damages, 13 Journal of Legal Studies 147, 149–50, 152–57, 159–67 (1984). Copyright © 1984 by The University of Chicago. Reprinted by permission.

loss. The penalty issue will arise when the clause turns out to be an inaccurate forecast of the actual loss. In this situation the damages are unreasonable ex post, but they may have been either reasonable or unreasonable ex ante. The economic explanation for a contract with unreasonable ex ante (and ex post) damages is very different from the explanation for reasonable ex ante and unreasonable ex post damages, and I will treat these two cases separately. . . .

Ex Ante Determination of Damages

Risk Aversion

If the seller is risk-neutral and the buyer is risk averse, it is beneficial to both parties for the seller to insure the buyer. The desired amount of insurance, ignoring incentives, will equal the actual loss when there are no nonpecuniary losses, as would be the case in a commercial setting. Damages in excess of full compensation would not appeal to a risk-averse buyer because he would face an unwanted variation in his income, depending on whether the product failed. He would not wish to pay a higher price for the product (to cover the added insurance cost) in order to increase risk.

If the buyer were risk preferring, a clause that specified damages in excess of losses would provide an opportunity to gamble. In the first edition of his text, Posner suggested that courts are taking a stand against gambling contracts when they refuse to enforce such clauses. Clarkson, Miller, and Muris point out that this motivation is historically flawed because wagers were enforced for a long time after courts ceased enforcing penalties. It seems likely that the parties to most contracts are risk averse or at least risk neutral. None of the commentators has pointed out a case in which gambling was the motivation for unreasonable damages, and it is unlikely that a preference for risk prevails in the commercial world.

Goetz and Scott argue that the court's unwillingness to compensate nonpecuniary losses following breach of contract has led courts to undercompensate victims of breach and has induced contracting parties to attempt to contract around the courts' rules. Goetz and Scott conclude that damage clauses calling for apparently excessive payments should be enforced in order to insure these losses. However, I have shown that it is usually irrational to insure such losses. The insurance decision involves transferring income from the state of the world in which the contract is not breached (a higher price will be paid for the contract) to the state of the world in which the contract is breached (damages will be received if breach occurs). If the loss is nonpecuniary, it is likely that the utility of the additional income in the breach state will fall short of the utility of the forgone income in the nonbreach state. Consequently nonpecuniary losses would not usually be fully compensated in an efficient contract. Therefore there will be few situations in which nonpecuniary losses provide an explanation for damages that are viewed as unreasonable by courts.

Often it is not realistic to assume that the seller is risk neutral. If the risk is caused by random production costs or random product failure, risk-averse sellers and buyers will lower the damages specified in the contract below the actual loss in order to share the risk. If the buyer were risk neutral and the seller were risk averse, the insurance motive, as opposed to the incentives motive, would suggest that there would be no damages. . . .

Moral Hazard

When the probability of a breach can be affected by either the buyer or the seller, the first-best contract would make the damages contingent on the appropriate precautions taken by each party. Given that it is frequently impossible to determine the extent of precautions taken, ex ante or ex post, there is a moral hazard problem. A seller who has to bear the full cost of breach will have an incentive to take efficient precautions, but the insured buyer may reduce his costly precautions in ways which cannot be detected. The rational response to this type of imperfect information in most situations is for the level of damages to be reduced. Given the lack of perfect information, the damages must be a compromise between incentives for the buyer, incentives for the seller, and allocation of risk between the parties.

Clarkson, Miller, and Muris (henceforth referred to as CMM) . . . claim that economic efficiency requires that the courts not enforce damage clauses in situations in which the nonbreaching party can influence the probability of breach and will benefit from breach. Their argument does not explain why damages would deliberately be set in excess of losses. Moral hazard, which arises because the buyer's precautions cannot be observed, will be taken into account at the time of the bargain and will lead to a reduction in the predetermined damage level below the actual damages. The CMM explanation is applicable to damages which are reasonable ex ante but unreasonable ex post (see below).

Imperfect Enforcement

Borrowing from the literature on enforcement of criminal laws, one can argue that damages should be punitive when the costs of detection or enforcement are such that breachers will be penalized with a probability less than one. For example, the breach may be costly to detect in franchise contracts. The franchisor is interested in maintaining quality of service, but it is costly to monitor continually the quality of all of the franchisees. Termination of the contract by the franchisor, imposing a substantial loss on the franchisee, may be an efficient method of encouraging quality. The same analysis would apply to termination-at-will employment contracts. Terminations of franchises and employment have not typically been challenged as penalties. It does not seem likely that breaches will go undetected in the kinds of cases which occur most frequently, such as delay in completion of a construction contract.

Penalties as a Signal

In many situations the promisor may know the probability of breach but the promisee may not. In the second edition of his text Posner points out that penalties promote efficient exchange by signaling a party's intention to honor his contracts. Although no one ever knows for sure that he will honor a contract, those who know that they are more likely to honor than others will find it less costly to agree to penalty clauses. If buyers cannot differentiate low risk from high risk sellers, a seller's acceptance of a penalty clause is a signal of a low probability of breach.

The signaling role of damage clauses offers some important insights into the process of contract formation but does not imply that signaling with damage clauses in excess of the actual loss is necessarily efficient and should be enforced. Penalty clauses are costly because they (1) induce excessive precautions by the promisor, (2) induce deficient precautions by the promisee, (3) overinsure the promisee, and (4) expose the promisor to additional risk. If the clause is used as a signal, it must generate benefits in the form of a more efficient match of parties in contracts. In a paper on creditors' remedies, I show that it may be Pareto efficient to ban the use of signaling clauses that use nonpecuniary penalties such as "arm-breaking" when there is default on a personal loan, but this type of contract is not prevalent in the penalty cases.

Even if signaling is the primary motive for damage clauses, there does not appear to be any reason a signal would involve a damage clause in excess of the actual loss. This can be demonstrated by considering a situation in which (a) there is only one type of customer with one type of loss, (b) there are numerous sellers who have different degrees of reliability, and (c) sellers can signal their reliability by offering guarantees or liquidated damages. If the damages equal the actual loss, the extent of reliability becomes irrelevant for the buyer. He will simply choose the lowest-cost seller offering full compensation in the event of breach. Sellers of less reliable products with the same manufacturing cost will not be able to compete in the market. The low-cost sellers, taking into account the buyers' losses, can drive the others out of the market without offering damages in excess of actual losses. . . .

Conclusion on Ex Ante Unreasonableness

What can the court conclude in a situation in which damages are ex ante unreasonably large? There are at least four possible explanations for such a conclusion. First, the excess damages may have been optimal. In light of the discussion above this seems unlikely.

A second explanation for a finding of unreasonable damages is that the court made an incorrect evaluation of the loss. Although it is tempting to conclude that the existence of the clause is prima facie evidence of the court's mistake, much of the law rests on the assumption that courts can measure losses with a sufficient degree of accuracy to justify the court's intervention. The penalty doctrine implicitly recognizes the possibility of court error by tending to enforce damage clauses when the loss is difficult to evaluate. Furthermore, damages not ordinarily awarded, such as attorney's fees, are recognized as losses when a damage clause is evaluated.

A third explanation is that there has been a procedural deficiency in the formation of the contract. For example, one party may not have realized the implications of the damage clause. Such a situation is more likely to arise when one party is an unsophisticated consumer than in a commercial setting. The doctrine of unconscionability could apply to this situation.

A fourth explanation for excess damages is that the parties, although knowledgeable, made a mistake about the nature of the contract, the possible loss, the probability of loss, or the implications of the damage clause. . . .

An overestimate of the damages is more likely to be a mutual mistake than a unilateral one. If the seller had overestimated the damages (a unilateral mistake), the buyer would have been willing to negotiate lower damages in return for a lower price. If the buyer alone had exaggerated the loss, the seller would have an incentive to inform him of his mistake, although the buyer might not find the assurances of the seller to be credible. The incentive will take two forms. First, if the seller is risk averse, his risk is lowered if predetermined damages are lowered. Second, the amount that the buyer would be willing to pay under the contract would be increased if he knew that the potential losses were lower. It seems reasonable to assume that the buyer usually has the best information on the size of the possible loss and that if he were mistaken, the seller would also be mistaken. The mutual mistake lowers the value of the contract by overinsuring the buyer, increasing the risk for the seller, and leading to excessive seller precautions. Ex ante the cost of the mistake will be borne by both parties. Nonenforcement of the excess portion of damages forces the party best able to acquire information on losses, the buyer, to acquire more information. If the buyer does not acquire the information, he not only bears his share of the cost of the inefficient contract, he also forgoes the excess insurance for which he paid an additional premium (built into the price of the contract).

In contrast to the situation of excess predetermined damages, there are many valid reasons for limitations on damages. Therefore, evidence of a limitation on damages is much less likely to indicate some unobservable mistake or unconscionability, and enforcement of limits on damages will lead to efficient treatment of most cases.

In summary, it appears that the penalty doctrine is not as anomalous as has been generally believed. The heart of the doctrine is that those damage clauses that were unreasonably large ex ante will not be enforced. A careful examination of the factors influencing predetermined contractual damages suggests that there are few instances in which excessive damages will be desired by the contracting parties. The courts are correct in viewing such clauses with suspicion. Their refusal to enforce the clauses when losses can be easily measured is consistent with the doctrines of mistake and unconscionability.

Damages Reasonable Ex Ante But Unreasonable Ex Post

The analysis in the previous section indicates that negotiated damages are unlikely to exceed the predicted losses, in the absence of a mistake or unconscionability. Nevertheless, negotiated damages will not exactly equal the predicted losses. . . . [T]he courts have been divided in their treatment of cases in which the damages were ex ante reasonable but excessive ex post. An economic perspective on the cases indicates that there are reasons for the courts' lack of unanimity.

There are two general reasons the predicted damages will turn out to be incorrect. First, changed circumstances may have altered the consequences of breach. The termination of war prior to the delivery of gun carriages is an example. Second, the size of loss may have been known to be random and the damages were set equal to the expected loss. The distinction between these two situations is based on foreseeability. In the first class of case, the parties did not foresee some event or type of breach that affected the extent of damages. In the second, they foresaw the event and its consequences, but chose to set damages equal to the expected loss. These two situations are very different from cases in which the damages were unreasonable ex ante. Within each category the economic implication of enforcing damages will depend on whether the damages became known before the breach.

Random Loss: Loss Not Known Until Breach Occurs

In some cases the actual loss resulting from breach is randomly distributed. If the parties know the distribution of losses at the time of contract formation, risk-neutral parties might choose to set the damages equal to the expected loss, guaranteeing that the decision to perform will reflect the expected cost of breach. This arrangement may substantially reduce the transaction costs. At the time of contracting the promisee has no incentive to exaggerate his expected losses because this will raise the cost of the contract without providing any net benefits to him. After a breach, he has an incentive to exaggerate his losses, and it becomes more costly to negotiate a settlement. If no settlement is reached, the parties must bear the costs of a trial. Obviously, the greater the difficulty in proving losses, the greater the relative advantage of predetermined damages. Although difficulty of measurement and ex ante uncertainty are analytically distinct, in practice the two are likely to be correlated.

Predetermined damages will appeal to risk-neutral parties but may not be the optimal risk-sharing arrangement for those who are risk averse. Moving from actual damages to predetermined damages will increase the risk faced by the promisee. The promisor will still face the risk associated with uncertain costs, alternative offers, or whatever else might cause him to breach.

If the court implements the penalty doctrine and at the same time enforces predetermined damages as a limit on the promisor's liability, the expected damages paid will fall short of the expected loss. Since this raises the probability of breach, there will be a tendency to raise the predetermined damages and, in the extreme, to eliminate predetermined damages. This will create inefficiency because of the added transaction cost and the allocation of risk objective which originally induced the parties to favor predetermined damages in the absence of the penalty doctrine.

Random Loss: Loss Known After Contract But Before Breach

When the loss becomes known before the breach decision, there will be efficient breach if the promisor bears the actual loss instead of the expected loss. We would not expect risk-neutral parties to be as willing to set predetermined damages in this situation as in the previous one, but risk aversion and the cost of measuring actual losses may still lead to predetermined damages in the absence of a penalty doctrine.

One can conclude that ex ante uncertainty of damages, whether known before or after breach, may lead to predetermined damages for allocating risk and for reducing the sum of ex ante and ex post negotiating costs. The ex post transaction costs are likely to be particularly high when losses are difficult to measure. It follows that the court's willingness to enforce predetermined damages when losses are hard to measure contributes to economic efficiency. Nonenforcement of damage clauses when damages were known to be random leads to inefficiency.

Unforeseen Change in Loss: Loss Not Known Before Breach

When the loss is unexpectedly large or small relative to the predetermined damages, the only argument for intervention is that the court can make the contract more efficient by supplying the allocation of unexpected losses that the parties would have agreed to had they considered the contingency that occurred. This would be true if the cost of anticipating that contingency and including it in the contract were higher than the cost of ex post monitoring of the contract by the court. If the parties are risk neutral, there is no advantage in altering the terms of the contract. If they are both risk averse, the loss should be shared by both parties, depending on the relative degree of risk aversion. It seems implausible that the relative risk aversion of the parties would lead to the pattern of risk allocation implicit in the current treatment of penalties and underliquidated damages. Consequently, the existing doctrine is not likely to lead to efficient risk allocation. Furthermore, courts are unlikely to have sufficient information on the relative risk aversion of the parties so that they are able to make contracts more efficient by altering the allocation of losses, unless circumstances are sufficiently altered that the doctrine of impossibility or frustration is applicable.

Unforeseen Change in Loss: Loss Known Before Breach

If the magnitude of loss becomes known before the time when breach might occur, the unforeseen change in loss will alter the efficient level of precautions against breach. It must be assumed that it is excessively costly to specify in the contract the actual amount of precautions taken by either party, otherwise this would have been done ex ante. If the precautions taken by either party can be observed ex post, the damages can be made contingent on such precautions. For example, if the promisee induces breach, the court can deny him damages, without calling the damages a penalty. The more likely situation is one in which the precautions actually taken cannot be observed at a reasonable cost. Consequently, predetermined damages are used to influence the behavior of each party, taking into account the degree of influence that each party is likely to have on the breach, the degree of risk aversion, and the types of losses (pecuniary or nonpecuniary) involved. When changed circumstances cause damages to diverge from those that were expected, it follows that the predetermined damages will no longer be optimal, and the parties will respond to inappropriate damages. When losses are lower than expected, the promisor's care will be excessive and the promisee's care will be deficient.

Clarkson, Miller, and Muris (CMM) argue that predetermined damages exceed losses because circumstances change between the time of negotiation and the breach of contract, reducing the losses that will result from breach. Under these conditions the nonbreaching party will gain if the other party breaches, and the former has an incentive to breach in undetectable ways. CMM argue that the court will contribute to efficiency if it identifies situations in which the promisee might have influenced the outcome, and they claim that courts refuse to enforce damage clauses in precisely these cases.

A major problem with the CMM explanation of penalties is that it ignores the effect of predetermined damages on the promisor's incentives. The promisor's efforts to avoid breach will be excessive if circumstances have changed so as to make the predetermined damages large in light of actual losses and deficient if the predetermined damages are small in light of the actual losses. Therefore, there are efficiency costs of inaccurate damages regardless of the promisee's possible ability to induce breach. Instead of looking only at the potential promisee influence on breach, courts that were concerned with incentives would have to identify cases in which the circumstances changed, the parties knew of the magnitude of the loss before breach, and one or both parties could effect breach. This exercise may be too costly to justify the efficiency gains. Furthermore, an explanation for the penalty doctrine based on breach incentives is not consistent with the enforcement of underliquidated damages.

One can conclude that the court will not be likely to contribute to efficiency if it uses the penalty doctrine in situations in which the losses were random and reasonable ex ante. When losses are altered as a result of some unforeseen event, the court may be able to make contracts more efficient by allocating risk according to the relative risk aversion of the parties, but the court can easily make mistakes in this allocation. The penalty doctrine may improve efficiency in situations in which the victim of breach can induce breach, but this only applies when there are unforeseen losses which become apparent before breach.

Notes on Liquidated Damages

1. The Clarkson, Miller, and Muris article referred to in the Rea excerpt is Kenneth W. Clarkson, Roger LeRoy Miller, and Timothy J. Muris, Liquidated Damages vs. Penalties: Sense or Nonsense?, 1978 Wisconsin Law Review 351. Other recent analyses of liquidated damage clauses include Paul H. Rubin, Unenforceable Contracts: Penalty Clauses and Specific Performance, 10 Journal of Legal Studies 237 (1981); Phillippe Aghion and Benjamin Hermalin, Contracts as a Barrier to Entry, 77 American Economics Review 388 (1987); Tai-Yeong Chung, On the Social Optimality of Liquidated Damage Clauses: An Economic Analysis, 8 Journal of Law, Economics, and Organization 280 (1992); and Lars A. Stole, The Economics of Liquidated Damage Clauses in Contractual Environments with Private Information, 8 Journal of Law, Economics, and Organization 582 (1992).

2. Both of the excerpts in this section point out that the enforceability of liquidated damage clauses is closely related to questions concerning the proper scope of freedom of contract and the proper uses of the unconscionability doctrine. The enforceability of such clauses also raises questions concerning the circumstances under which parties should be released from contracts (or from particular clauses) due to mutual mistake, or to a change in circumstances after the contract was signed. This latter issue will be discussed in Chapter 3 in connection with the doctrines of mistake and impracticability. Issues relating to unconscionability and freedom of contract will be discussed in Chapter 5.

3. When Rea concludes that it may be appropriate for courts to refuse to enforce liquidated damage clauses whose amounts appeared unreasonable ex ante, he asserts that courts can determine "with a sufficient degree of accuracy" what damage amounts would have been reasonable ex ante (supra at page 109). Otherwise, courts would be unable to recognize those cases where the stipulated amount was *un*reasonable ex ante.

For a more skeptical view of courts' ability to judge the reasonableness of stipulated damages, see Alan Schwartz, The Myth that Promisees Prefer Supracompensatory Remedies: An Analysis of Contracting for Damage Measures, 100 Yale Law Journal 369, 383–87 (1990). Schwartz agrees with Rea that promisees would rarely if ever want liquidated damage clauses set at an amount that was supracompensatory ex ante. However, he argues that—absent a more specific showing of some form of unconscionability—the existence of an apparently unreasonable clause should be taken as evidence that the courts must have erred in their assessment of what reasonable damages would have been, rather than as evidence that the parties must have made a mistake. As Schwartz puts it (id. at page 384):

> [E]xcept for the liquidated damage clause itself, courts do not ask whether the parties' agreement rested on predictions that were objectively reasonable given the evidence that the parties had before them when they signed the contract. This restraint follows from courts' correct belief that they are not as good at drafting contracts as business people are. The liquidated damage rule, however, permits enforcement only of damage predictions that are reasonable ex ante and so directs courts to evaluate the parties' predictions. Courts should be expected to do this job badly.

As this debate should suggest, the relative likelihood of errors by the court and errors by the parties will also be relevant to Chapter 3's discussion of the doctrines of mistake, frustration, and impracticability; and to Chapter 5's discussion of unconscionability.

4. When liquidated damage clauses are set at an amount that exceeds the measure of expectation damages that would be awarded in the absence of such a clause, enforcement of the liquidated damage clause would have many of the same effects as an award of specific performance. For example, enforcement of the higher measure of damages (a) would put more pressure on the promisor to perform, (b) would give the promisee a stronger negotiating position if the promisor for some reason wanted to buy his or her release from the contract, and (c) would give the promisor an incentive to take more precautions against accidents that might lead to either of the two disadvantages just described. Thus, it should not be surprising that many of the arguments in these two excerpts are similar to those raised in the readings on specific performance (supra at pages 93–100), and even earlier in the readings on expectation damages (supra at pages 42–55).

5. Liquidated damage clauses could also have the effect of optimizing the promisee's incentive to choose an optimal level of reliance, and to mitigate damages efficiently after a breach, as long as the amount of damages recoverable under the clause was fixed at an amount that was invariant with respect to the level of reliance or mitigating behavior actually engaged in by the promisee. This aspect of liquidated damage clauses was discussed in the earlier excerpts by Cooter (supra at pages 63–64) and Epstein (supra at pages 69–71).

§ 2.3.3 **Rescission and Restitution**

The Mitigation Principle

CHARLES J. GOETZ AND ROBERT E. SCOTT

The Perfect Tender Rule

Under the common-law perfect tender rule, an obligee has no duty to accept defective performance or offers to remedy the defect following a tender of nonconforming goods. Assume, for example, that Seller tenders a compressor in due course on the 15th of September. On examination, Buyer discovers that the aluminum casing used is Type B rather than Type A as specified in the contract. Seller concedes that Type B casing is a slightly lighter alloy and costs $3,000 less than the contract-specified variety. Furthermore, although the lighter casing will increase the risk of heat dispersion to some extent, an insurance policy can be purchased for $2,500 to guard against this contingency. Nonetheless, Buyer decides to reject the "nonconforming" tender. Seller then offers Buyer $3,000 if he retracts his rejection and accepts the compressor with the Type B casing. Must Buyer accept Seller's offer? The common-law perfect tender rule offers Buyer a choice: accept the deficient performance and recover the reduction in value as damages or reject the defective tender and recover the difference between the contract price and the market price at the time of tender.

The discretion offered by the perfect tender rule is not inconsistent with the mitigation principle. Even without legal compulsion, Buyer would have incentives to elect the cheapest readjustment option to reduce his exposure as a creditor seeking reimbursement of the damage bill. Buyer may also earn goodwill and a reputation as an efficient mitigator that will be reflected in the terms of subsequent transactions. Only a counterbalancing opportunity for opportunistic gains, therefore, would divert Buyer from the appropriate course, and such gains are unlikely to present themselves in a developed market. For example, the cost of potential adjustments or a corresponding unwillingness to accept a compensatory payment in place of rejection can be easily tested by the market. If Buyer's refusal is idiosyncratic or strategic, Seller simply offers the compressor elsewhere at a competitive price. The existence of many and close substitute performances reduces not only the bargaining range within which strategic claims can be made, but also the

Charles J. Goetz & Robert E. Scott, The Mitigation Principle: Toward a General Theory of Contractual Obligation, 69 Virginia Law Review 967, 995–1000, 1009–10 (1983). Copyright © 1983 by the Virginia Law Review Association. Reprinted by permission.

uncertainty that encourages them. Furthermore, Seller has an incentive to reveal the cost of his various adjustment options because ultimately he bears these costs anyway.

Suppose, however, that a "double lightning bolt" causes both Seller and Buyer to experience regret. This could happen, for example, if the market price for a conforming compressor declined, but a temporary shortage increased the cost of Type A aluminum casing. Seller would have an incentive, in such a situation, to substitute the less costly Type B casing to protect his expected profits. Buyer, on the other hand, would now prefer to escape the contract—by rejecting Seller's tender as "nonconforming"—and purchase a compressor at the lower market price. However uncommon this result might be, the drafters of the Code, and Professor Llewellyn in particular, feared that in such a case the common-law perfect tender rule would not yield optimal results. During the years in which the Code was in draft form, Llewellyn and other scholars expressed concern about the phenomenon of "surprise rejections" and the imposition of unnecessary costs on the breaching party. Although they framed the argument in terms of "economic waste"—wastefully forcing a seller to take goods back—they feared strategic moves by the buyer to exploit a minor defalcation for his own purposes.

Section 2–508(2), the result of this concern, permits a seller to "cure" a defective tender. If our Seller reasonably believed that Buyer would find a compressor with a lighter casing acceptable with or without a money allowance, then section 2–508(2) grants Seller an additional period of time to substitute a "conforming" tender. Determining when such a right of cure should be available, however, presents a problem. The conventional surprise rejection illustration suggests that, if a seller were under an obligation to tender 1,000 widgets and tendered only 999, rejection by the buyer would trigger the right of cure under section 2–508(2). One cannot reconcile this analysis, however, with the Code's definition of "conforming" goods. If the parties reasonably believed such a tender would be acceptable because of prior dealings or prevailing usage of trade, then the contract would incorporate this contextual understanding within the section 2–106 definition of "conforming goods." The surprise rejection illustration typically used as an example of section 2–508(2) thus merely illustrates a seller's legitimate surprise at the rejection of *conforming* goods. Such a rejection would be wrongful ab initio and never require the application of section 2–508.

Circumstances in which goods require adjustment before they are in good working order also provide inappropriate illustrations of section 2–508(2). If adjustment is regarded as incidental to the tender of such goods, then the need to adjust does not render the goods nonconforming. A buyer's attempted rejection would again be wrongful.

To invoke section 2–508(2) properly, one must imagine a case in which the tender is clearly defective, but the seller nonetheless anticipates that the buyer will accept the tender as a legitimate readjustment option. These requirements are satisfied where one can evaluate the deficiency on the market and correct it with a monetary payment accompanying the tender. What surprises

Seller in our hypothetical is not the rejection of his defective performance, but the rejection of the defective performance accompanied by the offer to send $3,000—more than enough money to insure against the consequences of the defect.

Granting a seller the right to cure under appropriate conditions serves two functions. First, section 2–508(2) encourages a buyer anticipating special losses from nonperformance to bargain for additional protection at the time of contracting. If a buyer attaches an idiosyncratic valuation to the seller's performance, the inadequacy of ordinary methods of cure necessitates such augmented protection. Even without section 2–508(2), however, the objective damages principle would encourage the early communication of idiosyncratic values. Second, and perhaps more importantly, the cure provision restrains opportunistic claims by Buyer. Unfortunately, it also invites evasion by Seller through the tender of inadequate substitutes as a "cure."

Even in a competitive market situation, however, a right of cure may be necessary to restrain an obligee's opportunistic behavior. Despite the initial existence of a market at the time of contracting, the double lightning bolt transforms the contract into a specialized relationship with few and imperfect substitutes. This species of bilateral monopoly raises legitimate fears of costly renegotiation following the defective tender. The doctrine of avoidable consequences, moreover, fails to restrain opportunism under these conditions. Although an obligee will be unable to recover avoidable costs and any expenses saved if he strategically rejects the obligor's tender, the exclusively defensive character of the rule fails to protect accrued contract rights that optimal mitigation could have preserved. Seller has valuable return rights in the favorable price shift, but possesses no effective remedy other than renegotiation if Buyer elects to reject the inadequately housed compressor and merely walks away from a now disfavored deal.

In short, once circumstances eliminate close substitutes, categorical standards of performance may no longer reliably serve to reduce costs. The common-law perfect tender rule encourages the nonbreacher in a bilateral monopoly to demand a premium to accept the retendered goods. Reducing readjustment costs under these circumstances requires a rule that reduces buyer opportunism by more than it increases seller evasion. The failure of the Code's cure provision thus lies in its ambiguity and generality. Only where the context clearly signals a need for more complex rules should the seller be empowered to demand the right of cure. . . .

Substantial Performance

. . . Unlike the perfect tender obligation imposed on the sale of goods, courts have implied only a "substantial performance" obligation on a defective performance in service or construction contracts. The rule of substantial performance—or material breach—assures the breacher of his accrued contractual gains whenever the tender is consistent with the overall scheme of the contract, although deficient in some particulars. The doctrine expands the duty to mitigate in specialized environments by requiring the mitigator to accept a deficient performance, together with objectively measured damages.

The substantial performance doctrine reduces opportunistic claims by softening the breacher-nonbreacher distinction, thereby removing opportunities to exploit inadvertent breaches. Such a rule is sensible in cases such as construction contracts where the circumstances suggest that renegotiation costs otherwise will be substantial. Once construction is substantially underway, the alternatives for both parties become inferior to the existing relationship, thus expanding the bargaining range within which the parties must reach agreement on post-breach adjustments. The obligor will regard removing the half-constructed object from the site and offering it to a third party as vastly inferior to acceptance plus damages because of the costs of removal. Without the substantial performance rule, the owner-obligee would have an incentive to exploit this situation. On the other hand, a minor deviation will not significantly increase the risk that a disappointed obligor can evade his responsibility to the obligee. So long as the performance is consistent with the general purpose or plan of the contract, objective damages will provide an adequate surrogate to most parties for any deficiencies. Furthermore, completion of the bulk of the performance reduces the risk of nonsatisfactory breach.

Good Faith in the Enforcement of Contracts

ERIC G. ANDERSEN

The Goals of Contract Enforcement

A brief review of traditional remedies for breach of contract reveals an attempt to accommodate two competing goals: (1) securing to the injured party the benefit of its bargain, (2) without imposing unnecessary costs on the breaching party. Two of the most important common-law remedies—money damages and cancellation for material breach—illustrate how this accommodation is made. . . .

Common Law Damages and Cancellation for Material Breach

A party who breaches a contract usually is liable to pay the other party enough money to put the other in the same position as if the contract had been

Eric G. Andersen, Good Faith in the Enforcement of Contracts, 73 Iowa Law Review 299, 306–12, 318–22, 345–47 (1988). Copyright © 1988 by The University of Iowa. Reprinted by permission.

performed without breach. The breaching party must compensate for injury to the other party's "expectation interest." A court will apply the mitigation principle, however, by calculating damages as if the injured party had taken reasonable steps to reduce the extent of its own injury and, thus, the breaching party's liability. For example, a court will calculate the damages awarded an employee who is discharged in breach of contract as if other, reasonably available employment had been taken. Similarly, a court will determine the damages awarded a seller of property whose buyer wrongfully failed to complete the transaction as if the property had been resold to someone else at the market price. These examples illustrate that, although the injured party's interest in having the benefit of its bargain has priority in contract enforcement by an award of damages, the breaching party's interest in minimizing its liability clearly matters. The enforcing party is not "expected to take steps that involve undue burden, risk, or humiliation," but is given an incentive to eliminate costs for the other side that are not necessary to protect the expectation interest.

The damages remedy often falls short of adequately protecting the expectation interest. The circumstances of the breach may indicate that the party committing the breach is unlikely to perform the remainder of its duties as promised. If the injured party has not yet begun performance or has performed only partially, and if the breaching party may become insolvent or impossible to find, recourse to damages alone is unlikely to provide the injured party with the monetary equivalent of the promised performance. For example, if a contractor breaches soon after beginning a building project, the owner is put at risk if the only available remedy is to pay the contract price and sue for damages. Moreover, even if the breaching party can be made to pay, recovery limitations based on the foreseeability of the injury caused by the breach and the certainty with which the extent of the injury can be determined may leave the injured party less than fully compensated. In any event, reducing a cause of action to judgment may be time consuming, disruptive, and expensive.

The common law therefore provides an additional, and in some respects more potent, device for protecting the expectation interest of a contracting party: constructive conditions of exchange. When a contract contains such a provision, performance as promised by one party is a condition of the other's duty to perform. When one party's breach constitutes the nonoccurrence of a constructive condition of exchange, the other party is entitled to end the contract and to be discharged of its own remaining performance, as well as enjoy a cause of action for damages in lieu of all future performance by the breaching party. Constructive conditions of exchange are powerful tools because, in addition to making damages available, they permit self-help by allowing the injured party simply to withhold its own performance without prior judicial intervention.

Constructive conditions of exchange protect the breaching party's interests as well as those of the injured party. Although, subject to the limitations mentioned above, damages always are available for breach, only material breaches discharge the aggrieved party and justify withholding the return performance. The definition of materiality, while far from clear in existing cases, may be characterized as permitting a discharge and a cause of action in lieu of future performance only when the drastic remedy of complete discharge is needed to protect the legitimate interests of the victim of the breach.

Moreover, many courts will allow a breaching party to recover in restitution the amount by which the net benefit conferred upon the other party by an incomplete or defective performance exceeds the damages to which that party is entitled. The common-law doctrines of money damages, cancellation for material breach, and restitution work in tandem to reduce the costs to the breaching party of the remedies available for breach of contract.

Agreed Damages Clauses

Contracting parties may agree to establish their own terms to accomplish the purposes of common-law remedies. When an agreed, remedial term is in issue, however, the policy of eliminating needless costs against the breaching party is not built into the operative rules, as with the mitigation doctrine, nor found within the definition of the remedy, as with material breach. That policy must be brought into play, if at all, as a rule or doctrine outside the agreed remedy itself.

Such a doctrine is well established relative to what is perhaps the most common agreed remedy, the liquidated damages clause. These clauses are useful because they allow parties to avoid the delays and uncertainty of judicial calculation of damages. The parties are restrained, however, in their power to create a damages remedy by agreement. Under familiar principles, a promise to pay an agreed sum of damages, even if bargained for at arm's length, will not be enforced if the amount constitutes a penalty—that is, if it is excessive relative to the actual injury anticipated to be caused by the breach. Thus, when a contract is enforced through the invocation of a liquidated damages clause, the law requires the same accommodation of the parties' interests that is made under the common-law damages remedy. The enforcing party's expectation interest will be protected, but only in a way and to an extent that eliminates unnecessary costs to the breaching party.

Other Agreed Enforcement Terms

Just as contracting parties may agree to a liquidated damages term in lieu of, or in addition to, the common-law damages remedy, they also may include express conditions in their agreement to replace or supplement constructive conditions of exchange. The need to accommodate the interests of the injured party and the breaching party also is present when the parties have agreed to use such express conditions to secure the expectation interest. For example, suppose a term in a mortgage loan agreement provides that the lender may accelerate the debt and terminate the agreement if the mortgagor is tardy in paying property taxes on the mortgaged property. One readily can imagine circumstances in which the mortgagor's injury from the acceleration of the debt would be greatly disproportionate to the lender's harm from the late payment of property taxes. Or consider a term in an installment contract for the sale of goods that entitles the seller to dispose of all future installments elsewhere if the buyer does not pay in cash upon delivery. If the market has risen substantially, the seller's invocation of the term in question might impose costs on the buyer that are not justified by the actual risks to the seller entailed by a delay in payment.

The same principles that undergird common-law remedies and agreed damages should apply here. There should be a way to protect the legitimate interests of the lender or the seller without visiting upon the other party to the contract the harshness that might result from enforcing the agreed term under some circumstances. Yet, the common law lacks a rule or clear statement of principle making that accommodation. No satisfactory counterparts to the penalty and mitigation doctrines exist when contract enforcement is accomplished by express conditions that do not operate directly through a liquidated payment obligation.

Although courts have not applied the principle of eliminating needless costs against a breaching party when express conditions are in issue, they often have recognized its relevance, perhaps intuitively. A satisfactory doctrinal framework for the principle's application, however, has not been established. Courts thus resort to a variety of other techniques or principles, such as interpretation, avoidance of forfeiture, waiver, or estoppel to accommodate the parties' interests. These techniques and principles often are ill-suited to accommodating the competing interests that are at stake and fail to establish a useful, general doctrinal framework for resolving future disputes. They commonly are manipulated to do justice, but the relevant conception of justice is left unclear. Therefore, the law in this area appears highly unpredictable. The remainder of this Article proposes a doctrine of good faith in enforcement that accommodates the competing goals of contract enforcement relative to agreed contract terms and provides a principled method for courts to decide future cases.

Good Faith in Enforcement

Enforcement Terms That Primarily Would Serve Purposes Other Than Those They Were Intended To Promote

An enforcement term may have a materially different effect upon the parties when it is invoked than the parties intended at the time of contract formation. If the different effect is attributable to a mistake of fact or to supervening events that the parties assumed would not occur, a number of doctrines exist to deal with the problems raised. In the absence of such factors, however, the question is best dealt with under the rubric of good faith. By definition, an enforcement term represents the parties' attempt to secure to at least one of them the expected benefit of the bargain—that is, it is intended to protect the expectation interest. If, in fact, the term primarily would accomplish some different end at a substantial cost to the other party, a court should decline to sanction or, at least, should limit the extent of the right or power granted by the enforcement term.

Holiday Inns of America, Inc. v. Knight illustrates this idea. [Holiday Inns contracted for an option to purchase real property, which could be extended for up to five years upon payment of $10,000 each year, but which provided that the option would terminate if Holiday Inns was tardy in making any of the annual payments. Holiday Inns was slightly tardy with the third payment, and the Knights declared the option terminated.] If the provision permitting cancellation without notice or opportunity for cure was an enforcement term, the court's refusal to permit cancellation can be justified under the good faith analysis. As noted above, the cancellation term's purpose was to prevent the optionee from speculating against the optionor longer than the agreed period of time. As Professor Farnsworth has pointed out:

> Holiday Inns was risking payment of only $10,000, to keep open an option on property worth 20 times that amount. It is unlikely that Holiday Inns would speculate by withholding this relatively small sum for a few days when so much [that is, the automatic cancellation of the right to purchase] was at stake. The importance to the Knights of strict observance of the July 1 deadline, to keep Holiday Inns from speculating, was therefore not great.

Thus, under the circumstances at the time of the parties' dispute, the effect of enforcement would not have been to prevent Holiday Inns from playing the market against the Knights, but to permit the Knights to take advantage of the property's greatly appreciated value by terminating the optionee's right to purchase. Invocation of the cancellation term under those circumstances would not be in good faith under the proposed test. . . .

When the Enforcing Party Could Have Taken Reasonable Steps To Make Invocation of the Enforcement Term Unnecessary

The second situation in which a term might be denied enforcement for lack of good faith arises when the enforcing party could have taken reasonable steps, without prejudicing its own position, that would have made invocation of the term unnecessary to advance the purpose for which the term was included in the agreement. This "mitigation" side of the analysis recognizes that, even if an enforcement term protects precisely the interests the parties had intended to secure, under particular circumstances, it might do so at an unnecessarily high cost. In such a case, a court should apply the same reasoning that permits a reduction in money damages to reflect the injury that would have resulted had the enforcing party taken reasonable steps to protect itself. The corresponding application of good faith in enforcement would limit the effect given to an enforcement term in light of the reasonable steps the enforcing party might have taken to secure the interest that term was intended to protect.

For example, consider an enforcement term in a contract for the sale of real property permitting the buyer to terminate if the seller fails to obtain necessary zoning approval by a certain date, it being understood that the buyer is under no duty to seek the approval. Suppose that the seller makes a genuine but unsuccessful effort to obtain the approval, but had the buyer responded to the seller's request to make itself available to the zoning board to provide information about its proposed use of the property, the required consent probably would have been granted. Invocation of the term would safeguard precisely the intended interest—it would protect the buyer against having to purchase land that could not be put to the intended use. It would do so, however, at an unnecessary cost to the seller: termination of the contract and complete loss of its benefits. Under those circumstances, invocation of the termination clause by the buyer would not be in good faith.

The two sides of the good faith analysis—the fit between intended and actual effect and mitigation—are not identical. The former tests for whether enforcement has gone awry because a term's effect would be substantially different than anticipated. The latter assumes that the effect on the enforcing party would be as intended, but questions whether something less severe might have accomplished the same result. The common denominator is the needlessness of the costs that would be imposed upon the party against whom the term is invoked. The elimination of such costs is, in large part, the purpose of established, remedial principles of contract law. The doctrine of good faith in enforcement can bring that policy to bear upon the entire range of enforcement terms.

Justification of the Good Faith in Enforcement Doctrine

The good faith in enforcement doctrine may appear to exert a limiting effect on the parties' agreement; it can prevent a party from taking advantage of an agreed enforcement term as written. The doctrine's ultimate purpose and effect, however, is to further the agreement. It does so by recognizing a hierarchy in the elements of a contract: a party's interest in performance—the expectation interest—has a higher priority than its interest in enforcement. By definition, enforcement is in the service of performance. The justification for the good faith in enforcement doctrine is precisely that it protects this priority.

Thus, good faith in enforcement is distinct in important respects from other doctrines that are grounded in, and justified by, public policies rather than the parties' agreement and therefore have a different reach and purpose than does good faith in enforcement. For purposes of this discussion, "public policy" refers to a broad range of societal interests that may restrict either private parties' power to form a contract or the legal effects of a contract they have formed. Public policy is a source of limitation on freedom of contract wholly external to the undertakings of the parties. It may preclude enforcement of a valid agreement, as illustrated by the growing number of cases in which termination of an at-will employee is found to have been wrongful. When a supervening event has rendered performance by one side extraordinarily burdensome or has greatly reduced the bargain's value to one of the parties, public policy may intervene by discharging an otherwise binding contractual obligation. It also may prevent enforcement of contracts infected with illegality in their manner of creation, particular terms, or basic purpose. In such cases, the propriety or efficacy of enforcement does not matter because the subject matter of the underlying promise itself runs afoul of the restriction. An important and relatively recent example of a restriction on contracting based on public policy is the doctrine of unconscionability, which denies effect to a contract in whole or in part because of procedural unfairness in its formation or substantive inequality in its terms.

All of these limitations take effect not because of what the parties intended or agreed, but in spite of it. Their focus is not on the parties' purpose in entering the agreement, but on accomplishing a social objective that may be unrelated to that purpose. Public policy is a means by which social or commercial morality is brought to bear in the contractual relationships of the market place.

By contrast, good faith in enforcement operates within the sphere of the parties' agreement. Its terms of reference are the ends the parties pursued in forming their contract—the benefits they bargained to receive—and the means they employed to help secure the receipt of those benefits. The doctrine recognizes that the means chosen may be stated with such breadth or imprecision that, if viewed and applied without reference to their purpose in an agreement, they could produce results quite different than what the parties contemplated. It responds to the possibility that, under the circumstances at the time of enforcement, the means selected will be unnecessarily costly because other steps easily might accomplish the same purpose. But it does not pass judgment on the validity of the expectation interest itself. That task is left to a variety of other doctrines.

Notes on Rescission and Restitution

1. In most of the scenarios discussed in these excerpts, the nonbreaching party prefers to exit the contract (and recover the market value of any benefits

he or she has conferred) rather than being put in the position he or she would have occupied if the contract had been performed. Typically, this means that the contract would have been a losing one for the nonbreacher, but the other party is the one who actually breaks the contract. (This is the "double lightning bolt" referred to by Goetz and Scott.) The result is that rescission and/or restitution is more favorable for the nonbreacher than expectation damages would have been.

Whenever the rescission or restitution remedy leaves the nonbreacher with more than his or her expectation interest, Goetz and Scott note that the threat of having to pay such a remedy could discourage an efficient breach. They also note that an efficient breach could still take place through renegotiation, in which the party who wanted out of the contract would have to share some of his or her gains with the party who wanted the contract enforced (supra at page 117). A similar analysis of the restitution measure and the incentives to renegotiate can be found in Henry Mather, Restitution as a Remedy for Breach of Contract: The Case of the Partially Performing Seller, 92 Yale Law Journal 14 (1982).

However, the possibility of renegotiation does not eliminate other possible effects of the rescission or restitution remedies. In this respect, restitution and rescission raise many of the same issues considered earlier in connection with other more favorable remedies such as specific performance or liquidated damages. For example, even if renegotiation is costless, the threat of having to share some of the gains from breach might lead the potential breacher to charge a higher price for entering into the contract initially, or to take too many precautions against unfavorable events that might lead him or her to want to get out of the contract. On the other hand, to the extent that the expectation measure (as actually applied by the courts) provided too little incentive for precautions, or less than the optimal amount of insurance, a measure such as restitution or rescission that gave the nonbreacher a little bit extra could provide a useful correction.

2. When the contract does not explicitly call for restitution or rescission and the courts must decide whether to make that remedy available, many of the relevant questions are the same ones that come up whenever the courts must select a remedial default rule. In particular, when the remedies of rescission or restitution would exceed the expectation measure, the issues raised are the same as those raised by other supercompensatory measures of damages, such as punitive damages (discussed infra at pages 127–37) or—in some cases—specific performance (discussed supra at pages 93–100).

In some cases, though—for example, in some cases involving express conditions on the contract—the contract explicitly provides for the rescission or restitution remedies if certain contingencies occur. In these cases, deciding whether to give effect to those conditions is similar to deciding whether to enforce a liquidated damage clause (discussed supra at pages 100–14), or a clause requiring specific performance as a remedy (supra at page 100). The

Andersen excerpt notes this parallel and points out that courts frequently refuse to enforce both express conditions and liquidated damage clauses. Are Andersen's arguments in defense of this pattern persuasive in light of the earlier readings on liquidated damage clauses?

3. When the broken contract would have been a favorable one for the nonbreacher—probably the more common case—the restitution and rescission remedies will usually leave the nonbreacher with *less* than his or her expectation interest. For example, if the only benefit the nonbreaching party has conferred on the other party is an advance payment of the contract price, the restitution remedy will simply require the breaching party to return the contract price. This is the form of the restitution measure that has most often been analyzed by economists. See, for example, A. Mitchell Polinsky, An Introduction to Law and Economics 33 (Boston: Little, Brown & Co., 2d ed. 1989); Steven Shavell, Damage Measures for Breach of Contract, 11 Bell Journal of Economics 466 (1980). In this special case when the restitution remedy consists of nothing more than the return of the contract price, the size of the remedy will not be affected by the extent of the nonbreacher's reliance and/or mitigation activities. As a consequence, when the restitution remedy consists merely of returning the contract price, it should not distort the nonbreacher's incentive to rely excessively. (Cf. the discussion of stipulated damage measures in the earlier excerpt by Cooter, supra at page 63.)

4. The restitution remedy could more frequently exceed the expectation measure if it were interpreted to require the breaching party to disgorge any extra profits he or she made as a result of the breach. For example, a seller who breaches a contract with B1 in order to sell to a higher-valuing B2 could conceivably be required to make good B1's expectation losses *and* to turn over to B1 the profits he or she made from the sale to B2. For discussions of this interpretation of the restitution principle, see Sidney W. DeLong, The Efficiency of Disgorgement as a Remedy for Breach of Contract, 22 Indiana Law Review 737 (1989); E. Allan Farnsworth, Your Loss or My Gain? The Dilemma of the Disgorgement Principle in Breach of Contract, 94 Yale Law Journal 1339 (1982); and Daniel Friedmann, Restitution of Benefits Obtained through the Appropriation of Property or the Commission of a Wrong, 80 Columbia Law Review 504 (1980).

5. When rescission of a contract requires that goods be returned to the seller, this will automatically affect one mitigation issue, for it will shift to the seller initial responsibility for finding an alternative use for the rejected goods. For analyses of this issue, see Alan Schwartz, Cure and Revocation for Quality Defects: The Utility of Bargains, 16 Boston College Law Review 543 (1975); George L. Priest, Breach and Remedy for the Tender of Non-Conforming Goods Under the Uniform Commercial Code: An Economic Analysis, 91 Harvard Law Review 960 (1978).

§ 2.3.4 **Punitive Damages**

Punitive Damages: Divergence in Search of a Rationale

BRUCE CHAPMAN AND MICHAEL TREBILCOCK

The Compensation Rationale

Types of Proscribed Conduct

[T]here is something closely analogous to genuine "punitive damages," namely "aggravated damages," which, under a compensatory rationale that normally would allow recovery for losses caused by mere negligence objectively considered, must *a fortiori* allow recovery for the special sense of dignitary loss a plaintiff feels when a defendant has harmed him or her "maliciously," "vindictively" or otherwise "outrageously." . . .

The kind of indignity or insult that is the source of aggravated damages under the compensatory rationale might tend to arise most often in cases of an intentional breach of a fiduciary duty. There is, after all, a special sense of wrong in being "used" by someone one trusts. An extreme example is marital rape. Ordinary rape might properly be thought of as the exclusive concern of criminal law, with an additional and separate private law remedy for damages confined to recovery for the severe sort of emotional and physical distress that commonly attaches to this kind of wrongdoing. But the wrong in "stranger" rape is in part to be explained by the idea that the victim is fungible (that is, any woman would have satisfied the criminal wrongdoer), something that, while it explains the deep emotional distress of the victim who is accustomed to being loved as a unique individual rather than being treated as a substitutable object of sexual gratification, does not require special damages for the added victimization of being wronged by someone one trusts. In marital rape, by contrast, there *is* this added sense of victimization which might properly be a source of aggravated damages.[1]

Less extreme forms of intentional breach of the sorts of obligations that

Bruce Chapman and Michael Trebilcock, Punitive Damages: Divergence in Search of a Rationale, 40 Alabama Law Review 741, 761, 764–72, 780–83, 786–87, 797–98, 806–09, 818–19 (1989). Copyright © 1989 by Bruce Chapman and Michael Trebilcock. Reprinted by permission.

1. We should not be thought of as suggesting that there is *no* dignitary loss in stranger rape, only in marital rape. Our point is only that there may be an *additional* sense of dignitary loss in marital rape that calls for *additional* compensation, namely aggravated damages.

arise out of special relationships, and which might properly provide grounds for a special sense of being "used" by someone one trusts, could be found in cases of wrongful dismissal from an employment relationship or breach of promise of marriage. In both contexts, the relationship that is intentionally breached is more than a merely contractual one for which the usual contract remedies would be adequate. Here the relationships are built on trust and dependency, and the victims of breach are particularly vulnerable to a sense of being "used." It should not be surprising, therefore, that punitive or aggravated damages have often attached to such cases. . . .

[T]he increasing number of American punitive damages cases in the context of insurance contracts might be such an instance. According to the American Bar Association's special report on punitive damages, the California courts first developed the rule, now followed in a majority of American jurisdictions, that an insured can sue in tort and recover punitive damages when an insurer breaches the "covenant of good faith" implied by law in the contract. Originally, these cases of "bad faith breach" involved third-party liability insurers who rejected settlement offers within policy limits, knowing that there was no reasonable basis for doing so, with the result that the insured was forced to pay personally a liability award in excess of the policy limits. Now the majority of courts hold the insurer liable in tort for bad faith breach, permitting the insured to recover the "extra-contractual" amount of liability as tort damages. While such a remedy can quite plausibly be explained as a deterrence rule aimed at a particular kind of principal-agent problem (namely, the limited liability of the insurer under the contract induces the insurer to take litigation risks that are excessive from the insured's point of view), it is also reasonable to suggest that the insured has suffered a special type of injury by virtue of the fiduciary nature of the relationship between insurer and insured, a relationship where the insurer occupies a position of trust and power and where the insured is dependent and vulnerable. The latter analysis would explain a comparable reluctance, now to be found in a literature critical of the idea of efficient breach, to allow other fiduciaries, such as banks and pension funds, to escape their contracts merely by paying damages.

In summary, therefore, the compensation rationale, while it focuses on the plaintiff, can nevertheless explain why in an action for punitive or, more accurately, aggravated damages there is a need to consider the nature of the defendant's conduct. There is a special kind of loss, namely a dignitary loss, in being maliciously targeted by the defendant as a victim. More often than not this particular targeting of the victim will be because there is an intentional breach of a special relationship of trust already existing between the defendant and the plaintiff, a relationship that makes the plaintiff feel especially vulnerable to being used and that helps to characterize the defendant's conduct, at least when it is intentional, as particularly outrageous. Moreover, since it is so often a special relationship between the two parties that is the source of upset, it makes sense for any remedy of that wrong to combine the two parties in a single action, and for the defendant in particular to pay damages for the wrong to the plaintiff. How these damages are to be quantified, however, has still to be considered.

Quantum of Punitive Damages

In the traditional compensatory theory of tort law, tort awards for damages are said to have the purpose of "making the victim whole"—of restoring (as far as is possible to do so with money) the victim to his or her pre-injury position. However, with the growth of law and economics scholarship, and a corresponding increased appreciation that the plaintiff-victim may end up paying for this level of compensation (for example, in a higher product price), this traditional full compensation view of tort awards has increasingly been replaced by something closer to an insurance approach. The question now is less what award would make the victim whole ex post, but rather what amount of insurance the average risk-averse victim would voluntarily have purchased ex ante.

Now for the relatively simple cases of single activity accidents where a potential injurer's activity only imposes a risk of *pecuniary* loss on a potential victim, a victim would typically insure for full compensation in the event of injury. Thus, there is no difference in quantum in treating the tort award from an ex post compensatory or ex ante insurance point of view. However, when the accidents or torts involve *nonpecuniary* losses, and in particular when the loss, given the accident, does not result in the victim putting any additional value on money than he or she would have done had the accident not occurred, then the insurance rationale for tort awards will imply a lower award than the compensatory rationale. This is because a nonpecuniary loss is the sort of loss for which a victim would not insure at all ex ante, but is nevertheless a real loss for which the victim might seek compensation ex post.

The sorts of dignitary loss referred to above that are fundamental to actions for punitive or aggravated damages are paradigmatic instances of nonpecuniary loss. There is no market for replacing lost dignity, nor could there ever be. Dignity simply cannot be bought and sold. Indeed, being "bought" often carries with it a loss of dignity. One can imagine, for example, being compensated, perhaps richly, for the loss of one's dignity ex post. But it is harder to think of insuring for this loss of dignity ex ante, since money is of so little use in replacing it.

This suggests that the insurance rationale would imply a zero quantum of aggravated or punitive damages. A full compensation rationale, by contrast, would imply a quantity sufficient to make the victim indifferent about the actual occurrence of the dignitary loss—quite possibly a very large amount. However, this contrast is overdrawn if one recognizes that awards under the compensatory rationale have a tendency to become reduced to those more appropriate to the insurance rationale, at least in cases where the two parties are already in a contractual relationship. In a products liability situation, for example, rich compensatory awards for dignitary loss, or aggravated damages awards, would tend to be passed on to the consumer-victims in higher prices.

But these higher prices would indicate premia for the kinds of losses for which consumers would rather not insure. Thus, in a perfectly functioning market one would expect consumers to have some incentive to contract out of any fully compensatory protection for dignitary loss; the proper award under the compensation rationale, therefore, would be reduced to the award implied by the insurance theory. This might explain why, until relatively recently, punitive damages were generally unavailable in actions for breach of contract.

However, one must be careful not to overplay the contractual forces that tend to reduce the compensatory awards for dignitary loss to the levels prescribed by the insurance theory. The one historical exception to the general rule of nonavailability of punitive damages in actions for breach of contract, namely actions for breach of promise to marry, is suggestive of the reasons why. Imagine what it would mean in pre-marital negotiations for a potential defendant to price the possibility of liability for dignitary loss into the marriage contract. Such behavior clearly signals to the other party a lack of reliability or, worse, bad intentions. It is hard to believe that such pricing behavior in prior negotiations would not endanger the contract's very existence.

The same holds true for comparable contractual negotiations in the other special relationships of trust referred to above. A potential defendant employer, for example, would find it difficult to price for the possibility that he or she might be held liable for imposing loss of dignity on a potential employee by way of some kind of outrageous or vindictive dismissal. Mere mention of the possibility again endangers the contract and the trust upon which the relationship depends. Moreover, since the source of the indignity in these types of losses is in the *intentional* breach of a special relationship of trust, other potential employers can easily avoid having to price for the possibility of liability for such losses if they have no such intentions. Thus, problems both of prior negotiation and effective competition should lead one to expect that contracts would not effectively reduce full compensatory awards for dignitary loss to the near zero level implied by the insurance argument. In the particular context of liability for intentional breach of trust leading to dignitary loss, the insurance argument for little or no quantum of damages simply has less bite than it usually does. This should not be surprising, since the insurance perspective holds best in contexts of strict liability where intentional breaches of a standard of behavior are not at issue.

The Retributive Rationale

. . . In the retributive view, the justification of any punishment is backward-looking and desert-based rather than forward-looking and consequentialist. Moreover, Kant, the paradigmatic retributive theorist, was adamant that even if punishment could be seen as advancing some other good extrinsic to the desert-based justification of punishment, the pursuit of this good must always be constrained by the requirement that punishment first be deserved:

> Judicial punishment can never be used merely as a means to promote some other good for the criminal himself or for civil society, but instead it must in all cases be imposed on him only on the ground that he has committed a crime; for a human being can never be manipulated merely as a means to the purposes of someone else and can never be confused with the objects of the Law of things . . . . He must first be found to be deserving of punishment before any consideration is given to the utility of this punishment for himself or for his fellow citizens. The law concerning punishment is a categorical imperative, and woe to him who rummages around in the winding paths of a theory of happiness looking for some advantage to be gained by releasing the criminal from punishment or by reducing the amount of it . . . .

It is, moreover, generally accepted among retributivists that the sort of criminal culpability that attracts punishment on a desert-based view is to be distinguished from that kind of fault that might attract liability in a private law tort action. The former involves intentional, or at least reckless, wrongdoing whereas the latter might only imply some kind of inadvertent transgression on the rights of others. Some examples should help to illustrate the general point and show why the former implicates punishment and state action in a way that the latter does not.

Consider, for example, the facts of *Regina v. Shymkowich*. A beachcomber, after removing two logs from a logging company's booming ground, was charged with theft. In his defence he claimed that he believed that the two logs had drifted into the boom and that, as drifting logs, he had a right to salvage them. If one accepts the beachcomber's story, it seems that the case only involves a mistake about entitlements, that is, a confusion as to where the line is drawn between the rights of the company and the rights of the beachcomber. The beachcomber by his action is not denying that the company's rights are relevant. Rather he accepts that the company has rights, but simply disputes that they extend to the logs in question. The appropriate response by the logging company is a private action against the beachcomber's conversion of the two logs.

However, if one does not accept the beachcomber's story as true, then the character of his transaction and our response to it is changed. Then it seems that he has intentionally stolen the logs and is rightly charged with, and convicted of, theft. His actions amount to more than a denial that the company has any rights to these logs; instead, they amount to a denial of the relevance of rights altogether. Since the infringement is of rights in general (or, as Kant or Hegel would have it, since the infringement is of the category of Right), correction of the transgression is more than just the private affair of those (no matter how many) whose particular rights have been infringed. The state, as guardian of the category of Right, and not just some private individual as a rights-holder, must take public action against the thief.

But recognition of the public law nature of the criminal law action can achieve a better understanding of the nature of criminal wrong itself. For a thief to deny the category of Right, he must engage that category *as a cate-*

gory, that is, conceptually. This means that his denial of Right must be cognitive, involving conscious or advertent indifference to the relevance of rights. Thus, the public form of the criminal action not only makes sense of, but more strongly positively requires, subjective mens rea on the part of the accused. Anything less, such as mere negligence or mistake, cannot explain why the state, as guardian of the category of Right, is a party to the action. That is, anything less cannot explain the public law nature of the criminal law. . . .

It might now be objected that there are some kinds of intentional or cognitive wrongdoing which are not the subject of state action against the wrongdoer. An intentional breach of contract, for example, is typically only a private law matter between the contracting parties. However, this objection fails to fully appreciate to what matter the wrongdoer's intention, or cognitive sense, must attach if the state, as guardian of the category of Right, is properly to respond in the form of a criminal law action. In the standard breach of contract case, and in particular in the intentional (often "efficient") breach of contract case, the breaching party is not really denying the relevance of rights to the dispute. Rather, the breaching party is at all times prepared to pay compensation, and thus to recognize the category of Right if a breach of the contract or damages flowing from that breach can be proved. In this respect, she is quite unlike the thief who, at the time the impugned transaction takes place, steals without any intention of paying compensation. Thus, the usual case of breach of contract, even a case of intentional breach of contract, is properly enforced by way of a private law action for damages. The thief, on the other hand, challenges the very idea of rights as protected spheres of autonomy or liberty. Moreover, in doing so, the thief negates any claim that he can make to liberty himself. The necessary implications of the thief's own conduct, therefore, is [sic] the denial of the thief's own liberty, a denial that ultimately manifests itself in the thief's own punishment by the state. . . .

[I]f we accept that the state must take action . . . there is still the question of whether the state's obligation is only a prima facie one that is capable of being "contracted out" to more efficient private enforcers. Here we argue that such contracting out is acceptable as long as certain requirements of the retributive rationale are not sacrificed to the demands of administrative efficiency. In particular, we shall argue that the proportionality and desert requirements set severe limits on the use of punitive damages as a method of privately enforcing the law against advertent wrongdoing. Additional methods of raising the level of enforcement, although subject to their own difficulties, might have to be considered.

It is easy to appreciate that a retributivist should be interested in the levels of punitive enforcement that are achieved under state action. After all, under the retributivist rationale individuals are supposed to get what they deserve, and failures of enforcement are failures in precisely that respect. Thus, the retributivist should be prepared to consider the possibility that private enforcers may bring more wrongdoers to justice. However, the greater efficiency of enforcement should never be allowed to overshadow the requirements of retributive justice. Retribution must not be replaced by privately motivated revenge, proportionality of punishment to offense must be preserved, and the administrative cost of prosecution should not be allowed as an excuse for consciously tolerating lower than possible levels of enforcement action against certain defendants. Thus, while the retributivist state might be prepared to delegate some enforcement of public sanctions to the private sector, it must at all times maintain its control over how that enforcement is carried out. . . .

One idea the retributivist will not entertain is the possibility of grossing up the penalty to reflect the fact that the probability of its enforcement is less than one. This is a common recommendation among deterrence theorists. However, for the retributivist, the promotion of any good extrinsic to punishment as desert, even the promotion of deterrence, is unjust in that it treats that individual who is caught and obliged to pay the penalty as a mere means for the good of others. The same sort of retributive concern would also preclude consciously choosing a low level of enforcement merely to avoid inconvenience or high administrative costs. . . .

Deterrence Rationales

. . . In the case of criminal prohibitions (e.g., murder, theft, and rape), in which the standards are clear and no social utility is attached to the conduct in question (i.e., the conduct is absolutely normatively proscribed), then initial analysis would suggest that it is immaterial how high the sanction, or f (for fine) is set. Unlike the compensatory rationale for punitive damages, where obviously demonstrable injury bounds f, or the retributive rationale for punitive damages, where moral notions of just desert and proportionality bound f, no similar constraints apply to the deterrence rationale.

However, further analysis suggests other issues that require resolution under the deterrence rationale. First, why is the imposition of sanctions for criminal wrongdoing not the exclusive preserve of the criminal law? In response, a case must be made out for private enforcement of criminal sanctions. The literature on private law enforcement suggests that the following kinds of consideration should be addressed. It might be argued, on the one hand, that for reasons canvassed in the previous section, public enforcement will often be incomplete. However, this may simply suggest that p [the probability of punishment] will often be sub-optimally low, the antidote to which is to raise f so that the *expected* costs of engaging in wrongdoing (namely, p $\times$ f) ensures compliance with the standard. The constraints on raising f in a public enforcement regime predicated on deterrence rationales are conventionally identified as follows:

1. the problem of marginal deterrence—the expected penalty for theft cannot be as high as armed robbery, otherwise there will be incentives for wrongdoers to substitute the more serious for the less serious offense; this sug-

gests some outside limit on how high f's for particular offenses can be raised without causing a convergence of expected penalties for disparate offenses;

2. the problem of liquidity limitations—f cannot be raised so high as to exceed a defendant's wealth, because obviously a penalty has no marginal deterrent effect beyond this point. At this point, alternative sanctions to fines—for example, imprisonment—may need to be invoked even though more costly to society. This may suggest that in a public enforcement context less wealthy violators may face imprisonment more frequently than more wealthy violators with respect to whom it is socially less costly to impose higher monetary penalties;

3. the problem of increased incentives for bribery—obviously the higher f is raised, the stronger the temptation for wrongdoers to attempt to bribe law enforcement officials to suppress wrongdoing and waive enforcement. Officials have correspondingly greater incentives to accept such bribes, given that they derive no personal benefits from the imposition of the formal penalty;

4. the problem of risk aversion or overdeterrence—Polinsky and Shavell have argued that the presence of risk aversion precludes lowering p and raising f beyond some range for offenses like parking violations without inducing socially inappropriate modifications of conduct (e.g., declining to park illegally in order to get a critically ill person to a hospital emergency ward). However, this tends to suggest that penalties for parking and like offenses should be seen as prices and not sanctions, and thus do not fall within the category of absolutely normatively proscribed conduct. . . .

Breach of Contract

Cooter argues that the proper way to conceive of the standard remedies for breach of contract (damages as the presumptive remedy; specific performance as an exceptional remedy) is that contract law has chosen to adopt a pricing regime rather than a sanctions regime. Subject to the standard categories of excuses (such as mistake or frustration), contractual liability also entails a form of strict liability. Cooter argues that this choice of regime is defensible within his framework because determining when it is or is not socially optimal to perform a contract, given the myriad of contingencies that may impact on contractual performance, would entail severe uncertainties, while damages (typically pecuniary) for breach are relatively easily assessed in the bulk of contractual settings.

This view of contractual liability (i.e., that a contractual obligation gives a promisor a right to elect between performing, on the one hand, or breaching and paying damages, on the other) has not gone unchallenged. However, it seems accurately to describe, for the most part, established contract law—that is to say, breaches of contract are not absolutely normatively proscribed, as are crimes, intentional torts, and torts of negligence, and may have positive social utility, at least in some cases (i.e., in cases of efficient breach). On this view of contractual liability as a pricing mechanism, the role for punitive damages would seem properly restricted to cases where low probability of enforcement (low p's) render it unlikely that the contract breacher will face ex ante the full expected social costs of his breach. Here, grossing-up damages in individual cases to offset low p's may discourage inefficient breaches (i.e., breaches that generate more costs for promisees than gains for promisors). The prime candidates for the award of punitive damages, on deterrent grounds, in this context would seem to be cases of mass consumer fraud or product defects where individual losses are insufficient to warrant the costs of suit ex post. This is also the class of case where class actions may offer enforcement efficiencies, but these, along with associated inefficiencies, would need to be weighed against the enforcement resources conserved by a high f, low p, private enforcement strategy along with its associated inefficiencies.

Notes on Punitive Damages

1. The scholarly literature on punitive damages has grown rapidly in recent years. Noteworthy collections of articles include the symposia published at 56 Southern California Law Review 1 (1982); and at 40 Alabama Law Review 3 (1989) (from which the Chapman and Trebilcock article is taken). The Cooter article referred to in the Chapman and Trebilcock excerpt is Robert Cooter, Prices and Sanctions, 84 Columbia Law Review 1523 (1984).

There is also a large economic literature on the effects of costly and imperfect enforcement, a topic whose significance extends far beyond punitive damages. Much of this literature is surveyed in Robert D. Cooter and Daniel L. Rubinfeld, Economic Analysis of Legal Disputes and Their Resolution, 27 Journal of Economic Literature 1067 (1989).

2. In a portion of the article not reprinted here, Chapman and Trebilcock describe the idea of "grossing up" as follows:

> By "grossing up" we mean multiplying the required penalty by the reciprocal of the probability of its enforcement. Thus, if the probability of a penalty X being enforced was only .5, then the penalty meted out in cases of actual enforcement would be 2X. This is done to insure that wrongdoers can expect ex ante to pay a penalty equal to X (since 2X discounted by a .5 probability of enforcement is equal to X). This ex ante perspective suggests why "grossing up" is more appropriate to the forward-looking concerns of deterrence than the backward-looking concerns of retribution.

Chapman and Trebilcock, 40 Alabama Law Review at 797, note 221. As they note, this multiplier concept has often been endorsed in the economic literature on punitive damages. For an example, see the Rea excerpt on liquidated damage clauses (supra at page 107).

In addition to the objections discussed in the Chapman and Trebilcock excerpt, other objections to a damage multiplier can also be raised. Calculation of the appropriate multiplier will rarely be easy, for it depends on the probability that the defendant will be sued, which is itself a function of the size of the multiplier (higher damage awards generally give rise to more lawsuits), thus introducing complicated feedback effects. Moreover, a multiplier of one over the probability of being sued will not always be appropriate from the standpoint of optimizing the defendant's precaution decision. When the probability that a defendant will be sued is less than one, but is higher for a defendant who breaches in a flagrant way than for a defendant who commits only a marginal breach (surely a plausible assumption), the damage multiplier that optimizes the defendant's incentives to take precautions will always be less than one over the probability of punishment. Depending on the exact relationship between the egregiousness of the defendant's breach and the probability of receiving punishment, it is even possible for exactly compensatory remedies to induce too many precautions, implying that the optimal damage multiplier should be something less than one. A multiplier may also be unattractive to risk-averse customers if it makes them pay an even larger price for a lottery with lower odds of winning but a larger prize for those who win. For discussions of these issues, see Alan Schwartz, The Myth that Promisees Prefer Supracompensatory Remedies: An Analysis of Contracting for Damage Measures 100 Yale Law Journal 369, 395–403 (1990); Richard Craswell, Contract Remedies, Renegotiation, and the Theory of Efficient Breach, 61 Southern California Law Review 629, 664–65 (1988); Jason S. Johnston, Punitive Liability: A New Paradigm of Efficiency in Tort Law, 87 Columbia Law Review 1385 (1987).

3. By definition, punitive damages give the promisee more than the expectation measure would. As a consequence, punitive damages raise many of the same arguments discussed earlier in connection with other remedies that are more favorable than the expectation measure, such as specific performance (supra at pages 93–100), liquidated damage clauses (supra at pages 100–14), and (in some cases) restitution and rescission (supra at pages 115–26). Thus, many of the arguments discussed in connection with those remedies could also be brought to bear on punitive damages.

For example, the Rea excerpt on liquidated damage clauses (supra at 105–13) suggested that most parties would not want to contract for damage amounts above the expectation measure, and that if they did so, it was probably the result of a mistake. If most parties would not want damage awards above the expectation level, what could justify courts in imposing such awards? Does the retributive rationale discussed by Chapman and Trebilcock justify punitive damages even if the parties would not have wanted punitive damages? For an expression of the contrary view, see Alan Schwartz, The Myth that Promisees Prefer Supracompensatory Remedies: An Analysis of Contracting for Damage Measures, 100 Yale Law Journal 369 (1990).

4. By the same token, many of the arguments applied to punitive damages by Chapman and Trebilcock could in theory be applied to each of the other supercompensatory remedies discussed above. For example, if the promisor's

breach was deliberate in a way that reflected a complete disregard for the promisee's rights, that might supply an argument for more liberal use of specific performance or cost-of-completion damages in addition to supplying an argument for punitive damages. This possibility was touched on earlier in the notes at page 92, in connection with measuring expectation damages by the cost of completion rather than by the diminution in value in those cases where the promisor's breach was "willful."

3

Defining the Performance Obligation

The readings in Chapter 2 discussed the remedies for breach. These remedies are relevant only if the contract has been breached, but determining whether a contract has been breached is not always easy. In many cases, there is a dispute about just what the contract requires—for example, is a contract for sale breached if the seller delivers a product that works, but does not work very well? Is a construction contract breached if unexpected soil conditions make the building impossible to construct? Questions like these are the subject of the readings in this chapter.

The readings in the first section discuss the "implied excuses" of impracticability, frustration, mistake, and nondisclosure. These doctrines *cut back* on the apparent content of the promise by treating the promisor as having implicitly disclaimed responsibility for certain unexpected problems, even though no such disclaimer was explicitly written into the contract. These doctrines overlap significantly, but the organization of the readings in the first major section of this chapter reflects a functional concern. Those in the first subsection—dealing primarily with impracticability, frustration, and mutual mistake—focus on risks that were not known to either party at the time of the contract. The readings therefore emphasize questions about which party should have gathered more information about the problem, or taken steps to prevent the problem from occurring, or purchased insurance against it. By contrast, the readings in the second subsection—dealing primarily with nondisclosure and unilateral mistake—deal with problems that were already known to one of the parties. Accordingly, these readings focus on a special kind of precaution: whether the informed party should be required to share his or her knowledge with the other party.

The second major section of this chapter deals with the doctrine of implied warranty. This doctrine *expands* the apparent content of the promise by treating the promisor as having assumed responsibility for certain problems, even though no such undertaking was explicitly written into the contract. As both the implied warranty and the implied excuse doctrines assign responsibility for risks, the doctrines discussed in these first two sections raise very similar issues.

The third section considers how the scope of the promise should be defined in "relational contracts." These are contracts that are expected to govern many facets of a contractual relationship over a relatively long period of time. Because the relationship can be quite complex, it is very common for issues to arise that were not explicitly provided for in the formal contract. The first set of readings analyzes these issues from the standpoint of economics (in the articles by Charles Goetz and Robert Scott and by Alan Schwartz) and sociology (in the article by Ian Macneil). The remaining readings, by William Klein and Gillian Hadfield, address the problems that arise when one party seeks to terminate a relationship after the other party has relied by making investments whose value would be lost outside the relationship.

Finally, the reading in the fourth section (by Aivazian, Trebilcock, and Penny) deals with the parties' efforts to redefine the scope of their obligations after the contract has been signed. Such efforts at contractual modification raise issues of consideration, duress, and contract formation (among others), so this reading could easily have been placed in Chapters 4 or 5. But these modifications also have the effect of reallocating the risk of unexpected contingencies, and thus raise issues very similar to those raised by implied excuses and implied warranties, as discussed in the earlier sections of this chapter.

§ 3.1 Implied Excuses

3.1.1 Impracticability, Frustration, and Mistake

Impossibility and Related Doctrines in Contract Law

RICHARD A. POSNER AND ANDREW M. ROSENFIELD

Ordinarily the failure of one party to a contract to fulfill the performance required of him constitutes a breach of contract for which he is liable in damages to the other party. But sometimes the failure to perform is excused and the contract is said to be discharged rather than breached. This study uses economic theory to investigate three closely related doctrines in the law of contracts that operate to discharge a contract: "impossibility," "impracticability," and "frustration." These are not the only excuses for nonperformance of a contract. Among other excuses, not discussed in this study, is the closely related doctrine of mutual mistake (sometimes called "antecedent impossibility"). Also related, and only incidentally discussed herein, is the doctrine of *Hadley v. Baxendale*, limiting the liability of the breaching party to the foreseeable damages of the breach. . . .

The Economics of Impossibility

The typical case in which impossibility or some related doctrine is invoked is one where, by reason of an unforeseen or at least unprovided-for event, performance by one of the parties of his obligations under the contract has become so much more costly than he foresaw at the time the contract was made as to be uneconomical (that is, the costs of performance would be greater than the benefits). The performance promised may have been delivery of a particular cargo by a specified delivery date—but the ship is trapped in the Suez Canal because of a war between Israel and Egypt. Or it may have been a piano recital by Gina Bachauer—and she dies between the signing of the contract and the date of the recital. The law could in each case treat the failure to perform as a breach of contract, thereby in effect assigning to the promisor the risk that war, or death, would prevent perfor-

Richard A. Posner & Andrew M. Rosenfield, Impossibility and Related Doctrines in Contract Law: An Economic Analysis, 6 Journal of Legal Studies 83, 83, 88–94 (1977). Copyright © 1977 by the University of Chicago. Reprinted by permission.

mance (or render it uneconomical). Alternatively, invoking impossibility or some related notion, the law could treat the failure to perform as excusable and discharge the contract, thereby in effect assigning the risk to the promisee.

From the standpoint of economics—and disregarding, but only momentarily, administrative costs—discharge should be allowed where the promisee is the superior risk bearer; if the promisor is the superior risk bearer, nonperformance should be treated as a breach of contract. "Superior risk bearer" is to be understood here as the party that is the more efficient bearer of the particular risk in question, in the particular circumstances of the transaction. Of course, if the parties have expressly assigned the risk to one of them, there is no occasion to inquire which is the superior risk bearer. The inquiry is merely an aid to interpretation.

A party can be a superior risk bearer for one of two reasons. First, he may be in a better position to prevent the risk from materializing. This resembles the economic criterion for assigning liability in tort cases. It is an important criterion in many contract settings, too, but not in this one. Discharge would be inefficient in any case where the promisor could prevent the risk from materializing at a lower cost than the expected cost of the risky event. In such a case efficiency would require that the promisor bear the loss resulting from the occurrence of the event, and hence that occurrence should be treated as precipitating a breach of contract.

But the converse is not necessarily true. It does not necessarily follow from the fact that the promisor could not at any reasonable cost have prevented the risk from materializing that he should be discharged from his contractual obligations. Prevention is only one way of dealing with risk; the other is insurance. The promisor may be the superior insurer. If so, his inability to prevent the risk from materializing should not operate to discharge him from the contract, any more than an insurance company's inability to prevent a fire on the premises of the insured should excuse it from its liability to make good the damage caused by the fire.

To understand how it is that one party to a contract may be the superior (more efficient) risk bearer even though he cannot prevent the risk from materializing, it is necessary to understand the fundamental concept of risk aversion. Compare a 100 percent chance of having to pay $10 with a one percent chance of having to pay $1,000. The expected cost is the same in both cases, yet not everyone would be indifferent as between the two alternatives. Most people would be willing to pay a substantial sum to avoid the uncertain alternative—for example, $15 to avoid having to take a one percent chance of having to pay $1,000. Such people are risk averse. The prevalence of insurance is powerful evidence that risk aversion is extremely common, for insurance is simply trading an uncertain for a certain cost. Because of the administrative expenses of insurance, the certain cost (that is, the insurance premium) is always higher, often much higher, than the uncertain cost that it avoids—the expected cost of the fire, of the automobile accident, or whatever. Only a risk-averse individual would pay more to avoid bearing risk than the expected cost of the risk.

The fact that people are willing to pay to avoid risk shows that risk is a cost. Accordingly, insurance is a method (alternative to prevention) of reducing the costs associated with the risk that performance of a contract may be more costly than anticipated. It is a particularly important method of cost avoidance in the impossibility context because the risks with which that doctrine is concerned are generally not preventable by the party charged with nonperformance. As mentioned, if they were, that would normally afford a compelling reason for treating nonperformance as a breach of contract. (Stated otherwise, a "moral hazard" problem would be created if the promisor were insured against a hazard that he could have prevented at reasonable cost.)

The factors relevant to determining which party to the contract is the cheaper insurer are (1) risk-appraisal costs and (2) transaction costs. The former comprise the costs of determining (a) the probability that the risk will materialize and (b) the magnitude of the loss if it does materialize. The amount of risk is the product of the probability of loss and of the magnitude of the loss if it occurs. Both elements—probability and magnitude—must be known in order for the insurer to know how much to ask from the other party to the contract as compensation for bearing the risk in question.

The relevant transaction costs are the costs involved in eliminating or minimizing the risk through pooling it with other uncertain events, that is, diversifying away the risk. This can be done either through self-insurance or through the purchase of an insurance policy (market insurance). To illustrate, a corporation's shareholders might eliminate the risk associated with some contract the corporation had made by holding a portfolio of securities in which their shares in the corporation were combined with shares in many other corporations whose earnings would not be (adversely) affected if this particular corporation were to default on the contract. This would be an example of self-insurance. Alternatively, the corporation might purchase business-loss or some other form of insurance that would protect it (and, more important, its shareholders) from the consequences of a default on the contract; this would be an example of market insurance. Where good opportunities for diversification exist, self-insurance will often be cheaper than market insurance.

The foregoing discussion indicates the factors that courts and legislatures might consider in devising efficient rules for the discharge of contract. An easy case for discharge would be one where (1) the promisor asking to be discharged could not reasonably have prevented the event rendering his performance uneconomical, and (2) the promisee could have insured against the occurrence of the event at lower cost than the promisor because the promisee (a) was in a better position to estimate both (i) the probability of the event's occurrence and (ii) the magnitude of the loss if it did occur, and (b) could have self-insured, whereas the promisor would have had to buy more costly market insurance. As we shall see, not all cases are this easy.

The Analysis Applied

. . . A, a manufacturer of printing machinery, contracts with B, a commercial printer, to sell and install a printing machine on B's premises. As B is aware, the machine will be custom-designed for B's needs and once the machine has been completed its value to any other printer will be very small. After the machine is completed, but before installation, a fire destroys B's premises and puts B out of business, precluding B from accepting delivery of the machine. The machine has no salvage value and A accordingly sues for the full price. B defends on the ground that the fire, which the fire marshal has found occurred without negligence on B's part—indeed (the same point, in an economic sense), which could not have been prevented by B at any reasonable cost—should operate to discharge B from its obligations under the contract.

The risk that has materialized, rendering completion of the contract uneconomical, is that a fire on B's premises would prevent B from taking delivery of the machine at a time when the machine was so far completed (to B's specifications) that it would have no value in an alternative use. The fact that the fire occurred in premises under B's control suggests that B had the superior ability to *prevent* the fire from occurring. This consideration is entitled to some weight even though the fire marshal found that B could not, in fact, have prevented the fire (economically); the fire marshal might be wrong. Certainly as between the parties B had the superior ability to prevent the fire. But in light of the fire marshal's finding, ability to prevent cannot weigh too heavily in the decision of the case.

Turning to the relative ability of the parties to *insure* against the machine's loss of value as a result of the fire, we note first that while B was in a better position to determine the probability that a fire would occur, A was in a better position to determine the magnitude of the relevant loss (the loss of the resources that went into making the machine) if the fire did occur. That loss depended not only on the salvage value of the machine if the fire occurred after its completion but also on its salvage value at various anterior stages. A knows better than B the stages of production of the machine and the salvage value at each stage.

Assuming the actuarial value of the risk has been computed, there remains the question which of the parties could have obtained insurance protection at lower cost. Depending on the volume of A's production and on A's prior experience with contingencies such as occurred in the contract with B, A may be able to eliminate the risk of such contingencies simply by charging a higher price—in effect, an insurance premium—to all of its customers; A may in short be able to self-insure. B is less likely to be able to do so: the magnitude of its potential liability to A in the event of a default may greatly exceed any amount it could hope to pass on to its customers in the form of higher prices. As for market insurance, it seems unlikely that B could obtain for a reasonable price a fire insurance policy that protected it not only against the damage to its premises (and possibly to its business) caused by a fire but also against its contractual liability to A which, as mentioned, depends on the stage in the production of the machine at which the fire occurs, a matter within the private knowledge of A.

We are inclined to view A as the superior risk bearer in these circumstances and thus to discharge B. This inclination would be strengthened if it turned out that A was a publicly held, and B a closely held, corporation, for then the owners of A could eliminate the risk of the loss of the machine's value simply by combining their shares in A with shares of other companies in a suitably diversified portfolio. It is generally more difficult for the owners of a closely held corporation to diversify away the risks associated with their holdings in the corporation, for often those holdings represent a large fraction of their net assets.

Contract as Promise

CHARLES FRIED

It is not the presence of risk or uncertainty that vitiates agreement, since contracts generally are a device for allocating risks. In a contract for future delivery the seller takes on himself the risk that the goods will rise in price or that for some other reason it will become more burdensome for him to perform, and the buyer assumes reciprocal risks. The language of mistake suggests that certainty is the paradigm, but in fact contracts are largely a deliberate attempt to deal with uncertainty. The parties might have allocated the risk of the king's illness [in *Krell v. Henry*] or of the fire [in *Taylor v. Caldwell*], but they had not done so. And parties may allocate risks as to existing as well as to future circumstances. The prospector who buys a claim is taking a risk about the presence of minerals. In the celebrated case of *Wood v. Boynton* the court upheld the sale for one dollar of an unidentified stone, which turned out to be a rough diamond, as involving in effect a gamble on the part of the buyer and seller as to the stone's value; while in *Sherwood v. Walker* the court found that the seller had not transferred nor had the buyer paid for the chance that an apparently barren prize cow was in fact pregnant.

That the presence of risk, which is common to these cases, is not the cause of the difficulty is illustrated by yet another celebrated case, *Raffles v. Wichelhaus,* in which a buyer refused to accept delivery of a shipment of cotton arriving in Liverpool from Bombay on the ship *Peerless,* as he had contracted to do, because the *Peerless* he had in mind had sailed in October, while the seller's ship *Peerless* had sailed in December. To speak of problems of risk, allocated or unallocated, in that case seems not a little strained. The straightforward point is that the two parties, though they seemed to have agreed, had not agreed in fact. And this—not mistake or risk—is at the heart of all of these cases: There just is no agreement as to what is or turns out to be an important aspect of the arrangement. In *Raffles* there was agreement to purchase cotton from India shipped on a ship named *Peerless*. As it turned out, there was no agreement at all on the crucial issue of which ship *Peerless*. Similarly in the coronation cases and *Taylor* there was agreement about some things, but none about risks (the king's illness, fire) that might have been covered and that turned out to be crucial. So risk comes into it, but only as one element about which the parties might have reached agreement but unfortunately did not. In all of these cases the court is forced to sort out the difficulties that result when parties think they have agreed but actually have not. The one basis on which these cases cannot be resolved is on the basis of the agreement—that is, of contract as promise. The court cannot enforce the will of the parties because there are no concordant wills. Judgment must therefore be based on principles external to the will of the parties. . . .

It would be irrational to ignore the gaps in contracts, to refuse to fill them. It would be irrational not to recognize contractual accidents and to refuse to make adjustments when they occur. The gaps cannot be filled, the adjustments cannot be governed, by the promise principle. We have already encountered the two competing residuary principles of civil obligation that take over when promise gives out: the tort principle to compensate for harm done, and the restitution principle for benefits conferred. Each of these has some application, but only a limited or puzzling one, to the cases that concern us. If a contracting party has knowingly concealed or negligently overlooked an eventuality that sharply alters the risk—for example, if the owner in the *Stees* case knew or easily could have learned of the difficult condition of his land—we may force him to bear the resulting loss for that reason. If a party has conferred benefits (built something, paid in advance) under a contract that subsequently fails because of frustration, the benefit should be paid for or returned. In the cases arising out of the cancellation of the coronation, for instance, the rooms overlooking the procession route were valuable just for that purpose and could be rented again when the coronation was rescheduled. To allow the owners to collect twice would have been preposterous. (But what if there had been no later procession?)

Unfortunately in many cases both parties are harmed, neither is at fault, neither benefits. The half-built house is destroyed by an earthquake. The half-built machine is rendered useless by a government regulation. The program printed for the canceled yacht race is of no interest to anyone. In such situations a distinct third principle for apportioning loss and gain comes into play: the principle of sharing. Consider these cases:

> III. A man and a woman spending the night together in an inn discover an
> envelope containing a large sum of money at the back of a bureau drawer. The

original owner cannot be traced. Should the owner of the inn, the man, or the woman, keep the money, or should they share it?

IV. In an unusually severe storm a freighter loses some but not all of the valuable cargo of several shippers. Should the loss lie where it falls, should it be borne by the owner of the vessel, or should it be shared among all the shippers and the owner of the vessel?

These cases point up the difference between sharing on one hand and the benefit and harm principles on the other. In benefit and harm the predicate for shifting a burden or an advantage is the responsible act of one of the parties. Such responsibility may arise out of culpability—including negligence—a voluntary act, or a prior assumption of responsibility, as by a contract. As cases III and IV illustrate, however, in some situations there may be no basis for holding the parties responsible or accountable to one another. Rather, persons in some relation, perhaps engaged in some common enterprise, suffer an unexpected loss or receive an unexpected gain. The sharing principle comes into play where no agreement obtains, no one in the relationship is at fault, and no one has conferred a benefit. Sharing applies where there are no rights to respect. It is the principle that would apply if a group of us were to land together on some new planet. It is peculiarly appropriate to filling the gaps in agreements, to picking up after contractual accidents. Applied to the collapsing house in *Stees* or the half-built machine now rendered useless, it says that the loss would not lie where it falls, but the parties would share that loss—which would mean perhaps that the owner or buyer would pay for half of the useless work that was done. And this is the direction in which courts are now moving.

Why, you might ask, should just the parties to the agreement share the benefits and burdens? In the case of the half-built machine why should not the government pay part of the cost? In the case of the house, why should not the neighbors chip in, since the earthquake damage might have happened to them? In the case of the money found in the inn, why should not the boon be spread more widely? The question necessarily leads into very general issues of political philosophy, and the hesitation to recognize and to fill gaps in contractual arrangements may have arisen as a result of worries at this most general level. Admitting a general obligation to share is rightly seen as a threat to the principles of autonomy and personal responsibility. If we must share the benefits and burdens of random contractual accidents, why not share all of life's benefits and burdens? Why not view good and bad luck in investments, choice of occupation, or market strategy as accidents to be evened out by sharing? And indeed why not view even variation in talents, character, or disposition as accidents? In the end there would be full sharing, and no one would enjoy the benefits or bear the responsibilities of his personal choices—indeed of his person. In such a system the concept of autonomy, which lies behind contract as promise, would be rendered meaningless.

Modern liberal democracies and liberal political theory have sought to develop a concept of sharing that yet leaves the person and his liberties intact. . . . [T]he accommodation is sought through the basic division of function in the modern welfare state between private market (contractual) autonomy and general redistributive welfare schemes. This accommodation assumes that the obligation to share is a general one in which all should participate by tax contributions. The system attempts to reduce the extreme disparities in overall wealth that undermine the possibility of community and at the same time to provide both fair opportunities for advancement and a guaranteed social minimum. Such a system is designed to provide a framework, a structure within which individuals may indeed exercise their autonomy and reap the benefits or suffer the consequences of their choices. Though the exact formula for what constitutes tolerable disparities and a decent social minimum must be subject to shifting, political judgments, I affirm that this structure and its purpose are in principle sound.

Does this mean that there is no room for the principle of sharing in the contractual domain, that its admission there would threaten to undermine the healthy compromise of autonomy and community implicit in liberal democracies? I think not. By engaging in a contractual relation A and B become no longer strangers to each other. They stand closer than those who are merely members of the same political community. Like the persons in my examples they are joined in a common enterprise, and therefore they have some obligation to share unexpected benefits and losses in the case of an accident in the course of that enterprise. Just as we do not say that C must come in to share those losses, so we do not say that A and B must share losses that are wholly outside the scope of their enterprise.

This appealing resolution has problems. Is not the contractual enterprise just the enterprise in which mutual obligations have been willingly undertaken; and yet do not contractual accidents occur precisely because no mutual engagements have been made; so that we are constructing a kind of nonconsensual penumbra around the consensual core? In terms of what do we construct this penumbra? Not in terms of the wills or promises of the parties. Obviously some standard of sharing *external* to the parties must control. But if the law is free to impose standards external to the intention of contractual parties to the end of fairly dividing up losses in the case of contractual accident, does this not show that principles of sharing are available potentially to allocate burdens and benefits among any set of citizens at all? If so, then not only is the contractual nexus unnecessary to create the focusing predicate *for* sharing, sharing itself may be seen as so powerful a principle as to overwhelm that nexus, reversing the effects of agreements even where there has been no accident.

We need not go this far. Those in concrete or personal relations must have a greater care for each other than those who stand to each other in the abstract relation of fellow citizens, or fellow man. By this principle family members and friends are owed and may engross a greater measure of our concern than abstract justice prescribes. By this principle too, direct or intentional harm constitutes a wrong. Making another person the object of your intention, a step along the way in your plans, particularizes that person and forms a concrete relation between you and him. Typically such concrete relations are freely chosen. Though some family ties are not chosen, ties of friendship are, and, as to harm, the intention to harm a specified other person (stranger or not) is quintessentially voluntary.

A contractual relation is a good example of a concrete relation that may give rise to a more focused duty to share another's good or ill fortune. The relation is, after all, freely chosen. Indeed this is the same idea as that the contractual parties are in a common enterprise—an enterprise they chose to enter. True, the bond does extend beyond the explicit terms of the contract, since the problems we are concerned with are by hypothesis not explicitly disposed of in the contract. The contract does, however, also imply a limit to the obligation. If I have furnished machinery to your factory over the years I am to some unspecified further degree involved in your manufacturing endeavor, but surely not in the misfortunes that befall you in some unrelated speculation, or in your family travails. If I have agreed to sell you a small standard part, an extension cord or a light bulb, my implication in your venture (and yours in mine) will have very little penumbra beyond the actual agreement of sale. Thus in filling the gaps it is natural to look to the agreement itself for some sense of the nature and extent of the common enterprise. Since actual intent is (by hypothesis) missing, a court respects the autonomy of the parties so far as possible by construing an allocation of burdens and benefits that reasonable persons would have made in this kind of arrangement. (It treats the contract as a kind of charter or constitution for the parties' relation.) This is, as I argued earlier, an inquiry with unavoidably normative elements: "reasonable" parties do not merely seek to accomplish rational objectives; they do so constrained by norms of fairness and honesty. Finally, this recourse to principles of sharing to fill the gaps does not threaten to overwhelm the promissory principle, for the simple reason that the parties are quite free to control the meaning and extent of their relation by the contract itself.

Mistake, Frustration, and the Windfall Principle of Contract Remedies

ANDREW KULL

[T]he characteristic and traditional response of our legal system to cases of mistaken and frustrated contracts is neither to relieve the disadvantaged party nor to assign the loss to the superior risk bearer, but to leave things alone. The party who has balked at performing will not be forced to proceed, but the completed exchange will not be recalled. Walker will not be forced to deliver to Sherwood a breeding cow sold for the price of beef; but neither will Wood be allowed to recover the yellow diamond, already delivered, unwittingly sold to Boynton for the price of a topaz. Where an exchange has been interrupted after part performance—the usual case in the context of frustration—courts following the traditional rule will not intervene to readjust the allocation of losses that chance and the parties' agreement have created. Persons hiring rooms from which to view a coronation procession will be excused by the king's indisposition from paying any balance of the price not yet due; but they will not be allowed to recover sums already paid at the time the ceremony is canceled.

The principle of inertia that frequently seems to guide the remedies for mutual mistake and frustration may seem harsh in some particular applications, but it is neither arbitrary nor illogical. Disparities between anticipation and realization in contractual exchange, the risk of which has not been allocated by the parties, are in the nature of "windfalls" (including those, carrying adverse consequences, that might more properly be described as "casualties"). The law will not act to enforce such windfalls—to compel an exchange on terms that were not bargained for—because its objective is limited to giving effect to the parties' agreement. But if the parties have not allocated the risk of a particular windfall or casualty to one of them, neither have they allocated it to the other. There is thus no basis in their bargain on which to justify a court's intervention to shift windfall benefits and burdens in either direction. As a matter of social utility, excluding for the moment considerations of fairness, it will ordinarily be a matter of indifference whether the windfall cost or benefit, once realized, falls to A or to B. Reallocation after the event thus involves significant administrative costs while achieving no compelling social advantage. The judicial disposition to let windfalls lie—to answer the claim of mistake or frustration by confirming the status quo—is here referred to as the "windfall principle." . . .

The Trouble With "Gap-Filling"

The Peculiar Intractability of Mistake and Frustration

. . . A common presumption sees a contract as "incomplete" if it fails to address specifically any ground of subsequent dispute. All contracts are potentially "incomplete" in this sense, a circumstance that may be explained in terms of transaction costs or simple human fallibility. The most common defense of judicial intervention asserts that a court-supplied clause, properly chosen, can rectify an omission by providing the term the parties would have negotiated if they had thought about it. So conceived, the judicial role comes very close to enforcing the parties' own agreement rather than writing one for them. The only problem is whether the question of what the parties "would have done" can be reliably answered.

The idea that incomplete contracts result from transaction costs leads to a related conception, that the court's role in dealing with gaps is to design suitable default rules (in the manner of U.C.C. provisions), to govern both the case at bar and future contracts unless varied by express agreement. The standard approach to the choice of default rules recommends that they be those the parties to a typical contract would be most likely to select for themselves. This view looks to the economies to be realized by contracting parties, who will negotiate fewer tailor-made provisions when they may, if they choose, incorporate off-the-rack terms by saying nothing at all. But this standard prescription for judicial intervention confronts distinctive difficulties when used to fill the most visible and interesting sort of contractual gaps, those that result in mistaken and frustrated contracts.

The suggestion that a court fill a gap in a contract by supplying the term the parties would have chosen themselves is feasible only if the unanswered question is one that most parties similarly situated would answer the same way most of the time. If the parties have not specified a time for performance, a court serves efficiency goals if, in the absence of contrary indications, it supplies a term calling for performance within a reasonable time. Default terms actually employed by the courts rarely depart from this comfortable level of generalization. But the commonsense reasoning that permits a court to supplement a contract in this manner is powerless to answer the question, incomparably more complex, of how the parties would have allocated the risk of a particular frustrating circumstance that they chose not to address in their agreement.

The seemingly insuperable difficulty of divining the parties' unexpressed ex-ante preferences might appear to be resolved if we could identify a universal principle by which all contracting parties, properly informed, would choose to allocate those risks of mistake and frustration not otherwise addressed by their agreement. Thus Posner and Rosenfield proposed, in a well known article, that judges allocate the risk of a frustrating circumstance by assigning it to the party determined to be the superior risk bearer. Because the superior risk bearer is the party who can bear the risk in question at lower cost, allocation of the risk to him is in the economic interest of both parties; it is therefore the informed choice of all parties for all risks.

The relationship between superior risk bearing and risk allocation provides an important insight, but commentators have not generally been clear about its implications. Where the difference in the parties' risk-bearing capacity is sufficiently pronounced, an identification of the superior risk bearer may enable us to reconstruct an allocation of risks that was an implicit (i.e., actual and intended) part of their agreement. Indeed, the idea that risk bearing will serve as a guide to the intent of the parties appears to have been the principal claim of Posner and Rosenfield. If the owner of a building has insurance covering the value of repairs or additions while work is in progress, and the builder's interest is uninsurable, a court might reasonably conclude that the risk of destruction prior to completion was implicitly allocated to the owner even if his contract with the builder says nothing about it. But the status of economic analysis in this example is merely that of any other circumstantial evidence as to the parties' intent; and if we can properly identify a *contractual* allocation of the risk, by inquiring into risk-bearing capacity or any other evidence, the contract is not frustrated.

The assertion that risk-bearing capacity should influence the choice of remedies for frustrated contracts necessarily implies, therefore, that a judicial allocation to the superior risk bearer can substitute for a contractual allocation the parties did *not* make for themselves. This contention overlooks the simple but critical fact that the contract comes before the court only after the risk of the allegedly frustrating disparity has materialized. The parties' neglected opportunity to make an ex-ante allocation of *risks* has been irretrievably lost. Instead, a court confronted with an allegedly frustrated contract (which it may enforce or discharge on whatever terms it chooses) is necessarily limited to an ex-post allocation of *losses*. . . .

The judicial assignment of frustration losses to superior risk bearers might nevertheless be thought desirable for its influence on future transactions. Because parties in a costless negotiation would assign every risk to the superior risk bearer, a default rule to this effect might appear to permit savings on transaction costs while encouraging the efficient allocation of risks. But compared to the alternatives—one possibility being the rule that the loss lies where it falls—a default rule based on an after-the-fact judicial assessment of risk-bearing capacity is singularly unappealing. Its effect would be to cause parties to assign every risk of a frustrating event, not otherwise allocated by their contract, to that party determined by a court (after the frustration of the contract) to have been the superior risk bearer with respect to that risk. This is an off-the-rack term that no rational contracting party would willingly adopt, given the near impossibility of predicting which party would later be found by the court to have been the superior risk bearer with respect to the risk in question.

Risk-bearing capacity is ordinarily understood to include, not merely the parties' relative aversion to risk (in itself a condition that real-world courts can scarcely ascertain), but also their relative capacity to avoid a given risk or to insure against it, by self-insurance or otherwise. These are conditions that depend on the nature of the risk in question, so that the superior risk bearer with respect to a given risk frequently cannot be identified, even theoretically, until the nature of the risk is known. But the function of a default term in this context is to allocate unidentified risks, many of them "unforeseeable" in common parlance. Moreover, even if the nature of certain classes of unidentified risks could be accurately identified in advance, the determination of superior risk-bearing capacity depends on so many additional variables, many of them difficult to establish conclusively, that it will normally be impossible for contracting parties to predict with confidence how a future court might decide the issue. Superior risk-bearing capacity *as determined after the fact by judges* is therefore a default term that conveys no usable information to the parties; and an uninformative default term cannot be the source of any economies from superior risk spreading.

The imposition of a default term whose effect is uncertain will not reduce transaction costs. On the contrary, it will increase them: parties will attempt to exclude by contract the added uncertainty of unpredictable judicial intervention. Where they fail to do so, the uncertainty of outcomes under any such legal rule will encourage litigation. Nor can a default term yielding unpredictable results produce more efficient risk spreading, since increased efficiencies can only be realized when each party knows which risks he will bear. But it is precisely where a risk-bearing advantage is clear enough to be recognized that the parties are most likely to allocate it as an express or implied term of the contract. Such risks will therefore not be assigned by the default rule, whatever it may be. While the sheer *in terrorem* effect of an unpredictable default rule would presumably encourage a greater investment by the parties in contractual negotiation and specific risk allocation, this result is not a distinctive advantage of the rule under consideration. . . .

The Displaced Private Allocation

A legal regime that leaves losses from frustration to lie where they fall . . . is one under which the parties to any contract allocate two levels of risks. To the extent they think it worthwhile, the parties allocate the risks of various frustrating circumstances, either specifically or through such broad-brush devices as force majeure clauses. Those specific risks that the parties either do not recognize or do not choose to allocate comprise the residual risk that the contract will eventually be frustrated for one reason or another. . . .

The ex-ante allocation of losses from frustration operates chiefly by adjusting the advantage or disadvantage to either party from an unexpected interruption of contractual performance. The basic mechanism for shifting such losses, employed to some extent in most contracts, is simply the payment term: payment in advance versus payment on delivery, with the intermediate possibilities of instalment payments, progress payments, and the like, all of which may be made subject to such further conditions as the parties choose. Expressed more generally, the incidence of losses from frustration will be a function of the timing and conditions of each party's obligations to perform. On the facts of *Fibrosa*, for example, the loss caused by the frustrating event that later discharged the contract had been clearly allocated between the parties as a function of the payment terms. Frustration loss early in the course of performance was assigned entirely to the buyer; frustration loss just before completion was assigned preponderantly to the seller. The allocation was somewhat crude, but that is presumably because the parties saw no advantage in negotiating a more complex one. The contractual devices for allocating such losses are intuitively understood and limited only by the energies of draftsmen. They can and will be refined as far as the parties deem appropriate.

The residual allocation of frustration losses produced by these devices in the context of a windfall rule is neither artificial nor arbitrary. It is a result chosen by the parties, not in the tautological sense that it is demonstrably a function of a privately chosen term but in a broader sense that allows us to describe its consequences as consciously self-imposed. In real-world negotiation, the possibility of frustration may be remote from the minds of the parties, but the risk of interrupted performance (presumably from the other party's default) is ever present. The parties are also well aware that the courts provide no effective relief for most contractual defaults. A's realistic expectation is therefore that, in the event B ceases to perform and the parties are unable to negotiate a settlement, A's best recourse will frequently be to walk away from the deal without seeking or obtaining judicial intervention. Should such an event come to pass, the circumstances of the parties will be precisely the same as if their contract had been interrupted by frustration under a legal regime denying judicial intervention. In everyday negotiation, therefore, each party seeks to shift the incidence of the same kinds of losses as those resulting from a frustrated contract, using devices designed to safeguard against the consequences of default. . . . The contractual allocation of potential frustration losses is not only theoretically possible, it is effectively being made all the time.

The possibility of a comprehensive ex-ante allocation of frustration risks— one that includes both the risk of various frustrating circumstances and the incidence of loss should the contract nevertheless be frustrated—depends on the courts' refusal to shift those losses when they occur. Conversely, it is only by holding out the promise of judicial intervention to adjust losses *ex aequo et bono* that courts relieve the parties of the need to address the distribution of these losses in the context of their agreement. A rule that losses lie where they fall actually compels the parties to allocate such losses ex-ante, including losses from risks described as "unforeseeable"; while a rule permitting judicial intervention and loss sharing makes it impossible for them to do so.

Relational Contracts in the Courts

ALAN SCHWARTZ

It is helpful to begin with an example. The parties make a five-year contract under which the seller will supply the buyer's requirements for goods at a constant price of $100 per unit. The seller expects to make a 20 percent gross profit per item, and the buyer expects to value each item at $120. If the spot-market price for the goods in the third year is $200, the seller's opportunity cost of performance—$100 per unit—exceeds her expected gain under the contract—$20; if the market price is $50, the buyer's opportunity cost of performance—$50 per unit—exceeds his expected gain under the contract— $20. The seller in the former case and the buyer in the latter prefer breach. The question is whether the parties could have escaped these outcomes with a better contract.

Alan Schwartz, Relational Contracts in the Courts: An Analysis of Incomplete Agreements and Judicial Strategies, 21 Journal of Legal Studies 271, 284–90 (1992). Copyright © 1992 by The University of Chicago. Reprinted by permission.

The seller's problem can be avoided by a cost-plus contract (price equals seller's current cost multiplied by 1.25), but buyers seldom will agree to condition price on cost. A seller's costs commonly are unobservable. The seller under a cost-plus contract thus has no incentive to minimize costs, but does have an incentive to manipulate the cost data. The buyer's problem can be avoided by a contract that links price to the demand that the buyer faces, but sellers seldom will agree to condition price on a particular buyer's downstream demand because demand commonly is unobservable or unverifiable. The buyer under a demand-based contract will have an incentive to understate demand to induce price reductions. The parties, therefore, have two contracting problems: (1) to keep contract gains above the opportunity costs of performance when cost and demand are noncontractible and (2) to allocate the risk that their solution to the first problem sometimes will fail.

Parties attempt to solve the former problem with index clauses. A demand-side index relates the contract price to the current market price. Section 2–305(1) of the UCC contains such an index: when the contract is silent as to price or when it directs the parties to agree on a price but they cannot, "the price is a reasonable index at the time for delivery." Another form of demand-side index links the contract price to the price that the buyer pays in later deals with other sellers; the prices in these deals reflect current market conditions. The parties use such indices because market and contract prices ordinarily are verifiable. The indices are also desirable because the current market price is likely to reflect both demand and cost factors. Nevertheless, demand-side indices are relatively uncommon. This apparently is because of product heterogeneity: when the contract product differs materially from other products, external prices may correlate poorly with the conditions that the parties face.

A supply-side index links the contract price to verifiable measures that correlate with the seller's costs. For example, a contract for machines will set a base price that escalates with changes in the Metals Price Index. Supply-side indices are unsatisfactory in two respects. Initially, they omit demand factors that could substantially affect the opportunity costs of performance. Second, the correlation between the verifiable external measure and a seller's actual costs may weaken over time. The widespread use of supply-side indices thus suggests that asymmetric information is a concern in many long-term relationships.

The parties may respond to the risk that an index will fail with risk-allocation and reopener provisions. As an example of the former, when the seller's opportunity cost of performance exceeds her contract gain, the buyer often can earn rents. In the spot-market example above, if the spot-market price rises to $200, the buyer can purchase at $100 and resell. A maximum-quantity clause reduces the seller's risk by restricting the buyer to amounts that he needs for his own business. Section 2–306(1) of the UCC contains such a quantity-allocation term, as do some private contracts. . . . Reopener provisions permit parties to compel price renegotiations. It is an open question why these terms are only occasionally used.

Verifiable Inability To Perform

Exogenous events may affect a performing party's physical ability to perform the contract. Examples include floods, fires, wars, embargoes, strikes, and government regulations. . . . The occurrence and effect of major exogenous events ordinarily is verifiable: the seller's factory burns down so she cannot produce; the farmer's land is flooded so nothing will grow. Also, there seldom is private information of the probability that such events may occur. Consequently, private contracts should allocate the risk of nonperformance.

The buyer will bear the risk under the following assumptions. (1) Performing parties cannot affect the probability or impact of the exogenous events that can preclude performance altogether. (2) Performing parties cannot observe the gains that paying parties will make from performance. (3) When substitutes for the subject of the contract exist, performing parties have no comparative advantage over paying parties or other market participants at procuring substitutes. (4) The class of paying parties is not significantly more risk averse than the class of performing parties.

The first assumption implies that excuse will not create moral hazard; imposing the risk on the buyer will increase neither the likelihood that performance-affecting events will occur nor the costs that such events create. The second assumption implies that paying parties are superior insurers; if paying parties were to bear the risk, each of them would purchase the amount of insurance he needed at the appropriate price rather than pay a premium (impounded in the product's price) that reflects the mean loss of a large, probably disparate class. On these assumptions, excuse is prima facie efficient; it would permit insurance economies to be realized without creating moral hazard.

The last two assumptions hold that the prima facie case seldom will be overcome. The third implies that enforcement would not create transaction-cost economies; in particular, paying parties have no incentive to require performing parties who cannot produce to become wholesalers. The fourth assumption implies that the parties' risk-sharing preferences do not outweigh the efficiency case for excuse. Therefore, the typical contract will excuse the performing party if an exogenous event made performance physically impossible. . . .

The Evidence

The parties' response to the risk that the opportunity cost of performance will come to exceed gains that could be earned under the contract is consistent with the theory presented immediately above. Indices seem the best coping method, and they are common in long-term relationships. Four demand-side indices are observed. Contracts sometimes contain most-favored-nation or nondiscrimination clauses. Under either term, the price is adjusted to reflect the prices that the buyer pays in later periods to others selling the same product. These clauses thus are variants of the paradigmatic demand-side index, which equates the contract price to the market price at the time for delivery. A third demand-side index ties the contract price to the verifiable price of another product that is affected by changes in cost and demand factors in the same way as the contract product. For example, natural-gas prices sometimes are adjusted to be a specified fraction of current crude-oil prices. Finally, in real-estate markets, rent often is a fraction of the lessee's gross revenues. These normally vary with demand at the lessee's business. As stated above, the parties more commonly use the supply-side indices, which link the contract price to verifiable proxies for the seller's current costs.

Long-term contracts respond in various ways to the risk that an index will become imperfect. They set maximum and minimum prices and maximum and minimum quantities. For example, if the indexed contract price substantially exceeds the current market price, the buyer will prefer to purchase on the market rather than order under the contract. A minimum-quantity clause constrains the buyer's ability to shift the entire risk of a failed index to the seller in this way. Take-or-pay clauses are minimum-quantity terms: the take-or-pay percentage is the fraction of anticipated contract quantity that the buyer must purchase. Natural-gas contracts sometimes contain "market-out" clauses. These excuse a buyer from performing if price in his resale markets falls far below the contract price to him. Contracts also may contain reopener provisions, which permit renegotiation when a supply-side index generates prices that correlate badly with actual producer costs. . . .

Long-term contracts also respond in the predicted way to the risk that the performing party will be physically unable to perform; they excuse sellers. Force majeure clauses are very common. Apart from index and risk-allocation clauses (such as a reopener provision), long-term contracts ignore the risk that a party's performance will become "impracticable." The factors that generally cause impracticability—high production costs and low demand—are unverifiable or unobservable. Theory predicts and data confirm that parties do not explicitly contract about such variables.

Notes on Impracticability, Frustration, and Mistake

1. The doctrines of impracticability and frustration, like the rule of *Hadley v. Baxendale,* are usually said to apply to "unforeseeable" risks. This raises a problem very similar to the level of description problem discussed earlier in the notes on consequential damages (supra at page 78): the likelihood of any given event depends critically on how that event is described. For example, the likelihood that the Suez Canal would have closed on the exact day that it did, at the exact time, was surely quite small. On the other hand, the likelihood that the Suez Canal would have closed at *some* time during the contract was probably higher. The likelihood that *something* would disrupt performance of the contract—if not the closure of the Suez Canal, then something else—could

have been higher still. (See also the distinction between a "mistake" and mere "conscious ignorance," discussed in the notes infra at pages 170–71).

2. As the preceding note should suggest, many of the issues raised by the doctrines of impracticability and mistake are similar to the issues raised by certain remedial rules analyzed earlier in Chapter 2. For example, the excerpt by Schwartz and the excerpt by Posner and Rosenfield discusses the effect of the impracticability doctrine on the promisor's incentives to take efficient precautions to reduce the risks. Excusing the promisor from liability could also affect the promisee's incentives to choose an optimal level of reliance, or to mitigate properly following a breach. (Analogous points were made in the earlier excerpts by Cooter, supra at pages 59–64, Epstein, supra at pages 67–72, and Craswell, supra at pages 49–52). On the other hand, in some contexts failing to excuse the promisor could lead to cross-subsidization, as discussed in the excerpt by Quillen (supra at pages 72–76). A similar problem is discussed in the excerpt by George Priest concerning implied warranties (infra at pages 174–79).

This similarity between the economic analysis of implied excuses and the economic analysis of remedies for breach is not coincidental. The economic analysis of remedies for breach asks when the promisor ought to be held responsible for *all* damages caused by his or her nonperformance, and when it would instead be better to reduce the promisor's liability (e.g., under the mitigation doctrine or the rule of *Hadley v. Baxendale*). Viewed from this economic perspective, the implied excuse doctrines are simply another set of rules that define the conditions under which the promisor's liability will be reduced. Since the bottom line (reduced promisor liability) is the same no matter which doctrines are used, the effect on the parties' incentives are very similar.

3. Can the noneconomic arguments advanced in the excerpts by Fried and Kull be similarly applied to the remedies for breach? Consider that most contracts do not say what measure of damages ought to be paid if either party breaches, thereby leaving a "gap" concerning the remedy for breach. Could the arguments in the Kull excerpt be used to show that courts should not try to fill this gap and should instead leave the losses wherever they have fallen? Or could the Fried excerpt be used to argue that this gap should be filled by the principle of sharing—meaning that the breaching party should only be liable for 50 percent of the nonbreacher's loss?

A possible response to this use of the Fried excerpt is to argue that a breach of contract is a wrongful act, thus taking such a case out of the purview of the sharing principle and placing it instead under the principle of liability for damages wrongfully inflicted on others. However, what about breaches that do not seem wrongful? For example, suppose an automobile company takes every feasible precaution on its assembly line, but occasionally (through no fault of its own) produces an automobile that is defective in a way that violates the company's standard warranty. If the company's contract didn't say how much they would pay in the event of a breach of warranty, could Fried's sharing principle then be used to argue that the losses from the defect ought to be split between the auto company and its customer? If this seems unattractive, is it because gaps involving damages are just "different" from other kinds of gaps? Or is it because a manufacturer's auto warranty ought to be interpreted as assuming responsibility for full expectation damages (if no other measure of damages is stated), so that this contract shouldn't be treated as having a gap at all?

4. Problems of impracticability and mistake are also closely related to issues of contract interpretation. For example, most people would not apply Fried's sharing principle to a breach of warranty case because most people (and certainly most courts) would interpret a warranty as assuming full responsibility for the legally applicable measure of damages. But if warranties can be interpreted as overriding Fried's sharing principle, why not interpret other contracts in this way, too? For example, why not interpret shipping contracts as assuming full responsibility for the risk that the Suez Canal might close? Once we introduce the possibility of interpreting the contract to assign the full risk to one party or the other, the principles we use to interpret the contract become critical. (Similar issues of interpretation will be central to all the issues discussed in this chapter).

5. Interpretation issues are also raised by the Kull excerpt's proposed "windfall principle," which would leave the losses where they fall in cases of mistake or frustration. The excerpt's first argument on behalf of this proposal is that such a rule will be more predictable than a rule that assigns losses on the basis of economic efficiency, fairness, or some other criteria whose application is difficult to predict in advance. However, this argument may be somewhat weakened by the excerpt's suggestion that courts could still apply an economic analysis (after the fact) "to reconstruct an allocation of risks that was an implicit (i.e., actual and intended) part of their agreement." For example, Kull suggests (supra at page 151) that

> If the owner of a building has insurance covering the value of repairs or additions while work is in progress, and the builder's interest is uninsurable, a court might reasonably conclude that the risk of destruction prior to completion was implicitly allocated to the owner even if his contract with the builder says nothing about it.

Is this use of economic reasoning to interpret the contract any more predictable than the use of similar economic reasoning to interpret the impracticability doctrine?

Compare the following discussion of the consequences of abolishing the impracticability doctrine:

> Moreover, some of the problems with the defense would simply resurface elsewhere. Contracting parties often include vague force majeure clauses in their contracts, for example, and the courts cannot avoid the need to construe them in the event of disputes. Likewise, even in the absence of an impracticability defense, contracting parties can always urge that contracts be construed to incorporate certain implied terms—for example, an implied term that personal

services contracts terminate on the death of the promisor. Thus, any curtail-
ment or abolition of the impracticability defense would to some extent simply
shift the battleground and confront the courts with the same issues in a different context.

Alan O. Sykes, The Doctrine of Commercial Impracticability in a Second-
Best World, 19 Journal of Legal Studies 43, 94 n. 154 (1990).

Can these problems of unpredictable interpretation be reduced by Kull's
suggestion that economic analysis be used as an aid to interpretation only
where the difference in the parties' risk-bearing capacity is "sufficiently pronounced" (supra at page 151)? For arguments in favor of a similar approach to
the interpretation of long-term supply contracts, see the excerpt by Alan
Schwartz infra at pages 187–93.

6. Other recent analyses of impracticability and mistake include Victor P.
Goldberg, Impossibility and Related Excuses, 144 Journal of Institutional and
Theoretical Economics 100 (1988); Michelle J. White, Contract Breach and
Contract Discharge Due to Impossibility: A Unified Theory, 17 Journal of
Legal Studies 353 (1988); Alan O. Sykes, The Doctrine of Commercial Im-
practicability in a Second-Best World, 19 Journal of Legal Studies 43 (1990);
Steven Walt, Expectations, Loss Distribution and Commercial Impracticability, 24 Indiana Law Review 65 (1991); George G. Triantis, Contractual Allocation of Unknown Risks: A Critique of the Doctrine of Commercial Impracticability, 42 University of Toronto Law Journal 450 (1992); and Ian Ayres and
Eric Rasmussen, Mutual and Unilateral Mistake in Contract Law, 22 Journal
of Legal Studies 309 (1993). The readings on long-term marketing and supply
contracts, infra at pages 181–201, are also relevant here.

3.1.2 Nondisclosure and Unilateral Mistake

Mistake, Disclosure, Information, and the Law of Contracts

ANTHONY T. KRONMAN

Every contractual agreement is predicated upon a number of factual assumptions about the world. Some of these assumptions are shared by the parties to
the contract and some are not. It is always possible that a particular factual

Anthony T. Kronman, Mistake, Disclosure, Information, and the Law of Contracts, 7 Journal of
Legal Studies 1, 1–10, 12–18 (1978). Copyright © 1978 by The University of Chicago. Reprinted
by permission.

assumption is mistaken. From an economic point of view, the risk of such a mistake (whether it be the mistake of only one party or both) represents a cost. It is a cost to the contracting parties themselves and to society as a whole since the actual occurrence of a mistake always (potentially) increases the resources which must be devoted to the process of allocating goods to their higher-valuing users.

There are basically two ways in which this particular cost can be reduced to an optimal level. First, one or both of the parties can take steps to prevent the mistake from occurring. Second, to the extent a mistake cannot be prevented, either party (or both) can insure against the risk of its occurrence by purchasing insurance from a professional insurer or by self-insuring.

In what follows, I shall be concerned exclusively with the prevention of mistakes. Although this limitation might appear arbitrary, it is warranted by the fact that most mistake cases involve errors which can be prevented at a reasonable cost. Where a risk cannot be prevented at a reasonable cost—which is true of many of the risks associated with what the law calls "supervening impossibilities"—insurance is the only effective means of risk reduction. (This is why the concept of insurance unavoidably plays a more prominent role in the treatment of impossibility than it does in the analysis of mistake.)

Information is the antidote to mistake. Although information is costly to produce, one individual may be able to obtain relevant information more cheaply than another. If the parties to a contract are acting rationally, they will minimize the joint costs of a potential mistake by assigning the risk of its occurrence to the party who is the better (cheaper) information-gatherer. Where the parties have actually assigned the risk—whether explicitly, or implicitly through their adherence to trade custom and past patterns of dealing—their own allocation must be respected. Where they have not—and there is a resulting gap in the contract—a court concerned with economic efficiency should impose the risk on the better information-gatherer. This is so for familiar reasons: by allocating the risk in this way, an efficiency-minded court reduces the transaction costs of the contracting process itself.

The most important doctrinal distinction in the law of mistake is the one drawn between "mutual" and "unilateral" mistakes. Traditionally, courts have been more reluctant to excuse a mistaken promisor where he alone is mistaken than where the other party is mistaken as to the same fact. Although relief for unilateral mistake has been liberalized during the last half-century (to the point where some commentators have questioned the utility of the distinction between unilateral and mutual mistake and a few have even urged its abolition), it is still "black-letter" law that a promisor whose mistake is not shared by the other party is less likely to be relieved of his duty to perform than a promisor whose mistake happens to be mutual.

Viewed broadly, the distinction between mutual and unilateral mistake makes sense from an economic point of view. Where both parties to a contract are mistaken about the same fact or state of affairs, deciding which of them would have been better able to prevent the mistake may well require a detailed inquiry regarding the nature of the mistake and the (economic) role or position of each of the parties involved. But where only one party is mistaken, it is reasonable to assume that he is in a better position than the other party to prevent his own error. As we shall see, this is not true in every case, but it provides a useful beginning point for analysis and helps to explain the generic difference between mutual and unilateral mistakes. . . .

In the past, it was often asserted that, absent fraud or misrepresentation, a unilateral mistake never justifies excusing the mistaken party from his duty to perform or pay damages. This is certainly no longer the law, and Corbin has demonstrated that in all probability it never was. One well-established exception protects the unilaterally mistaken promisor whose error is known or reasonably should be known to the other party. Relief has long been available in this case despite the fact that the promisor's mistake is not shared by the other party to the contract.

For example, if a bidder submits a bid containing a clerical error or miscalculation, and the mistake is either evident on the face of the bid or may reasonably be inferred from a discrepancy between it and other bids, the bidder will typically be permitted to withdraw the bid without having to pay damages (even after the bid has been accepted and in some cases relied upon by the other party). Or, to take another example, suppose that A submits a proposed contract in writing to B and knows that B has misread the document. If B accepts the proposed contract, upon discovering his error, he may avoid his obligations under the contract and has no duty to compensate A for A's lost expectation. A closely related situation involves the offer which is "too good to be true." One receiving such an offer cannot "snap it up"; if he does so, the offeror may withdraw the offer despite the fact that it has been accepted.

In each of the cases just described, one party is mistaken and the other has actual knowledge or reason to know of his mistake. The mistaken party in each case is excused from meeting any contractual obligations owed to the party with knowledge.

A rule of this sort is a sensible one. While it is true that in each of the cases just described the mistaken party is likely to be the one best able to prevent the mistake from occurring (by exercising care in preparing his bid or in reading the proposed contract which has been submitted to him), the other party may be able to rectify the mistake more cheaply in the interim between its occurrence and the formation of the contract. At one moment in time the mistaken party is the better mistake-preventer (information-gatherer). At some subsequent moment, however, the other party may be the better preventer because of his superior access to relevant information that will disclose the mistake and thus allow its correction. This may be so, for example, if he has other bids to compare with the mistaken one since this will provide him with information which the bidder himself lacks. Of course, if the mistake is one which cannot reasonably be known by the non-mistaken party (that is, if he would have to incur substantial costs in order to discover it), there is no reason to assume that the non-mistaken party is the better (more efficient) mistake-preventer at the time the contract is executed. But if the mistake is actually known or could be discovered at a very slight cost, the principle of efficiency is best served by a compound liability rule which imposes initial responsibility for the mistake on the mistaken party but shifts liability to the other party if he has actual knowledge or reason to know of the error. Compound liability rules of this sort are familiar in other areas of the law: the tort doctrine of "last clear chance" is one example.

The cases in which relief is granted to a unilaterally mistaken promisor on the grounds that his mistake was known or reasonably knowable by the other party appear, however, to conflict sharply with another line of cases. These cases deal with the related problems of fraud and disclosure: if one party to a contract knows that the other is mistaken as to some material fact, is it fraud for the party with knowledge to fail to disclose the error and may the mistaken party avoid the contract on the theory that he was owed a duty of disclosure? This question is not always answered in the same way. In some cases, courts typically find a duty to disclose and in others they do not. It is the latter group of cases—those not requiring disclosure—which appear to conflict with the rule that a unilateral mistake will excuse if the other party knows or has reason to know of its existence. . . .

The Production of Information and the Duty To Disclose

It is appropriate to begin a discussion of fraud and nondisclosure in contract law with the celebrated case of *Laidlaw v. Organ*. Organ was a New Orleans commission merchant engaged in the purchase and sale of tobacco. Early on the morning of February 19, 1815, he was informed by a Mr. Shepherd that a peace treaty had been signed at Ghent by American and British officers, formally ending the War of 1812. Mr. Shepherd (who was himself interested in the profits of the transaction involved in *Laidlaw v. Organ*) had obtained information regarding the treaty from his brother who, along with two other gentlemen, brought the news from the British fleet. (What Shepherd's brother and his companions were doing with the British fleet is not disclosed.)

Knowledge of the treaty was made public in a handbill circulated around eight o'clock on the morning of the nineteenth. However, before the treaty's existence had been publicized ("soon after sunrise" according to the reported version of the case), Organ, knowing of the treaty, called on a representative of the Laidlaw firm and entered into a contract for the purchase of 111 hogsheads of tobacco. Before agreeing to sell the tobacco, the Laidlaw representative "asked if there was any news which was calculated to enhance the price or value of the article about to be purchased." It is unclear what response, if any, Organ made to this inquiry.

As a result of the news of the treaty—which signalled an end to the naval blockade of New Orleans—the market price of tobacco quickly rose by 30 to 50 percent. Laidlaw refused to deliver the tobacco as he had originally promised. Organ subsequently brought suit to recover damages and to block Laidlaw from otherwise disposing of the goods in controversy. . . .

From a social point of view, it is desirable that information which reveals a change in circumstances affecting the relative value of commodities reach the market as quickly as possible (or put differently, that the time between the change itself and its comprehension and assessment be minimized). If a farmer who would have planted tobacco had he known of the change plants peanuts instead, he will have to choose between either uprooting one crop and substituting another (which may be prohibitively expensive and will in any case be costly), or devoting his land to a nonoptimal use. In either case, both the individual farmer and society as a whole will be worse off than if he had planted tobacco to begin with. The sooner information of the change reaches the farmer, the less likely it is that social resources will be wasted.

Consider another (and perhaps more realistic) illustration of the same point. A is a shipowner who normally transports goods between New Orleans and various other ports. However, because of the naval blockade, he is unable to enter the New Orleans harbor. Some time after the treaty is signed, but before its existence is publicized, A enters a contract to ship cotton from Savannah to New York City. After news of the treaty reaches New Orleans, a tobacco merchant of that city offers A a "bonus" if he will agree to deliver a shipment of tobacco to Baltimore. If we assume that the offer is sufficiently attractive to induce A to breach his first contract and pay damages, although his ship will be properly allocated to its highest-valuing user, the cost of allocating it will be greater than it would have been had information of the treaty reached A before he entered his first contract. Resources will be consumed by A in transacting out of the first contract; from a social point of view, their consumption represents a pure waste.

Allocative efficiency is promoted by getting information of changed circumstances to the market as quickly as possible. Of course, the information doesn't just "get" there. Like everything else, it is supplied by individuals (either directly, by being publicized, or indirectly, when it is signalled by an individual's market behavior).

In some cases, the individuals who supply information have obtained it by a deliberate search; in other cases, their information has been acquired casually. A securities analyst, for example, acquires information about a particular corporation in a deliberate fashion. By contrast, a businessman who acquires a valuable piece of information when he accidentally overhears a conversation on a bus acquires the information casually.

As it is used here, the term "deliberately acquired information" means information whose acquisition entails costs which would not have been incurred but for the likelihood, however great, that the information would actually be produced. These costs may include, of course, not only direct search costs (the costs of examining the corporation's annual statement) but the costs of developing an initial expertise as well (for example, the cost of attending business school). If the costs incurred in acquiring the information (the cost of the bus ticket in the second example) would have been incurred in any case—that is, whether or not the information was forthcoming—the information may be said to have been casually acquired. The distinction between deliberately and casually acquired information is a shorthand way of express-ing this economic difference. Although in reality it may be difficult to deter-mine whether any particular item of information has been acquired in one way or the other, the distinction between these two types of information has—as I hope to show—considerable analytical usefulness.

If information has been deliberately acquired (in the sense defined above), and its possessor is denied the benefits of having and using it, he will have an incentive to reduce (or curtail entirely) his production of such information in the future. This is in fact merely a consequence of defining deliberately ac-quired information in the way that I have, since one who acquires information of this sort will by definition have incurred costs which he would have avoided had it not been for the prospect of the benefits he has now been denied. By being denied the same benefits, one who has casually acquired information will not be discouraged from doing what—for independent reasons—he would have done in any case. . . .

One effective way of insuring that an individual will benefit from the possession of information (or anything else for that matter) is to assign him a property right in the information itself—a right or entitlement to invoke the coercive machinery of the state in order to exclude others from its use and enjoyment. The benefits of possession become secure only when the state transforms the possessor of information into an owner by investing him with a legally enforceable property right of some sort or other. The assignment of property rights in information is a familiar feature of our legal system. The legal protection accorded patented inventions and certain trade secrets are two obvious examples.

One (seldom noticed) way in which the legal system can establish property rights in information is by permitting an informed party to enter—and enforce—contracts which his information suggests are profitable, without dis-closing the information to the other party. Imposing a duty to disclose upon the knowledgeable party deprives him of a private advantage which the informa-tion would otherwise afford. A duty to disclose is tantamount to a requirement that the benefit of the information be publicly shared and is thus antithetical to the notion of a property right which—whatever else it may entail—always requires the legal protection of private appropriation.

Of course, different sorts of property rights may be better suited for pro-tecting possessory interests in different sorts of information. It is unlikely, for example, that information of the kind involved in *Laidlaw v. Organ* could be effectively protected by a patent system. The only feasible way of assigning property rights in short-lived market information is to permit those with such information to contract freely without disclosing what they know. . . .

If we assume that the courts can easily discriminate between those who have acquired information casually and those who have acquired it deliberately, plausible economic considerations might well justify imposing a duty to disclose on a case-by-case basis (imposing it where the information has been casually acquired, refusing to impose it where the information is the fruit of a deliberate search). A party who has casually acquired information is, at the time of the transaction, likely to be a better (cheaper) mistake-preventer than the mistaken party with whom he deals—regardless of the fact that both parties initially had equal access to the information in question. One who has deliberately acquired information is also in a position to prevent the other party's error. But in determining the cost to the knowledgeable party of preventing the mistake (by disclosure), we must include whatever investment he has made in acquiring the information in the first place. This investment will represent a loss to him if the other party can avoid the contract on the grounds that the party with the information owes him a duty of disclosure. . . .

A rule which calls for case-by-case application of a disclosure requirement is likely, however, to involve factual issues that will be difficult (and expensive) to resolve. *Laidlaw* itself illustrates this point nicely. On the facts of the case, as we have them, it is impossible to determine whether the buyer actually made a deliberate investment in acquiring information regarding the treaty. The cost of administering a disclosure requirement on a case-by-case basis is likely to be substantial.

As an alternative, one might uniformly apply a blanket rule (of disclosure or nondisclosure) across each class of cases involving the same sort of information (for example, information about market conditions or about defects in property held for sale). In determining the appropriate blanket rule for a particular class of cases, it would first be necessary to decide whether the kind of information involved is (on the whole) more likely to be generated by chance or by deliberate searching. The greater the likelihood that such information will be deliberately produced rather than casually discovered, the more plausible the assumption becomes that a blanket rule permitting nondisclosure will have benefits that outweigh its costs.

Legal Secrets

KIM LANE SCHEPPELE

Sometimes the targets of the secret know about the potential existence of a secret, even though they do not know the content of the secret itself. Laidlaw's agent, for example, clearly suspected that there might be a treaty [in *Laidlaw v. Organ*]; otherwise he would not have asked Organ whether he had heard any news. Laidlaw's agent just did not know whether a treaty had been signed. When the target suspects that there might be a secret, we find *shallow secrets*. When the target is completely in the dark, never imagining

Kim Lane Scheppele, Legal Secrets: Equality and Efficiency in the Common Law (Chicago: University of Chicago Press, 1988), pp. 21–22, 70–72, 75–76, 78–79. Copyright © 1988 by The University of Chicago. Reprinted by permission.

that relevant information might be had, we find *deep secrets*. . . . The depth of a secret affects the sorts of justifications that can be made by those left out of the secret.

If someone knows that a secret exists, making this a shallow secret, she can ask for the information directly and claim that her direct questions should be answered. As in the card game Go Fish, where players ask whether their opponents have particular cards in their hands and where they can capture these cards if they guess correctly, people who ask direct questions can claim that the whole negotiation process, like the game, is undermined if their skill in asking questions is not rewarded. With shallow secrets, however, the secret-keeper can claim that the targets at least know that they are deciding with the risk of ignorance. Keepers of shallow secrets may be able to claim that the targets should search for the information rather than expect to have it handed to them outright.

The deep secret allows the targets to make stronger justifications against the secret-keepers. If the targets have no idea that the information exists, let alone what the information consists of, then the targets have a more forceful case that the secret amounts to fraud. The targets cannot protect themselves against information they cannot imagine, and so the secret-keeper can always gain advantage at the expense of the target. But the deep-secret-keepers may claim that they have property rights in the information and do not have to reveal it to those who do not know. . . .

A Contractarian Theory of Law

In direct secrets, one person withholds information from another and that other claims the right to know. To figure out what the optimal rule would be in cases like these, we should note that sometimes A will want to withhold information from B and that sometimes B will want to withhold information from A. Both A and B, then, will have some interest in choosing rules that would not work to their detriment in either situation.

In order to choose these rules, individual decision makers must first have some sense of who they would be and what they would want. But in order to minimize the overt play of narrow self-interest, a motivation unlikely to lead to the choice of fair rules, we need to restrict the ability of these individuals to foresee their own particular fortunes. As a first matter, they cannot know whether they are going to be secret-keepers or targets in any individual cases or even in the aggregate. This is a minimum condition designed to keep individual decision makers from rigging the rules for their own individual benefit. They cannot know how they personally would fare under any particular regime of rules and so they cannot produce completely self-serving rules. They must make some attempt to produce rules that incorporate more impartial standards.

But what can they be expected to know about themselves and how they will act in the face of uncertainty in this ideal world? First, we will assume, like Rawls, that the individuals deciding in advance on a system of rules will be rational. Rational actors will generally consider only their own self-interest in a particular decision. The fact that these individuals cannot know what will benefit them personally (because they do not know what position they will be in when the time comes) will lead them to set out to construct a set of ground rules for improving their collective lot. A set of ground rules would be agreed on in advance only if they improved the chances that each individual would be able to better her condition more reliably by following the rules than by living under a regime where odds of a good outcome were determined by chance. These rational individuals, deciding in advance of knowing their particular fates, would act not simply to promote narrow self-interest in the individual case (because they could not if they tried) but rather to promote self-interest in a broader perspective.

One may think that such rational individuals would choose rules that would make the outcome under the rules the same for all, but this is only true if these individuals are extremely risk averse or willing to live with a particular vision of equality *on average*. It seems unrealistic to assume that individuals at our hypothetical decision point are completely risk averse because here, under the piecemeal approach, the only thing at stake is the particular rule about secrecy. Individuals are not choosing once and for all the complete regime of institutions that will control their entire lives, a situation in which it would make more sense to be extremely risk averse. Much of the fear of bad consequences in these sorts of contractarian choices comes from not considering what sorts of *other* institutional arrangements the society has for protecting people harmed by particular decisions or rules. If a society has a generous welfare system, for example, a group of people deciding on what sort of tort-law system to adopt may reasonably agree on rules that fail to compensate people for particular kinds of losses because they know that injured persons will be able to get aid elsewhere. There is no reason to expect the sort of extreme risk aversion in piecemeal contractarianism that we find in more holistic forms if the background institutions are basically just.

There is also no reason to expect that people choosing rules will, as the utilitarians argue, opt to maximize utility *on average,* for the reasons we examined above. People do not live life as persons on average; and, if one receives a bad lot in an allocation of scarce resources, one can hardly be expected to be comforted by the knowledge that, on average, things worked out to be the best they could be. It is no justification for someone standing with one hand in a pot of scalding water and the other in a pan of ice to know that, on average, her hands are a comfortable temperature. Experience is not felt on average; the variance matters a great deal. Being a person means living a particular life. And so we might expect that persons designing a system of rules will want to make sure that the variance in outcomes does not demand more than people, living particular lives, can reasonably be expected to tolerate. . . .

What chances will people be willing to take, once they are protected from catastrophic losses? Individuals, deciding before they know which position they are going to be in when disputing over disclosure of a secret, would be willing to tolerate some uncertainty—but only if they know that they have *some* nontrivial chance to win. They have to have some sense of what the odds are and how those odds are stacked in order to fit their calculations about what to do into a rational framework; without this, they are helpless in the face of hidden knowledge. But more importantly, since the individuals whom we are describing are people who want to make lives for themselves by using their powers and talents, they will want to be able to *do* something to improve their chances of coming out ahead. With the exception of small-stakes lotteries (which are blind gambles), people do not generally like being at the mercy of forces out of their control, especially when the stakes are large (as they would be in the secrecy cases). If there is nothing that a person can do to better her lot while there may be something that a secret-keeper can do to make the situation quite dreadful for her, then potential targets of secrets will want protection. When targets are at the mercy of secret-keepers and do not have the information that they would need to be able to protect themselves against the secret-keeper's wiles, they have lost critical control over the things that matter to them. And in such situations, knowing that something is a gamble and that people often take gambles is no comfort. It is the *choice* in taking a gamble that people want, not simply the gambles themselves; and any legal rule that would force people into such gambles would be objectionable. Secrets to which people would consent must be secrets that are open to efforts at discovery.

This would lead rational decisionmakers to distinguish . . . between deep and shallow secrets. Deep secrets, those secrets that those kept in the dark do not know about at all, are unresponsive to effort. One can be clever and still not learn about deep secrets because one does not even know that there might be information out there that matters. One cannot rationally choose to search for information that, from the searcher's perspective, does not exist. Shallow secrets, on the other hand, are secrets about which the target has at least some shadowy sense. One knows enough of their existence to be able to decide whether a search will be worth the effort. Shallow secrets will be responsive to effort, in the sense that rational individuals can make decisions to search or not search for information. There is still risk involved (one may miss the information in a search or underestimate how valuable the knowledge actually is), but at least the decision problem is not completely intractable. As long as the secret is shallow, one has some chance of being able to triumph. . . .

With shallow secrets, however, rational individuals of the sort described above would want to take their chances on winning, as long as they were buffered from failing too badly if losses incurred in trying to win got too big. When people indicate that they are willing to tolerate variance in outcomes, they are saying that they are willing to take a fair gamble. . . . With the possibility of winning open to effort, they might be able to gain. If they had to disclose all their information, they would never be able to win. With a system of rules providing incentives for searching in those cases where the ignorant parties are capable of doing so, we might expect that more information would be discovered than would be the case under a regime where all information had to be immediately disclosed. Allowing shallow secrets enables the presence of enough incentives to encourage rational searches and to promote the discovery of information.

Not all shallow secrets ought to be acceptable, however. If A and B begin their search for information at radically different starting lines, then the person beginning with the disadvantage would be taking a much bigger risk of losing—and rational individuals choosing rules would wish to ensure against being at such a disadvantage. That would make the gamble they are taking seem not fair. People facing this situation would want a rule permitting shallow secrets only under those circumstances where neither A nor B started with a large advantage. Moreover, A and B would want to be protected if either one of them turned out to be incompetent and unable to make a rational choice. Shallow secrets allow the search for information to be open to effort, thus permitting individuals to make gains at least some of the time; but individuals who start off hobbled at a disadvantageous starting point would want to be able to recover when they fail. Having a floor constraint here means that such individuals should be allowed to recover in law not only when the stakes are high but also where the probabilities of suffering medium-sized losses are very high.

If the law reflected this sort of rational decision making in a strategic structure, then one would expect the law to always require disclosure of secrets that would seriously injure someone. And one would expect the law to always require disclosure of deep secrets, except when the stakes were very small. The law should *not* generally require disclosure of shallow secrets, unless the two parties in the transaction had started under such different circumstances that the information was not equally open to their efforts. Finally, in all these transactions, the law should protect from their own bad judgment those individuals unable to make rational decisions at all.

Notes on Nondisclosure and Unilateral Mistake

1. The Scheppele excerpt distinguishes between "deep" secrets (those whose existence the other party does not even suspect) and "shallow" secrets (those where the other party knows something is being concealed, even if he or she does not know the secret's exact content). Can this distinction be precisely defined? Even parties who never consciously considered some particular risk—say, the risk that the Suez Canal might close—may nonetheless have been aware of a more general risk, such as the risk that *something* would prevent the contract from being performed. Similarly, even a buyer who doesn't consciously consider the possibility that a house might have termites might nonetheless realize there was a chance that there would be *something* wrong with the house. Moreover, a buyer who knew there might be *something* wrong with the house could improve his or her information by hiring an expert investigator, even without knowing what specific problems the investigator

might be able to uncover. Thus, the information about the termites could be regarded as either a deep or a shallow secret, depending on whether the information is described as "information about termites" or as "information that something might be wrong with the house." (A similar problem was discussed in the notes on unforeseeable damages, supra at page 78, and in the notes on impracticability and frustration, supra at page 157.)

2. Even if we accept the distinction between deep and shallow secrets, is Scheppele correct in her conclusion that rational decision makers choosing behind a veil of ignorance—that is, decision makers who didn't know whether they would be the ones possessing a secret or trying to discover a secret— would want to be protected against all deep secrets? If the Kronman analysis is correct, sometimes a rule forcing the disclosure of deep secrets can make both the secret holder and the uninformed party worse off.

For example, suppose that a mining company might be able to find out whether there are minerals under a farmer's land. Suppose further that the farmer does not even suspect the presence of minerals: this is a "deep secret" as far as the farmer is concerned. If the mining company is not allowed to keep the information secret, they will not even bother to acquire the informa- tion (according to Kronman), so neither the mining company nor the farmer will benefit. If the mining company is allowed to keep the information secret, however, they will acquire that information and will receive a very large profit. The farmer will also receive a very small profit as the result of the mining company's information because the farmer will be able to sell the farm at a favorable price. (The farmer can always refuse to sell the farm, so the mining company will have to offer a price that the farmer considers favor- able.) This means that neither party will be worse off with a rule allowing the mining company to keep the information secret, and both parties will be made at least slightly better off by such a rule. If a rational decision maker didn't know whether he or she would be a mining company or a farmer, wouldn't such a decision maker still favor the rule that would benefit both parties?

3. The Scheppele excerpt also argued that rational decision makers would want the right to be told about shallow secrets *if* the consequences of not knowing the secret would be extremely severe. The desire to be protected from extremely large losses can be seen as a form of risk-aversion. If decision makers were not risk-averse at all, and instead were completely risk-neutral, they would be perfectly happy to maximize the *expected* or *average* outcome without being unduly bothered by the risk of large losses. If people are suffi- ciently risk-averse, though, they would be willing to incur some efficiency costs in order to be protected against large losses. For example, risk-averse decision makers might be willing to give up the gains that would result if the ability to keep secrets encouraged people to gather information in exchange for protection against the large losses those secrets might inflict on the other party to the transaction.

Do you think all potential contracting parties are risk-averse in this way? What if the party who was unaware of the secret (and who now seeks protection from the courts) is a large corporation whose shares are spread over thousands of shareholders? (See the discussion of risk-aversion and corporations in the excerpt by Posner and Rosenfield, supra at page 145.)

4. The Kronman excerpt focuses on information that would yield large gains to its possessor. Kronman's argument that society should encourage the production and use of such information, by letting the possessor keep the information secret, has been criticized for neglecting the distinction between information that is socially useful and information that has only private value. Information that an apparently barren cow is in fact fertile may have social value if it prevents a valuable breeding animal from being slaughtered for beef. By contrast, information about an impending rise in real estate prices acquired before the public announcement of a planned shopping center may have large private value but very little social value. Such information has private value because the party who obtains such information can profit from it by buying the surrounding land at a low price before the information becomes generally known. If these purchases merely shift the potential profit from the former owners to the purchaser without doing anything to alter the use of the land, it might be argued that the private gain to the purchaser is a mere transfer, which produces no new social value. See Robert Cooter and Thomas Ulen, Law and Economics (Homewood, Illinois: Scott, Foresman and Company, 1988), pp. 258–61; Jules L. Coleman, Douglas D. Heckathorn, and Steven M. Maser, A Bargaining Theory Approach to Default Provisions and Disclosure Rules in Contract Law, 12 Harvard Journal of Law and Public Policy 639, 693–96 (1989); Steven Shavell, Acquisition and Disclosure of Information Prior to Sale, 25 Rand Journal of Economics 20 (1994).

Should the law follow these critiques of Kronman and refuse to protect buyers or sellers who keep silent about information that would not produce any social value? Is it possible to draw a workable line between information that alters the use of resources in a socially valuable way and information whose only effect is to redistribute existing gains? For example, what if the rise in market prices caused by the informed party's purchases led to more efficient resource decisions by *other* landowners who were affected by the increase in price—would this make the informed party's information socially useful once again? For a further discussion of this possibility, see Randy E. Barnett, Rational Bargaining Theory and Contract: Default Rules, Hypothetical Consent, the Duty to Disclose, and Fraud, 15 Harvard Journal of Law and Public Policy 783 (1992).

5. Suppose that Kronman is right, and that (in at least some cases) it is better to allow the party with better information to keep that information secret to preserve the incentive to acquire the information in the first place. What if the other party directly asks the party with the better information if he or she knows anything about the secret? Should the party with the better information be allowed to lie in order to keep his or her information private?

For example, suppose that a mining company learns that there are valuable minerals under a farmer's land and attempts to buy the farm without disclosing its information. Suppose the farmer then asks the company's agents whether they have done any geological studies, or whether they know anything about the minerals under the land. (Assume that the farmer is smart enough to spot any evasion or equivocation, and keeps on putting the question directly.) If the company's agent answers honestly, the company will then have lost their chance to profit from their information, thus reducing their incentive to acquire the information in the first place. Flatly refusing to answer the questions may not be much better, for such a refusal would probably suggest to the farmer that the company *did* know something about the presence of minerals under the land. Thus, if we really want to allow the company to profit from their information, it may be necessary not only to free them from any duty to disclose their information, but also to license them to lie in response to direct inquiries from the other party.

6. When a seller knows information that would reflect unfavorably on his or her product, some economists have suggested that market forces may lead sellers to disclose that information on their own if buyers are sufficiently rational. Sophisticated buyers may be able to reason as follows: "If the seller knew that the undisclosed information would make the product more attractive, he or she would surely disclose it to me. Because the seller has not disclosed this information to me, he or she must know that the missing information would not in fact make this product more attractive. I should therefore assume the worst about the missing information."

For example, suppose that John sells oranges by the box. Each box has a removable lid and holds up to thirty oranges, but John has no legal obligation to disclose how many oranges are in each box. The market price for oranges is 50¢ per orange. If John refuses to say how many oranges are in the box, how much should Mary be willing to pay for the box? The clever answer is 50¢. If the box contains only one orange, 50¢ is the correct price; if it contained more than one orange, John would then have an incentive to take off the lid and prove to Mary that the box was worth more than 50¢. In economic terms, Mary's offer to pay 50¢ for the box is an "information forcing" proposal, which gives all reputable sellers an incentive to disclose the relevant information even without a legal duty to disclose. For a more formal discussion of this effect, see Sanford Grossman, The Informational Role of Warranties and Private Disclosure of Product Quality, 24 Journal of Law and Economics 461 (1981).

An important limitation pointed out by Grossman is that this incentive will only work if sellers who disclose that their products are not the worst must tell the truth. If there are no limits on deception by sellers, every seller (including those whose boxes contain no oranges) can claim that their products are worth more than the bare minimum—and because every seller can make such a claim, buyers will have no incentive to believe anyone. Sometimes it is easy for buyers to tell whether sellers are telling the truth, as in the simple example where Mary can count the number of oranges for herself. In other markets, where buyers have a harder time judging the truth for themselves, anti-deception laws and other regulations (e.g., federal inspection and grading of agricultural products) help make sellers' claims more credible to consumers.

Another limit identified by Grossman is that buyers must be sophisticated enough to realize that there is information that they don't know in order to realize they should assume the worst. Is this condition another way of stating Scheppele's distinction between deep and shallow secrets?

§ 3.2 Implied Warranties

A Theory of the Consumer Product Warranty

GEORGE L. PRIEST

Most discussions of product defects in the economics literature and most legal decisions regarding warranties regard the probability of loss from a defect as inherent in the nature of the product and independent of actions of the consumer. According to this approach, allocative investments by a consumer which serve to reduce the probability of losses are nonexistent, and the only relevant consumer and public policy choice is between consumer self-insurance and manufacturer liability, whether leading to an allocative investment or insurance. Often these analyses are qualified by a reference to consumer behavior, although seldom by more than an acknowledgment that in some cases a consumer may actively misuse a product. The implicit conclusion of each of these discussions is that allocative investments by a consumer are empirically unimportant to the optimization of the productive life of the good.

Of course, there is no theoretical justification for disregarding allocative investments by consumers. A more important implication of the theory, however, is that there is no meaningful way to consider a product defect independently of a consumer's allocative investments. Investments to reduce the probability of losses may take very subtle forms. I have alluded earlier to the control of children and the placement of an appliance—as it affects the number of times the appliance is scratched or jarred—as representing allocative investments. As a more general proposition, however, two forms of investment by consumers will affect the likelihood of defects in any consumer

George L. Priest, A Theory of the Consumer Product Warranty. Copyright © 1981 by The Yale Law Journal Company. Reprinted by permission of The Yale Law Journal and Fred B. Rothman & Company from The Yale Law Journal, volume 90, pages 1311–19.

product. The first is the consumer's selection of a product suitable for his expected needs. Warranty claims are likely to be more frequent, for example, where a washing machine is undersized or a vacuum cleaner underpowered, or where there occurs some unexpected increase in the demands that the consumer makes on the product. If the consumer accurately anticipates his uses, and if he selects a product designed most appropriately for those uses, the productive capacity of the good is more likely to be preserved. The second form of investment is the consumer's decision about the extent to which he will use the product. A consumer who operates an appliance infrequently may be said to be preserving the life of the product by choosing to store rather than to use it.

Initially, this conception may seem foreign because it is common to infer from personal experience some "normal" use of a product. Indeed, the law requires judges and juries to make inferences of "normal" use by implying in product sales a warranty of merchantability that a product is "of fair average quality" and is "fit for the ordinary purposes for which such goods are used." If it were possible to infer some "normal" use of a good, then the decision of individual consumer to use or not to use the good would be analytically irrelevant.

But preferences regarding the frequency of use of a product differ among consumers. The preferences of the particular set of consumers for whom the product has been designed in order to optimize sales cannot be determined by inference. Where the dominant set of purchasers operates the good infrequently and, thus, where the "normal" use of the good is storage, the level of the consumers' allocative investments in preservation of the product is high. As a consequence, the level of the manufacturer's allocative investment in product design or in insurance that optimizes productive services may be very low. In such a case, the design or manner of production of the product may be optimal even though the product appears grossly defective when operated with greater frequency, which is to say, when operated with lower allocative investments in care by consumers.

A warranty in this view is the instrument that expresses consumer preferences for allocative or insurance investments. It is a contract that divides responsibility for allocative investments and insurance between the consumer and the manufacturer. The content of the contract is determined by the respective costs to the two parties of allocative investments or insurance. According to this approach, a manufacturer makes investments to prolong product life up to the point at which the marginal cost of such investments equals the marginal benefit. A manufacturer, then, offers market insurance for those losses or items of service for which market insurance is less costly than insurance or allocative investments by the consumer himself.

To the extent that a manufacturer disclaims liability or excludes or limits warranty coverage, however, it shifts to the consumer the obligation to make allocative investments to preserve the product or to self-insure for its loss. A disclaimer or an exclusion of coverage is the functional equivalent of provisions, common in other contracts, that explicitly require one of the parties to take certain actions to prevent breach or to insure for losses from uncertain events. The theory predicts that disclaimers of liability and exclusions of coverage will be observed in consumer product warranties for those specific allocative or insurance investments that the consumer can provide more cheaply than the manufacturer. In this view, disclaimers and exclusions can be said to be demanded by consumers because of the relative cheapness of consumer allocative investments or of self-insurance. . . .

Reducing Differences in Risks

The task of defining optimal warranty provisions resembles the task of defining optimal rate classes in insurance contracts. In all insurance contexts, it is advantageous for an insurer to segregate applicants according to the level of risks added to the insurance pool. If the risk of loss of an individual can be predicted, then the insurance premium can be tailored to reflect the likelihood of future payouts. In particular, insurance coverage can be offered at a lower premium to an individual for whom the risk of loss is relatively low.

For most types of insurance, of course, it is prohibitively costly either to predict exactly the risk that an individual brings to a pool or to charge individual premiums. As a consequence, an insurer is forced to lump individuals into separate classes or, sometimes, into a single class. The premiums charged each member of the class must reflect the average level of risk of the class. Thus, the premium undercharges relatively high-risk individuals and overcharges relatively low-risk individuals. At the margin, some low-risk individuals are likely to find that the cost of market insurance exceeds the benefit, and will shift to allocative investments that reduce the likelihood of the loss or to self-insurance. In the context of consumer products, these individuals will shift their purchases to products sold without, or with less, warranty coverage. The more precisely the insurer is able to construct classes comprising individuals with relatively similar levels of risk, however, the smaller the discrepancy will be between the premium and the value of insurance to the lower risk members of the pool. Thus, the lower risk members become less inclined to substitute self-insurance for market insurance. As a general proposition, therefore, discrimination that reduces differences in risk between members of a given insurance class optimizes the sale of insurance.

It is common for life, medical, accident, and home insurers to obtain information about applicants prior to making contracts in order to place applicants in appropriate insurance classes. Insurers routinely solicit information about age, sex, property location and value, as well as medical records and driving histories in order to construct rate classes. Some insurers make it possible for individuals with characteristics that tend to be correlated with low levels of risk, such as abstemious smoking and drinking habits, to identify themselves in order to qualify for lower premiums. Analogues to these methods of discrimination, however, are not immediately apparent in the context of consumer product insurance. Typically, insurance polices for consumer product losses are tied to the sale of the product itself, so that the insurance pool invariably consists of all consumers who have purchased the product.

Consumers may differ in two general ways with respect to risk under a product warranty. First, the amount of use of a product during the period of warranty coverage may vary considerably between consumers. Compare, for example, the expected service costs to a washing machine manufacturer from a family with many children and from a family with only a single child. The costs of service to the large family will almost certainly be greater. If the manufacturer could define warranty coverage in terms of number of washloads, however, as an automobile manufacturer defines coverage in terms of mileage, then the expected costs from the two families to the manufacturer might be similar. But for washing machines, as well as for most other consumer appliances, the least costly measure of use appears to be duration of ownership. As a consequence, no matter what the period of coverage, the amount of use of the machine by the two families is likely to differ greatly. The insurance premium must be set to cover all expected costs of service. Thus, smaller families at the margin may find warranty protection to be worth less than its cost.

Second, the risk of loss may differ between consumers with respect to what I will call the "intensity" of product use. Compare now for the large and small families, the expected service costs to a television manufacturer. The amount of use of the television—that is, the number of viewing hours—might be identical for the two families. Nonetheless, the probability of a warranty claim is likely to be higher for the larger family, because of the greater number of individuals operating the set, because of the greater frequency of channel changes, and because of the greater risk in a large family that the set will be jostled, that the antenna will be struck, or that the machine will otherwise be treated roughly.

I define "intensity" of use as inversely related to the marginal cost to the consumer of "care" for the machine, that is, the cost of allocative investments to reduce the probability of a loss. The cost of monitoring the activities of children is likely to increase as the number of children increases. Thus, the family with many children is more likely than the family with a single child to substitute recovery under a warranty for allocative investments in care of the machine. As a consequence, the cost to the manufacturer of warranty coverage will be greater for the machine sold to the larger family. Again, at the margin, consumers with smaller families may find it advantageous to shift their purchases to machines sold without, or with less, warranty coverage.

Although product insurers do not directly acquire information about consumers prior to sale, a variety of subtle methods can enable them to segregate consumers. For example, a manufacturer can develop models of a product that differ with respect to characteristics related to differences in intensities or amounts of consumer use. A manufacturer of washing machines may produce models that differ in motor size or washbasket volume that are differentially convenient to families of different sizes. If these product characteristics segregate consumers according to the extent or intensity of use, then the manufacturer can offer, for each individual model, different allocative investments and levels of warranty coverage determined by the expected warranty claims for each model.

This technique, however, may achieve only partial success. The advantage to a manufacturer of culling out higher intensity or higher volume consumers from a particular insurance pool is to enable it to offer warranty coverage at a relatively lower premium, or greater coverage at the same premium, of models designed for lower volume or intensity uses. A lower premium or more attractive to all consumers, including those who expect to use the machine with greater intensity or in higher volume. At the margin, some of these consumers can be expected to purchase machines undersized for their needs. Such purchases substitute the extended warranty coverage of the lower volume machine for the mechanical superiority of the higher volume machine. This adverse selection by higher volume or intensity consumers will force manufacturers to reduce the extended coverage of the lower volume machine or to charge a higher premium for it. Either reaction will reduce the attractiveness of the lower volume machine to the lower risk members of the pool.

A separate but closely related method of segregating consumers is to offer warranty contracts with different terms at different premiums in conjunction with the sale of a given product. Recently, the domestic automobile manufacturers have introduced insurance policies for separate fees extending coverage for periods beyond the basic twelve-month warranty. The optional service contracts of many appliances is similar. These contracts segregate consumers according to the amount of insurance coverage they wish to buy. The warranty provides a term of basic coverage demanded by the lowest risk members of the pool. Those consumers for whom the risk is greater, however, can purchase more extensive coverage. Because relatively high-risk consumers are more likely to select such contracts, their premiums are likely to be proportionally higher for a given duration of coverage than the premiums of the basic warranty included in the sale price. . . .

Finally, a manufacturer may segregate consumers by means of explicit contractual provisions in the warranty. A manufacturer, for example, may exclude warranty coverage for a particular use of a product or specific class of consumers for which the volume or intensity of use is relatively high. The common provision that excludes coverage of commercial use is an obvious example. This provision narrows the class of those insured to domestic users of the product and may be incorporated to enforce a manufacturer's segregation of domestic and commercial purchasers by model design.

Some elements of product loss, however, may be excluded from coverage in the warranties of all product models. A common example is the exclusion of liability for consequential damages. The unavailability of any coverage of some loss, nevertheless, may be related to the reduction of differences in risk between members of the insurance pool. Where consumers differ substantially in the incidence or magnitude of a loss, such as consequential damages, there may be no single premium attractive to a sufficient number to justify offering coverage. Put another way, the increase in the premium required for coverage of such losses may be greater than the benefit of coverage to large numbers of consumers. If so, the sale of product insurance may be optimized by excluding coverage altogether.

Notes on Implied Warranties

1. This excerpt discusses the effect of warranty liability on the buyer's incentives to take an efficient level of precautions, or to rely too heavily by, for example, using a washing machine more heavily than it was designed to be used. This parallels the earlier discussion of the nonbreacher's precautions and reliance in Chapter 2 in the excerpts by Cooter (supra at pages 59–64) and Epstein (supra at pages 67–72). Those excerpts were considering the proper measure of damages for breach, not the proper scope of an implied warranty. However, in any case where expectation damages would not be an efficient remedy for breach—say, because it would give the buyer an incentive to rely too heavily or to take too few precautions—it will also be efficient to treat the seller as not having assumed liability for expectation damages, at least as long as the seller's contract does not explicitly provide otherwise. Refusing to imply a warranty is, of course, one way to treat the seller as not having assumed liability for full expectation damages. From an economic standpoint, then there is a good deal of similarity between the analysis of (a) how much a party ought to be liable for in cases where that party has unquestionably promised to do something; and (b) whether a party should be treated as having promised to do something in the first place, in cases where that party's promise is somewhat ambiguous.

2. The analysis in the Priest excerpt also raises several of the themes discussed earlier in this chapter in connection with impracticability, frustration, and mistake. For example, the excerpts by Posner and Rosenfield (supra at pages 141–45) also discussed the effect of liability on promisees' incentives to take an efficient level of precautions. The reason for this similarity is that an implied warranty is, in effect, an implied excuse viewed from the other side. That is, a judicial finding that the seller implicitly warranted some product is equivalent to a finding that the seller was *not* implicitly excused because of impracticability, frustration, or mistake.

Technically, the implied excuse doctrines start with the assumption that the promisor did promise to do something and then ask whether the promisor ought to be excused, while the implied warranty doctrine starts with the assumption that the seller did not explicitly promise anything and then asks whether such a promise ought to be implied. Still, both doctrines ultimately answer the question of whether a party should or should not be held liable for some problem that has arisen. Thus, it is not surprising that the relevant policy arguments should be the same.

3. An implied warranty is also similar to a misrepresentation—especially if the misrepresentation is not explicit, but is inferred from vague words or actions. For example, a seller who sells a house infested by termites will be held liable if the court decides either (a) that such sales should be interpreted as including an implied warranty, or (b) that the appearance of the house, or the way that it was described in the advertisements, constituted an actionable representation that implied that the house was termite-free.

To be sure, an implied warranty creates contractual liability, while misrepresentation is usually regarded as a tort (or as a defense to contractual liability). However, both forms of liability are "voluntary" in the sense that they can generally be disclaimed. That is, the seller can usually disclaim warranty liability by explicitly providing that the house is being sold "as is" with no warranty against termites. The seller can also usually disclaim liability for misrepresentation by clearly stating that he or she is not representing anything one way or the other about the soundness of the house, and encouraging the buyer to rely on an independent inspection.

The chief practical difference between warranty and misrepresentation concerns the measure of damages. As warranty is a contract action, the seller is usually liable for the buyer's expectation damages, while damages for the tort of misrepresentation are sometimes (though not always) measured by the buyer's reliance losses. Can this difference in remedies be justified? Or does the justification run the other way—that is, *if* expectation damages are deemed the superior remedy, the seller should be treated as having warranted the product, whereas if reliance damages are superior, the seller should only be held liable for misrepresentation? For a further discussion of the similarities and differences between warranty and misrepresentation, see P. S. Atiyah, Misrepresentation, Warranty and Estoppel, in Essays on Contract (Oxford: Clarendon Press, 1986), pp. 275–328.

4. Finally, the analysis of implied warranties is also relevant to many unconscionability cases—when a seller attempts to limit its warranty liability but does so in a way that may not be legally enforceable. Most sellers do attempt to limit their warranty liability in at least some respect—most often, to attempt to escape liability for all consequential damages that might follow a breach of warranty—and courts sometimes view such attempts with suspicion. Unconscionability and the other doctrines that can override the express language of a contract are discussed infra in Chapter 5.

§3.3 **Relational Contracts**

§3.3.1 **Marketing and Supply Contracts**

Principles of Relational Contracts

CHARLES J. GOETZ AND ROBERT E. SCOTT

[A] significant proportion of private contracts do not easily fit the presuppositions of classical legal analysis. One reason for this is the pivotal role played in conventional legal theory by the concept of the complete contingent contract. Parties in a bargaining situation are presumed able, at minimal cost, to allocate explicitly the risks that future contingencies may cause one or the other to regret having entered into an executory agreement. Under these conditions, the role of legal regulation can be defined quite precisely. Once the underlying rules policing the bargaining process have been specified, contract rules serve as standard or common risk allocations that can be varied by the individual agreement of particular parties. These rules serve the important purpose of saving most bargainers the cost of negotiating a tailor-made arrangement. If the basic risk allocation provided by a legal rule fails to suit the purposes of particular parties, then bargainers are free to negotiate an alternative allocation of risks. All relevant risks thus can be assigned optimally—either by legal rule or through individualized agreement—because future contingencies are not only known and understood at the time the bargain is struck, but can also be addressed by efficacious contractual responses.

In a complex society, however, many contractual arrangements diverge so markedly from the classical model that they require separate treatment. Parties frequently enter into continuing, highly interactive contractual arrangements. For these parties, a complete contingent contract may not be a feasible contracting mechanism. Where the future contingencies are peculiarly intricate or uncertain, practical difficulties arise that impede the contracting parties' efforts to allocate optimally all risks at the time of contracting. Not surprisingly, parties who find it advantageous to enter into such cooperative exchange relationships seek specially adapted contractual devices. The resulting "relational contracts" encompass most generic agency relationships, including distributorships, franchises, joint ventures, and employment contracts.

Although a certain ambiguity has always existed, there has been a ten-

Charles J. Goetz & Robert E. Scott, Principles of Relational Contracts, 67 Virginia Law Review 1089, 1089–91, 1095–98, 1111–19 (1981). Copyright © 1981 by the Virginia Law Review Association. Reprinted by permission.

dency to equate the term "relational contract" with long-term contractual involvements. We here adopt a very specific construction of the term that is based more precisely on a contrast with the classical contingent contract. A contract is relational to the extent that the parties are incapable of reducing important terms of the arrangement to well-defined obligations. Such definitive obligations may be impractical because of inability to identify uncertain future conditions or because of inability to characterize complex adaptations adequately even when the contingencies themselves can be identified in advance. As the discussion below illustrates, long-term contracts are more likely than short-term agreements to fit this conceptualization, but temporal extension per se is not the defining characteristic. . . .

Optimal Contractual Performance Under Certainty

Consider the common situation where one party is the distributor of a product supplied, at least in part, by the manufacturer. As an initial hypothetical case, assume that adjustments in distribution efforts are the only dimension of production influencing output on the margin. . . . For expository simplicity, we shall regard "distribution efforts" as units of product distributed. Although product volume has the virtue of being a clear and concrete conceptualization, the same analysis would apply if distribution efforts were reinterpreted to take the form of any other volume or quantitative adjustment, including, for example, quality level, advertising, or any other activity that affects joint profits. Assume also in this initial situation that the parties each know their own costs (but not necessarily the other party's costs) and know the external market conditions during the effective period of the proposed contract.

In Figure 3.1, IMC represents the aggregate of the marginal costs of "manufacturing" (MM) and of the marginal "distributing" costs (the vertical distance between IMC and MM) where a firm carries on both manufacturing and distribution as an integrated process. If MR is the marginal revenue curve, then the intersection of MR and IMC at b determines the output volume Q that maximizes profits. In other words, Q and its associated profit level is the best that the integrated firm can do by both manufacturing and distributing. The agent presumably enters the picture because he enjoys a productive advantage over the manufacturer, such as lower marginal distribution costs as exemplified by MD in Figure 3.1. The joint marginal cost curve JMC would result if the parties were able to treat their separate costs as a single entity.

The potential savings from separate performance of their respective functions now provide the manufacturer and distributor with an economic incentive to enter into a distributorship agreement. Two predictions can be made as to the terms of an agreement designed to exploit that potential. First, as Figure 3.1 suggests, the parties will not have exploited all of the potential gains from trade in the situation unless their agreement somehow calls for the manufacture and distribution of quantity Q^*, where the sum of the marginal

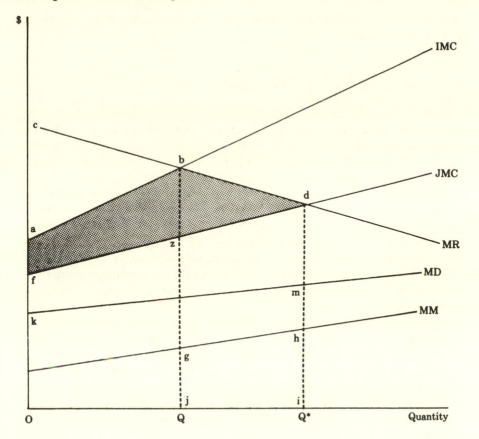

Figure 3.1.

costs to the joint producers equals the marginal revenues from sales. From any output other than Q*, a movement to Q* will increase the combined profits of the parties. If, therefore, there are no special impediments in the form of bargaining or other transaction costs, one would expect to find contract terms facilitating this "joint maximization" quantity outcome.

Second, certain limits can be placed on the minimum and maximum amounts that the distributor will pay to the manufacturer for the predicted Q* units of the product. Each party must be at least as well off under the contractual arrangement as it would have been otherwise. Consequently, the manufacturer must receive at least its additional manufacturing costs under the higher-volume distributorship agreement plus the "go-it-alone" profits that it would otherwise earn as an integrated firm. Graphically, these sums correspond, respectively, to area ghij and triangle *abc* in Figure 3.1. The distributor would, in turn, be willing to pay a maximum of the total revenue from sales minus its distribution costs (area kcdm). A range of indeterminacy exists because the gains from trade (cost savings fabz + profits on the expanded output zbd) must be divided through bargaining between the parties. These potential net gains from the distributorship arrangement are represented by the shaded area in Figure 3.1. . . .

Defining the Standard of Performance

Perhaps the most poorly understood class of relational contracts is that involving agreements wherein one party explicitly, or even implicitly, undertakes the contractual duty of using its "best efforts" to carry on an activity beneficial to the other. Some of the most common illustrations of such best efforts agreements are found in agency, licensing, franchising, and other distributorship arrangements. Notwithstanding the frequency with which such terms are observed empirically, the precise legal meaning to be attached to a best efforts requirement is not at all clear, either from a consideration of the case law or from theoretical discussions in standard legal scholarship. Nevertheless, there appears to be a relatively straightforward and persuasive definition that emerges from the preceding economic conceptualization of the problem faced by two parties who are attempting to set a contractual volume in which they have joint interests. . . .

Figure 3.2 illustrates a contract in which the distributor faces a marginal cost curve (MC) composed of his own marginal distribution costs (MD) plus the marginal "price" (R) negotiated with the manufacturer. In addition, it is assumed that the payment takes the form of a 50-percent royalty on gross sales. Absent any other information, one might expect that the distributor would then be entitled to choose volume Q where his marginal costs (MC = MD + R) are equal to marginal revenues (MR). This is the point at which the distributor's own profits are maximized. If he were required to sell an additional quantity beyond Q, the distributor's profits would be reduced, as exemplified by the shaded triangle A in Figure 3.2. Suppose, however, that there were some way in which he could oblige himself to adjust to the joint maximization output that we have previously identified at Q*. In exchange for such an undertaking, which at the time it is accepted represents a loss to the distributor, the manufacturer should be willing to agree in advance to a compensatory contractual concession through which the two parties can split the additional profits generated by the higher volume. These profits are represented by the cross-hatched triangle B in Figure 3.2.

Optimal Volume Definition

The obligation to produce at the joint maximization volume is the meaning that we propose for the best efforts term in commercial contracts. This interpretation of the best efforts provision has a great deal of theoretical attractiveness because, absent the specification of an alternative construction by the parties, it directs the outcome that maximizes the net gains that parties could achieve from their contractual relationship. In sum, it is a plausible means of identifying a goal presumably desired by most parties,

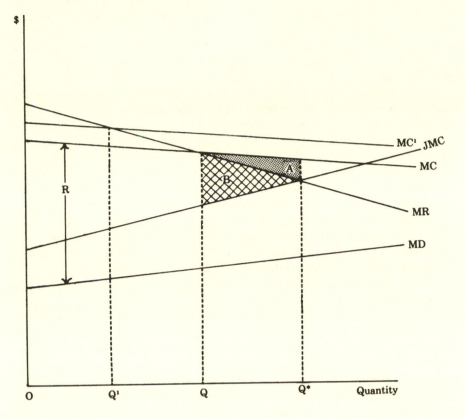

Figure 3.2.

albeit not always well articulated. In any case, business people need not be regarded as thinking explicitly in terms of the precise marginal conditions and other terminology of economic theory.

In addition, the definition suggested above is consistent with a "fairness" obligation of the kind formulated by distributive justice theorists. Under this conception, the distributor is required to treat the manufacturer "fairly," giving the manufacturer's interests (profits) equal weight with his own when output decisions are made. Moreover, such special consideration presumably has been paid for in advance by the manufacturer in the form of some compensatory commission.

In any specific fact situation, some retreat from the rigorous definition suggested above may be entirely appropriate. For instance, the duty of the best efforts promisor to take into consideration the other party's interests should be limited by the promisor's reasonable ability to foresee the extent of those interests. Thus, a failure by the distributor to account fully for the manufacturer's idiosyncratic accounting methods that unexpectedly reduce joint marginal costs and increase the additional volume necessary to reach Q* would not establish a breach of the best efforts obligation. This limitation is in the same spirit as the damage limitation rule of *Hadley v. Baxendale,* because it compels a party with unanticipatable interests to supply the information necessary for economically efficient behavior. Those parties with atypical or idiosyncratic requirements remain free to negotiate an individually tailored understanding of the best efforts obligation.

Unfortunately, a best efforts obligation, as defined above, inherently implies a serious monitoring problem. This is illustrated in Figure 3.2 where shaded triangle A represents the reduction in profits suffered by the distributor because he is obligated to produce at Q* rather than Q. Hence, the best efforts promisor will generally have an incentive to "chisel" on the obligation. In a world of cost-free information, such breaches of the best efforts requirement would be easily detected, and the behavior restrained through the legal damages imposed for breach of contract. In a real-world situation, however, the requisite information for proof of liability or quantification of damages may be prohibitively expensive to obtain, especially when the plaintiff bears the burden of such proof. Hence, the standard legal mechanism may not be a viable one for enforcement of this kind of contract provision.

Where recourse to the courts is not an attractive option, these economic considerations suggest that a best efforts promisee—such as the manufacturer—will attempt to contract for other means of controlling the standard of performance. Presumably, the self-interest of both contracting parties will induce them to seek out that combination of monitoring or bonding arrangements that represents the optimal tradeoff between expected costs of contractual governance and profits forgone because the ideal output of Q* is not enforced perfectly. As the cost of contract-specific monitoring strategies increases, the price of contracting to the best efforts promisor similarly increases. The best efforts promisor has an incentive, therefore, to propose cost-effective bonding agreements that reduce the costs of contractual control, thereby lowering the contract "price" paid by him. The "price" reductions might take the form of a reduction in the initial license payment required by the manufacturer, or a reduction in the royalty paid to the manufacturer on the contractual product or, indeed, any other adjustment in contract terms favorable to the distributor.

A commonly observed form of bonding is a termination privilege that could be invoked by the manufacturer if he detects a breach of the best efforts obligation. Moreover, the parties might be expected to negotiate a termination clause that granted the manufacturer considerable discretion as to the circumstances under which termination would be permitted. If, instead, the termination clause were only triggered by specific events, any attempt to exercise it might create precisely those problems of proof that the clause originally was designed to circumvent. A limited right of termination embodies less reassurance of contractual performance and would presumably induce some compensatory increase in the contract "price" paid by the distributor. . . . [A] discretionary termination clause is not an ideal safeguard. Often, however, it is a mutually beneficial adaptation to the inevitable conflict of interest generated by a best efforts agreement.

Although the best efforts result Q* is in theory a clearly optimal result for the parties, the realities of enforcement, especially when coupled with the inherent chiseling incentive, may dim the practical attractiveness of such agreements. Nevertheless, the problems arising in legal regulation of such agreements should not be viewed as dispositive. Many contractual provisions are honored even when there is no effective legal sanction for their breach. In some circumstances, this is due to the existence of informal, extra-legal sanctions, including a sense of commercial ethics. Notwithstanding practical difficulties of securing legal enforcement, therefore, a contractual provision also has value simply as a communication of understanding between the parties as to their mutual rights and duties. Hence, the inclusion of a best efforts term may, at a minimum, serve as a signal alerting good faith bargainers that the proposed contractual relationship is one in which special concerns are to be considered.

Where courts are compelled to attach a meaning to otherwise ambiguous contractual terms, it is sensible to look to the likely intent of the parties or the goal the parties might reasonably be deemed to have sought. The "optimal-output" definition of best efforts is, we argue, the single most plausible interpretation of the underlying economic motivations involved. . . .

Relational Contracts in the Courts

ALAN SCHWARTZ

[C]ourts pursue three strategies, broadly speaking, when deciding contracts cases: they protect process values, interpret language, and supply terms when the parties' contract fails to provide for the dispute that divides them. The third task is often referred to as "gap filling": the contract lacks a relevant term and, so, has a "gap"; the court fills the gap with a rule specifying how to resolve the dispute at issue. The court is said to supply a term because contract rules become implied terms in later agreements unless parties explicitly change them.

Courts also may (and sometimes do) supply terms when contracts apparently *lack* gaps. A sale between a manufacturer and a wholesaler illustrates how courts come to have this discretion. Suppose the wholesaler will face only two states of demand in his market, high or low. then consider two possible contracts that the parties could make: (a) the manufacturer's price for each of five units is $20 if the wholesaler faces a high demand and $10 per unit for a

Alan Schwartz, Relational Contracts in the Courts: An Analysis of Incomplete Agreements and Judicial Strategies, 21 Journal of Legal Studies 271, 272–74, 295–98, 301–03 (1992). Copyright © 1992 by The University of Chicago. Reprinted by permission.

low demand; (b) the price is $20 per unit for five units. The second contract apparently is complete; it sets a price for five units. The second wholesaler, however, never has an incentive to breach contract a because the price will be appropriate to either contingency. In contrast, he has an incentive to breach contract b if demand turns out to be low: the price then will be too high. This example suggests that the phrase "incomplete contract" should include more than the gap case.

The definition used here (which is now popular among economists) holds either than an incomplete contract has a true gap—for example, no price term—or that it partitions future states or potential contracting partners "too coarsely." Regarding the additional aspect of the definition, parties to sales transactions may face a large number of possible future states: demand in the wholesaler's market may be extremely high, very high, moderately high, average, and so forth. The initial contract a is complete under the definition because (it is assumed) only two contingencies could materialize and the contract has a two-state partition: it sets prices for the high- and low-demand states. Contract b is incomplete because it has a one-state partition in a two-state world: that is, the contract sets one price and, so, necessarily fails to treat one of the two cases that could arise.

To understand the relevance of this new definition to legal issues, one must recall that the wholesaler has an incentive to breach the incomplete contract. If he were to do so, then he likely would claim in a lawsuit that contract b is only superficially complete: it does not treat the low-demand state. Contract b nevertheless is sufficiently detailed to be the basis for a legal remedy: the seller could be awarded the difference between the contract price and the market price for five units. The court thus has a choice: it can supply a term that governs when the contract price arguably is inappropriate to the ex post state, or it can enforce the contract as written. The courts' rhetorical strategies sometimes conceal the existence of this discretion. A court that wants to excuse the buyer will supply a term and stress the parties' failure to consider the situation at hand—that is, the court will stress the contract's incompleteness. A court that prefers to enforce will give the seller damages and stress the contract's (apparent) completeness; the judge will recite the maxim that the courts do not make contracts for the parties. Recognizing that contract b is incomplete despite what some courts *say* permits this article to rephrase the question asked above in a more illuminating way: why do courts complete some incomplete contracts but not others?

The answer lies in why the parties fail to complete contracts themselves. When parties are asymmetrically informed about important factors, such as the buyer's downstream demand, they will inevitably write incomplete contracts. In the example above, the parties will write contract b if the seller could not observe demand in the wholesaler's market except at prohibitive cost. If contract a were to govern, then the wholesaler always would claim he faced low demand so he could pay the low price. Certain contracts are incomplete not because the parties overlooked or failed to understand the issue that came to divide them but because they are unwilling to bear the strategic-behavior risk that a complete contract would create. Relational contracts commonly are incomplete for this reason. Also, the process that produces relational contracts is often fair. A court therefore cannot pursue the three standard strategies when deciding the typical relational contract case: there are no process violations to cure; there is no ambiguous language to interpret; and the court can complete the contract no better than the sophisticated parties can on their own.

Courts respond to such cases by pursuing a "passive" judicial strategy that permits them to avoid making substantive determinations. In the example above, a passive court will not complete contract b with a term that conditions enforcement on the price being appropriate to the state of demand ex post; rather, the court will enforce the contract "as written." Courts that pursue such a passive strategy often appear to engage in simpleminded literalness (as in this example) or to defer unduly to strong parties, but, in fact, they are doing neither of these things. Rather, their response is the product of a traditional institutional constraint: courts would rather be passive than active when faced with problems they cannot solve. . . .

Quantity Cases

Contracts sometimes set a price or create a pricing formula but do not specify quantities. Disputes arise when a party offers or demands more or less than the other party believes is appropriate. . . . Courts decide these cases as the analysis here predicts: they regulate quantity when the requisite term can condition on verifiable actions or states; otherwise, they behave passively by giving almost total discretion over quantity to the party whom the contract authorizes to choose quantity.

Quantity disputes arise in connection with both requirements and output contracts. Under the former, the buyer agrees to purchase his requirements from the seller; under the latter, the buyer agrees to purchase the seller's output. These contracts do not specify the amount that a party will take or produce but sometimes do contain quantity estimates. More often, a quantity range is anticipated: the buyer is to have *some* requirements; the seller is to produce *something*.

The reason why parties use these contracts is not fully understood, but certain factors that affect contract content are identifiable. The amount that a party will find profitable to supply or demand is partly a function of economic factors that can vary with time. In this circumstance, the parties face a trade-off between incurring the cost of observing these factors again and recontracting or incurring the cost of acting on obsolete information. To understand why parties may omit a quantity term, suppose that the economic factors that affect the parties are more time variant for one than the other and that the significant time-variant factors are observable by the party whom they primarily affect but are not observable by the other party. Then, granting the discretion to choose quantity each period to the party whom the time-variant factors affect has two virtues. First, it saves the transaction costs of recontracting when the factors vary. Second, it also avoids the possibility that recontracting negotiations will fail: when parties are asymmetrically informed about relevant payoff matters—here, the time-variant factors—there may be a bargaining breakdown. The contract should grant discretion to a seller when her output is more affected by time-variant factors and to a buyer when his requirements are the more affected.

The opportunity cost of performing under such "open-quantity" contracts can come to exceed the gain for two reasons, both of which have the same behavioral consequence: the party with discretion supplies or demands an amount in the current period that varies substantially from quantities that she supplied or demanded in earlier periods. The first reason is that a party will vary quantity as a result of factors that affect her much more than they affect otherwise similarly situated firms. For example, an output seller reduces supply substantially because certain key employees leave and she no longer can produce profitably at the contract price with her remaining labor force. The second reason is that a party will vary quantity as a result of factors that affect all firms similarly. For example, systemic factors cause the market price to rise substantially above the contract price; the seller then prefers to sell output on the market rather than under the contract.

The factors that exclusively affect a single party often are noncontractible. As stated above, contracts seldom condition on sellers' production costs or a buyer's demand. In contract, the quantity traded and the contract and market prices usually are verifiable. Thus, a contract term or legal rule could provide that supplying output to the buyer is a breach when the market price had doubled and the seller is supplying to other firms. The remedy for breach would be to anchor damages to a contract estimate or to the mean of quantities supplied in past periods.

This explanation of open-quantity contracts has not been directly tested. The explanation generates six predictions about the cases; if the evidence supports the predictions, the explanation is, at least, plausible for the parties' strategies and for judicial performance. The first three predictions concern requirements contracts. (1) The requirements buyer cannot increase orders substantially over contract estimates or the mean of previous orders when his primary purpose is to resell. Buyers seldom can expand their physical plant in a short period; thus, the desire to resell on a rising market is the primary motive for a dramatic increase in orders. Requirements contracts commonly are meant to satisfy production needs, so purchase for resale contradicts the parties' intentions. More significantly, courts can verify when the buyer has entered the wholesale business. (2) The buyer cannot decrease orders substantially if he is purchasing the same amount on the market as he had ordered under the contract and the market price has fallen. (3) The buyer has the discretion to reduce requirements in any amount so long as he does not act as described in prediction 2. In this situation, the decrease in the requirements is commonly a function of factors peculiar to the buyer, such as input cost changes or changes in firm-level demand.

The analysis generates similar predictions for output contracts. (4) Specifically, sellers have the discretion to expand output substantially. A seller will increase production under a fixed-price contract only when production costs decline. Neither the parties nor the courts will condition quantity on a seller's costs. (5) Sellers are not permitted to reduce output under the contract in order to sell on the market when price has risen. (6) In all cases not described in prediction 5, sellers have the discretion to reduce output in any amount. . . .

Exclusive Distributorships

Persons with exclusive rights to make and sell products sometimes contract with others for production and distribution. The example used here involves publishing. An author owns the right to his book and contracts with a publisher to produce, promote, and distribute it. The publisher commonly is given exclusive rights. These contracts may be described in the language used above: an author/buyer purchases a quantity of distributorship services from a publisher/seller. The publishers also distribute books by other authors. Distribution contracts seldom specify the quantity of services the seller will provide.

There is a question as to why parties make these contracts rather than sell book rights. To understand the probable answer, let books come in two types, good and bad; a good book sells well. When parties contract, neither knows what type of book the author has written. Book types are revealed ex post to the publisher, who observes demand directly (she deals with retailers, for example). An author cannot observe demand directly. On these assumptions, publishers sell distribution services rather than buy rights because rights cannot be priced; their value is a function of a book's type, which is unknown at the time of the contract. (Veteran authors sometimes get advances or agreements to print a minimum number of copies because a book's type can correlate with previous book's types.) The question is how to price the publisher's services.

There are three possibilities. Publisher efforts, total sales, and publisher profits all increase or decrease with a book's type: the greater a book's potential sales, the greater the promotion effort the publisher will make, and so forth. Thus, authors can be paid a percentage of gross revenues or a percentage of profits, or the publisher can be paid on the basis of the promotional effort she supplies. For reasons that are now clear, distributorship contracts do not condition on profits. They also do not condition on effort. Authors seldom can observe effort, which commonly is unverifiable—that is, most authors would find that the costs of proving in court that a publisher made an unduly low effort would exceed the gains. In addition, the appropriate amount of effort (in an economic sense) is partly a function of demand; high efforts to promote bad books are wasted. Again, authors cannot directly observe demand, which is, in any event, hard to verify. Therefore, the common-exclusive distributorship contract pays the author a percentage of gross revenues from book sales.

Disputes can arise under these contracts when a publisher decides that a book's type is bad (and so supplies few services). Even sales of bad books can increase with publisher efforts, so authors always want publishers to do more. The publisher's problem, however, is to allocate effort optimally across all books. Thus, her concern is with the relative marginal productivity of effort. Would an additional dollar of effort increase sales of book x more than the author's book y? Should the publisher take a new book, y', on the same subject as book y? That is, would an additional dollar of effort increase the sales of y' by more than that dollar would increase sales of y? Authors may sue when the answers to these questions go against them. This analysis generates two predictions that are similar to those discussed above.

1. Courts permit publishers to supply any quantity of effort above zero to particular contract authors. A court can regulate effort more closely than this in two ways. First, it could adopt a rule that publishers have a legal obligation to allocate effort optimally across product lines (including the introduction of new products). Then the court can review a publisher's particular allocation decision to see whether she has made a mistake. Second, a court could adopt a rule that publishers have a legal duty to supply a fair amount of services. Then it can review a publisher's effort to see whether it was unfairly small. The theory presented here predicts that courts will reject the first strategy because they lack business expertise and because the information relevant to review— the state of demand and the actual effort exerted—is unverifiable. The theory also predicts that courts will reject the second strategy because the information relevant to review is unverifiable and because no norms tell just what level of effort is fair. Therefore, courts should give publishers discretion to choose the effort level.

2. Publishers cannot exert zero effort. This rule prevents forfeiture. A distributor will do nothing when the marginal cost of even a slight effort exceeds the gain. The distributor nevertheless may not cancel because matters can change. The rule that the distributor must do something thus forces the distributor to cancel unless the author waives his legal right; the author thus has the choice of whether to try elsewhere with another publisher or wait and see. The rule also is administrable because "doing nothing" is verifiable. This analysis generates the related prediction that suits will arise claiming that the distributor did nothing when the contract creates a barrier to cancellation.

The cases sometimes are difficult to understand, but, in general, they support these predictions. The UCC and contract law both require an exclusive distributor to use "best efforts to promote" contract goods but omit criteria by which to assess the quantity of effort.

Regarding the first situation in prediction 1, a publisher can reduce services for a particular book in two ways: she can supply more services to other books, or she can introduce a new book that competes with the original book. The cases give the publisher the discretion whether to supply more services to other books. There is an interesting Georgia case on an analogous issue that elucidates the reason why. There, the defendant was given the exclusive right to manage the plaintiffs' pecan groves and to sell the output. The plaintiffs sued, making two claims: the defendant improperly charged expenses to the plaintiffs (for example, it charged the plaintiffs for paint while doing no painting); and the defendant's labor charges were excessive—that is, it used more labor than the job required. The court gave the former set of claims to the jury but withheld the latter, explaining: "[d]etermining the amount of labor necessary to produce, harvest, and market pecans was a judgment Gold Kist was entitled to exercise under the contract and absent bad faith in deciding the amount of labor necessary, there is no issue for jury resolution. There was no testimony, other than the opinion testimony of the Flynns [plaintiffs], that the charges for labor were excessive. Gold Kist enjoyed discretion under the contract as to this issue." Courts can verify amounts charged and whether a distributor supplied any services at all. But information relevant to the question of whether a distributor supplied the appropriate amount of services is unverifiable. Courts act consistently with this distinction.

Economic Analysis of Contractual Relations

IAN MACNEIL

Discrete transactions differ from contractual relations respecting many key characteristics. Among those most important to economic analysis are: (1) commencement, duration, and termination; (2) measurement and specificity; (3) planning; (4) sharing vs. dividing benefits and burdens; (5) interdependence, future cooperation, and solidarity; (6) personal relations among, and numbers of, participants; and (7) power.

Understanding of these characteristics may be enhanced by putting them into a particular context: how a smelting operation—Smelter—might secure the coal needed for its operations. Among the possibilities:

A. Spot purchases from a stranger of 500 tons of coal in a market of many sellers, Seller's agents delivering the coal by truck dumped at Smelter's yard, cash being paid on delivery of each load.

B. Same as Illustration A, except that Seller is one from whom Smelter usually buys its coal, terms are thirty days credit, and Seller is tolerant of arrearages for fairly lengthy periods.

C. Once or twice a year, the Seller in Illustration B buys on speculation as much as 5,000 tons at a good price, thinking that Smelter will probably be

Ian Macneil, Economic Analysis of Contractual Relations: Its Shortfalls and the Need for a "Rich" Classificatory Apparatus. Copyright © 1981 by Northwestern University School of Law. Reprinted by special permission of Northwestern University School of Law, volume 75, issue 6, Northwestern University Law Review (volume 75, issue 6)—1018, 1025–39.

interested in stockpiling coal at a price lower than most of the market; so far Smelter has always bought, as it does this time.

D. Smelter enters a firm forward contract with Seller for 500 tons of coal a week for eight weeks, at a fixed price.

E. Smelter contracts with Coal Mine to buy all the coal it requires during one year; the specified price is subject to a quarterly escalator clause based on a designated market.

F. Same as Illustration E, except that in addition to the escalator clause there is a provision: "Should a party become dissatisfied with the price, the parties agree to negotiate a new price, and, in the absence of agreement, to refer the matter to X as arbitrator, to determine a fair and equitable price."

G. Same as Illustration F, except that Smelter and Coal Mine have had similar annual contracts for ten years.

H. Same as Illustration G, except that the latest contract entered into this year is for twenty years rather than one, requires Coal Mine periodically to provide Smelter with extensive cost information, to allow Smelter's experts to monitor mining operations, and to receive from Smelter recommendations respecting new equipment, improved methods of management, and the like. Smelter and Coal Mine also agree to build and operate a conveyor belt system from minehead to smelting plant, sharing the capital costs equally and operating the conveyor system jointly. As part of the deal Smelter gives Coal Mine a five-year loan to cover part of Coal Mine's costs of the conveyor system, and, in order to satisfy other lenders, guarantees Coal Mine's half of a twenty-year mortgage loan on the conveyor system.

I. Same as Illustration H, except that the payment by Smelter to Coal Mine is in return for 20 percent of Coal Mine's shares rather than a loan; Smelter is guaranteed two seats on Coal Mine's Board of Directors.

J. Same as Illustration I, except that it is now ten years later.

K. Same as Illustration I, except that Smelter's percentage of shares is 51 percent, and it obtains a majority on the Board. Coal Mine's old management is retained and it is allowed to operate as a largely independent division of Smelter.

L. Instead of any of the foregoing, Smelter merges with Coal Mine by mutual agreement. Smelter-Coal retains the same management and work force in the various divisions of the business, and hires a new engineering manager to supervise the building and operation of a conveyor system between minehead and smelter.

M. Same as Illustration L, except that the merger occurs when Smelter acquires the controlling shares of a deceased Coal Mine stockholder, all Coal Mine management is promptly fired, and tight centralized control of all operations is established.

N. Same as Illustration M, except that it is now forty years later.

Illustrations L through N are examples of the penultimate relational pattern in modern contracts: the large firm itself. The overwhelming variety of modern contracts, only sketchily revealed by our route from spot markets to firms, suggests the need for a "rich classificatory apparatus" such as the following.

Commencement, Duration, and Termination

Discrete transactions start sharply, are short-lived, and end sharply, either by clear performance or clear breach. Illustration A is the clearest example of this. In Illustration B—a spot purchase on thirty days' credit—these character- istics have begun (but only that) to erode; continuing relations between Smelter and Seller blur the sharpness of both beginning and ending, a point emphasized by the occasional arrearages. Duration is not just thirty days, but longer at both ends because of past and anticipated deals.

In contrast to Illustration A, or even Illustration B, Illustrations J and N, for example, show well how large-scale erosion of these discrete characteris- tics occurs in relations. In J (ten years into a twenty-year requirements con- tract, joint conveyor, acquisition of 20 percent of Coal Mine shares), the commencement itself of the relation was necessarily extended since it is not possible to plan and agree to such relations overnight. Moreover, the annual contracts preceding the present twenty-year contract are sensibly viewed as part of the commencement of that contract, since they undoubtedly provided much of the information, contract, and other considerations leading to it. The expired duration of the contract is ten years, with at least ten to go. And, so far as can be told, the relation has an indefinite future, no sharp date of termination being in prospect, renewal or other forms of continuance being likely. . . .

Measurement and Specificity

Discreteness calls for measurement and specificity, as demonstrated by Illus- trations A and B, the spot market purchases. Price and quantity must be precisely defined, along with the specific product, coal (undoubtedly of specified type, size, quality, etc.). In contractual relations, however, some aspects which eventually must become very specific—such as the amount of coal actually delivered in the thirty-eighth week of a one-year contract— may have been very *non*specific at the start of the relation, for example, "requirements." (This aspect will be treated in the next section on plan- ning.) Furthermore, in modern contractual relations, while much either starts out or finishes measured and specified, much does not. For example, many kinds of labor beggar any effort at precise specificity or measurement *at any time;* the output of any managerial or teaching job is illustrative. So, too, much of what individuals receive or pay in contractual relations—social and other psychic satisfactions and costs, such as prestige, power, discom- fort, and companionship, to name a few—remain nonspecific and immeasurable not only when initially planned, but later when actually acquired or incurred . . . .

Planning

Planning in the discrete transaction focuses on the substance of the exchange, for example, on the subject of sale, quantity, price, and payment terms, as in Illustrations A and B. But in relational contract, the planning of processes for conducting exchanges and other aspects of performance in the future, as well as conducting further planning, assume equal or even greater importance. Illustration F provides a simple example in the form of an agreement to negotiate price and use an arbitrator if negotiation fails. In Illustration M (merger after Smelter acquires majority control of Coal Mine) much of the initial planning, such as the new articles of incorporation and the bylaws, concerns processes for conducting further planning. . . . As the relation proceeds, substantive planning occurs regularly within these named frameworks, and within others, like collective bargaining, as will further process planning.

As the spot market illustrations suggest, the specificity and measurement characteristics of discrete contracts result in very complete and certain planning. On the other hand, when the subject is long-term requirements (one year, formally, in E, F, and G, and twenty in H through K), flexibility must be introduced. If planning is equated with certainty, or even with carefully calculated risk allocation, this flexibility reduces the completeness of planning. While planning of processes, such as the corporate governing structure, may be relatively complete for long periods, planning of the substance of exchanges, such as salaries, is necessarily relatively complete for only short periods. The very planning of processes in lieu of the substance of exchanges is a confession of incompleteness respecting planning of the latter. . . .

Sharing Versus Dividing Benefits and Burdens

Illustration A, the first spot market transaction, sharply divides benefits and burdens, assigning each kind explicitly to one party. For example, market risks of owning coal pass to buyer when the contract is made, as do physical risks when it is delivered, and the trucking is entirely the responsibility of the seller. This is still true in Illustration B, where credit is extended. But B is already a bit down the road towards a sharing of benefits and burdens, since Seller's quid pro quo for transferring the coal—the price—is now less absolutely assured, and Seller shares with Smelter some of the risks of Smelter's financial condition, as is demonstrated by the occurrence of arrearages. . . .

The requirements contracts, beginning with Illustration E, show a further mutualization of interests. Should the demand for Smelter's product decrease, resulting in lower production and hence smaller coal requirements, some of the costs will be born by Smelter, and some by Seller. Similarly, higher

Smelter production and sales will usually benefit both parties. The large sharing element in the joint building and operation of the conveyor system in Illustration H is enhanced further in Illustration I where Smelter also acquires a direct interest in Coal Mine's profits through share ownership. . . .

Interdependence, Future Cooperation, and Solidarity

The interdependence of the discrete transaction is so short-lived as to be easily overlooked. This is especially so in any analysis assuming the existence of markets; a participant in a market exchange is not dependent on the other participants to the exchange, but is only dependent on the existence of a market—a large number of possible participants willing to supply the goods or services desired. Thus, in Illustration A, for only a brief period between the placing of the order and performance of the deal on both sides, are Smelter and Seller interdependent. Even here the presentiation concepts likely to dominate thinking about such a transaction tend to obscure that interdependence.

The interdependence between exchangers becomes more obvious in more relational patterns. In Illustration D, a firm forward contract for a fixed quantity at a fixed price, for example, the parties have created an eight-week period during which Smelter is dependent on Seller for coal, and Seller is dependent on Smelter to pay for coal as promised.

Increased interdependence in contractual relations tends to be accompanied by a need for increased future cooperation and increased complexity of such cooperation. For example, in Illustration A, the only cooperation required of Seller is its performance: loading the coal onto trucks, driving to Smelter's yard and dumping. And the only cooperation required of Smelter is opening the gates, directing traffic in its yard, and payment. But a requirements contract with an escalation clause (Illustration E) requires a good bit more, such as notice of quantities and delivery time, and passing of information about price changes, to say nothing of dealing with the snags that inevitably develop in a long term contract. Moreover, as we move into more relational spheres, the cooperation among the parties becomes more complex and interwoven. The cooperation required in Illustration A is mainly that of producing and accepting an independent performance at the appropriate time. But the cooperation necessary in Illustration H, involving a jointly built and operated conveyor system, will be far more complex, and the activities of the parties will be far more interwoven.

One aspect of the foregoing is the need in relational contract to cooperate in planning the future, planning that will involve exercises of choice. For example, the joint conveyor system will require great amounts of day-to-day joint planning, even after it is built: work schedules, accounting programs, maintenance programs, to mention some. That Smelter and Coal Mine are likely to create a jointly owned organization to do this for them in a unitary administrative manner simply emphasizes this point. Indeed, all the illustrations from Illustration I through to the mergers demonstrate increasing intensification of the amount and complexity of cooperation necessary to secure the benefits of interdependence and reduce its risks and other costs. . . .

Personal Relations and Numbers

The essence of personal relations in the discrete transaction is what sociologists call nonprimary relations. These involve only a very small part of the personality, are very limited in scope and are nonunique; that is, it does not matter *who* the other party to the relation is. The contracting and deliveries in the spot market illustrations (A and B) demonstrate how relatively free of primary relations part of a complex economic activity can be organized.

It is unnecessary to examine situations as relational as are mergers to find primary relations in which "the participants interact as unique and total individuals" responding to "many aspects of" each others' "character and background." It is unlikely, for example, that Smelter and Coal Mine would be entering extensive requirements contracts, such as those in Illustrations E through K, especially repeatedly, without extensive primary relations developing between some sales and purchasing personnel, delivery and receiving employees, attorneys, accountants, and probably others. And certainly from Illustration H (joint conveyor system) on the more extensive relational patterns, countless primary relations are inevitable in accomplishing the basic task of securing coal from the minehead to Smelter. Modern internal corporate life may only seldom develop primary relations as close as those in family life, but they can be very close and very extensive. . . .

Power: Unilateral and Bilateral

Power is here defined as the ability to impose one's will on others irrespective of their wishes. Two kinds of power, unilateral and bilateral, may usefully be distinguished in analyzing contract and the economic model.

Unilateral power is any capacity a person has to subject another to some particular effect without the other's consent. Members of any society have unilateral power arising, inter alia, from the existence of property and other rights (such as liberty) conferring on the powerholders the ability to impose sanctions on others interfering with those rights. (This is, essentially, the "power to be let alone" with one's property and oneself.)

Bilateral power arises when the possibility of exchange exists by which two parties can release each other from some of the restraints imposed by their respective unilateral powers. For example, a potential employee may consider giving up some of his liberty in exchange for a potential employer's giving up some of its property rights in money. Bilateral power is thus held over different subjects (body and money) by different holders (potential employee and employer), but with an interparty nexus through potential exchange. Bilateral power is *exercised* through actual exchange. . . .

A complication concerning power in contractual relations is its common lack of specificity, if not in theory, then in fact. The power of Smelter in Illustration E to vary its requirements for coal is only vaguely delimited in law and even less delimited in practice. And in the merger illustrations, the diffusion of power occurring in fact, in theory, or in both *within* the firm, inevitably creates a lack of specificity respecting power.

Contractual relations introduce complicating new dimensions to bilateral power as well as to unilateral power. Bilateral power thus becomes a major contributing factor in any ongoing contractual relation. Illustration I (joint conveyor system, twenty-year requirements contract, Smelter as minority shareholder) shows this particularly vividly. Smelter's rights to monitor Coal Mine's operations, to make recommendations about equipment, management, and the like, and to have directors on Coal Mine's board are meaningless except as components in negotiational and agreement processes. So too with respect to cooperation in the building of the conveyor, and even more so in its operation over at least twenty years. . . .

In sum, contractual relations constantly generate changing power balances. In a discrete transaction, with its single exercise of bilateral power, the power status quo derives from outside the transaction. Thus, in Illustration A, the extent of Seller's power to extract a price for the coal from Smelter will depend on Smelter's alternatives in an actual market occurring in the existing status quo of property rights. The discrete transaction itself generates no force affecting this power. But in Illustration L (joint conveyor system, requirements contract, complex financial and ownership linkings, after ten years of operation) the relation itself will have become a major determinant of both unilateral and bilateral power positions at any given moment. This will be true not only of the now much interlinked "entities"—Smelter and Coal Mine—but also of individuals and of organizational sub-units. Contractual relations thus continually generate their own status quo affecting both bilateral and unilateral power.

Notes on Marketing and Supply Contracts

1. The illustrations at the beginning of the Macneil excerpt (supra at page 193) demonstrate that all the issues that arise between contracting partners can also arise within a single firm. This similarity between decisions within a large firm and decisions between two independent contracting partners has often been analyzed in the economics literature. One of the earliest discussions of this similarity was R. H. Coase, The Nature of the Firm, 4 Economica 386 (1937). The more recent economic literature is surveyed in Oliver Hart, An Economist's Perspective on the Theory of the Firm, 89 Columbia Law Review 1757 (1989).

2. One question not resolved, by either the legal or the economic literature, is whether the proper legal response to long-term or complex "relational" contracts is fundamentally different from the proper legal response to simpler contracts. It may be true that relational contracts are incomplete, as

all three of the excerpts in this section point out. However, even very simple contracts are often incomplete in at least some respects. For example, a simple contract of sale may fail to specify the measure of damages one party must pay if the contract is breached. It may also fail to specify the extent to which the seller warrants the quality of his or her products, or the conditions under which a mutual mistake would rescind the contract. Indeed, all of the legal doctrines discussed in this chapter and the preceding one are designed to fill just this sort of gap, even in contracts that are otherwise very simple.

Does this mean that the task of courts in interpreting "relational" contracts is no different from the task facing courts when they fill gaps in any contract? If so, all of the readings in Chapter 1 discussing the selection of default rules to fill gaps (supra at pages 16–30) are equally relevant here. From this perspective, it should not be surprising that the excerpts by Schwartz and by Goetz and Scott both suggest that courts should fill the gap with whatever rule would be most efficient in the sense of maximizing the total benefit to the parties. The two excerpts disagree about which rule would be most efficient, but both accept this efficiency approach (or the "what the parties would have agreed to" approach) to filling the gap.

Significantly, though, none of these excerpts suggests that the gaps in relational contracts should be filled with a penalty default rule of the sort discussed in the excerpt by Ayres and Gertner in Chapter 1 (supra at pages 22–27). This may indicate one significant difference between complex relational contracts and simpler ones. That is, in complex relational contracts it may not make sense for the law to try to encourage the parties to spell out all their rights and duties with greater precision in advance. On topics where the law would not use a penalty default rule anyway, though, this distinction has little practical significance.

3. Whenever the courts do not fill gaps with penalty default rules, the courts must instead select default rules that are most efficient, or most fair, or most desirable according to some other normative criterion. As noted above, the Goetz and Scott and the Schwartz excerpts both endorse economic efficiency as the appropriate normative criterion.

Notice, though, that in at least some cases the economic and noneconomic criteria may coincide. For example, Goetz and Scott argue that a rule requiring one party to maximize the total net benefits produced by a contractual relationship can also be defended on the grounds of fairness (supra at page 185), for it requires each party to treat the other party's interests on an equal footing with his or her own. Thus, you should not assume that efficiency and fairness considerations necessarily work in different directions in these cases.

4. The Schwartz excerpt asks whether courts are competent to identify the allocation of gains and losses that would be most efficient. This issue, too, was raised in connection with impracticability and frustration where the Kull excerpt questioned courts' ability to identify the best risk bearer (supra at page 152). The Schwartz excerpt adds to Kull's critique by pointing out that some contingencies are effectively "noncontractible" because they cannot be observed by courts. For example, if sellers can claim that their costs have gone

up and courts are in no position to verify such claims, it might seem to make little sense to have a legal rule under which either party's rights turned on whether the seller's costs had in fact gone up.

However, observability and verifiability are generally matters of degree, rather than all-or-nothing choices. For example, courts may be unable to measure sellers' costs without a large chance of error, but the size of the expected error will be low in some kinds of contracts (those where sellers have relatively simple cost structures) and high in others. The relevant question, then, concerns the point at which the expected judicial error will be sufficiently great to make it preferable to entrust complete discretion to one or the other of the parties under the rules discussed in the Schwartz excerpt.

5. Does the Macneil excerpt offer a rival normative theory of how gaps in relational contracts ought to be filled? Professor Macneil's views were further developed in a book—Ian Macneil, The New Social Contract: An Inquiry Into Modern Contractual Relations (New Haven: Yale University Press, 1980)—and a series of articles, most recently in Ian Macneil, Values in Contract: Internal and External, 78 Northwestern University Law Review 340 (1983). See also Robert W. Gordon, Macaulay, Macneil and the Discovery of Solidarity and Power in Contract Law, 1985 Wisconsin Law Review 565 (1985); Jay M. Feinman, Relational Contract and Default Rules, 3 Southern California Interdisciplinary Law Journal 43 (1993); and the excerpt by Gillian K. Hadfield (infra at pages 204–09). More critical reviews include Randy E. Barnett, Conflicting Visions: A Critique of Ian Macneil's Relational Theory of Contract, 78 Virginia Law Review 1175 (1992); and Richard Craswell, The Relational Move: Some Questions from Law and Economics, 3 Southern California Interdisciplinary Law Journal 91 (1993).

§3.3.2 Termination Decisions

Transaction Cost Determinants of "Unfair" Contractual Arrangements

BENJAMIN KLEIN

Most actual contractual arrangements consist of a combination of explicit and implicit-enforcement mechanisms. Some elements of performance will be specified and enforced by third-party sanctions. The residual elements of

Benjamin Klein, Transaction Cost Determinants of "Unfair" Contractual Arrangements, 70 American Economic Association Papers & Proceedings 356, 356–60 (1980). Copyright © 1980 by the American Economic Association. Reprinted by permission.

performance will be enforced without invoking the power of some outside party to the transaction but merely by the threat of termination of the transactional relationship. The details of any particular contract will consist of forms of these general elements chosen to minimize transaction costs (for example, hiring lawyers to discover contingencies and draft explicit terms, paying quality-assurance premiums, and investing in nonsalvageable "brand name" assets) and may imply the existence of what appear to be unfair contract terms.

Consider, for example, the initial capital requirements and termination provisions common in most franchise contractual arrangements. These apparently one-sided terms may be crucial elements of minimum-cost quality-policing arrangements. Given the difficulty of explicitly specifying and enforcing contractually every element of quality to be supplied by a franchisee, there is an incentive for an individual opportunistic franchisee to cheat the franchisor by supplying a lower quality of product than contracted for. Because the franchisee uses a common trademark, this behavior depreciates the reputation and hence the future profit stream of the franchisor.

The franchisor knows, given his direct policing and monitoring expenditures, the expected profit that a franchisee can obtain by cheating. For example, given the number of inspectors hired, he knows the expected time to detect a cheater; given the costs of low-quality inputs he knows the expected extra short-run cheating profit that can be earned. Therefore, the franchisor may require an initial lump sum payment from the franchisee equal to this estimated short-run gain from cheating. This is equivalent to a collateral bond forfeitable at the will of the franchisor. The franchisee will earn a normal rate of return on that bond if he does not cheat, but it will be forfeited if he does cheat and is terminated. . . .

It is important to recognize that franchise termination, if it is to assure quality compliance on the part of franchisees, must be unfair in the sense that the capital cost imposed on the franchisee that will optimally prevent cheating must be larger than the gain to the franchisee from cheating. Given that less than infinite resources are spent by the franchisor to monitor quality, there is some probability that franchisee cheating will go undetected. Therefore termination must become equivalent to a criminal-type sanction. Rather than the usually analyzed case of costlessly detected and policed contract breach, where the remedy of making the breaching party pay the cost of the damages of his specific breach makes economic sense, the sanction here must be large enough to make the expected net gain from cheating equal to zero. The transacting parties contractually agree upon a penalty-type sanction for breach as a means of economizing on direct policing costs. Because contract enforcement costs (including litigation costs which generally are not collectable by the innocent party in the United States) are not zero, this analysis provides a rationale against the common law prohibition of penalty clauses.

The obvious concern with such seemingly unfair contractual arrangements is the possibility that the franchisor may engage in opportunistic behavior by terminating a franchisee without cause, claiming the franchise fee and purchasing the initial franchise investment at a distress price. Such behavior may be prevented by the depreciation of the franchisor's brand name and therefore decreased future demand by potential franchisees to join the arrangement. However, this protective mechanism is limited by the relative importance of new franchise sales compared to the continuing franchising operation, that is, by the "maturity" of the franchise chain.

More importantly, what limits reverse cheating by franchisors is the possible increased cost of operating the chain through an employee operation compared to a franchise operation when such cheating is communicated among franchisees. As long as the implicit collateral bond put up by the franchisee is less than the present discounted value of this cost difference, franchisor cheating will be deterred. Although explicit bonds and price premium payments cannot simultaneously be made by both the franchisee and the franchisor, the discounted value of the cost difference has the effect of a collateral bond put up by the franchisor to assure his noncheating behavior. This explains why the franchisor does not increase the initial franchise fee to an arbitrarily high level and correspondingly decrease its direct policing expenditures and the probability of detecting franchisee cheating. While such offsetting changes could continue to optimally deter franchisee cheating and save the real resource cost of direct policing, the profit from and hence the incentive for reverse franchisor cheating would become too great for the arrangement to be stable.

Franchisees voluntarily signing these agreements obviously understand the termination-at-will clause separate from the legal consequences of that term to mean nonopportunistic franchisor termination. But this does not imply that the court should judge each termination on these unwritten but understood contract terms and attempt to determine if franchisor cheating has occurred. Franchisees also must recognize that by signing these agreements they are relying on the implicit market-enforcement mechanism outlined above, and not the court to prevent franchisor cheating. It is costly to use the court to regulate these terminations because elements of performance are difficult to contractually specify and to measure. In addition, litigation is costly and time consuming, during which the brand name of the franchisor can be depreciated further. If these costs were not large and the court could cheaply and quickly determine when franchisor cheating had occurred, the competitive process regarding the establishment of contract terms would lead transactors to settle on explicit governmentally enforceable contracts rather than rely on this implicit market-enforcement mechanism.

The potential error here is, after recognizing the importance of transaction costs and the incomplete "relational" nature of most real world contracts, to rely too strongly on the government as a regulator of unspecified terms . . . . While it is important for economic theory to handle significant contract costs and incomplete explicit contractual arrangements, such complexity does not imply a broad role for government. Rather, all that is implied is a role for brand names and the corresponding implicit market enforcement mechanism I have outlined.

Problematic Relations: Franchising and the Law of Incomplete Contracts

GILLIAN K. HADFIELD

To properly "interpret" the franchise contract one must read the written document in the context of the franchise "relation." Faced with the silence of the agreement, a court has only two options: Either it can refuse to decide the dispute, finding the contract invalid and unenforceable, or it can resolve the case, filling in the contract's silences. Early in the development of franchising, courts routinely took the first, "classical," route. Today, however, almost all courts choose the second.

Having taken this second route, a court has no choice but to look beyond the document and identify a configuration of commitments patterned not in the words of the contract but in the underlying relation itself. If a court is to decide that an auto dealership warrants termination by refusing to build a new showroom, it can only do so by discerning in the relationship an obligation on the part of the dealer to comply with a manufacturer's request for a new showroom. In deciding that a fast-food franchisor incurs no liability to potential franchisees when it fails to disclose its plans to dismantle the system, it must see within the franchise relationship no franchisor obligation to share with franchisees relevant available information necessary to make successful business decisions. Courts, in resolving franchising disputes, do exactly this: they supply a theory of the relationship in order to identify the commitments they must enforce. But the theory they supply is, I will argue, inappropriate.

The Business Judgment Approach

Essentially, the courts get half of the story right. They understand well the franchisor's interests in quality control and design and the vulnerability of these interests to franchisee failures. Thus, they see the franchisor's problem in enforcing the franchisee's commitment to maintain quality and adhere to the franchisor's system. Consequently, in the paradigmatic franchise dispute in which the franchisee has, for example, failed to observe practices set out in an operations manual, and where there is no evidence of improper franchisor motive, the courts easily find the obligation that makes such behavior a contract violation and a basis for termination.

Where the courts go astray, however, is in treating the franchisor's interest as if it represented the entirety of the relation. As one court expressed this view:

Gillian K. Hadfield, Problematic Relations: Franchising and the Law of Incomplete Contracts, 42 Stanford Law Review 927, 979–81, 983–90 (1990). Copyright © 1990 by the Board of Trustees of the Leland Stanford Junior University. Reprinted by permission.

> [T]he substantiality of a franchisee's noncompliance, as a legal concept, must be gauged in light of its effect upon or potential to affect the franchisor's trade name, trademark, good will and image which, after all, is the heart and substance of the franchising method of doing business.

Although correctly recognizing the effect of franchisee noncompliance on the franchisor's trademark, the court quoted above errs in adopting the view that this effect constitutes the "heart and substance" of the franchise. Even calling franchising a "method of doing business," and thus portraying it as the franchisor's choice of a distribution method, obscures the basic exchange aspects of the arrangement. Franchisors and franchisees alike perceive franchises as packages of services that promote the profitability of the fledgling small businessperson. It would be just as erroneous to portray this aspect of franchising as the entirety of franchising. The point remains, however, that both franchisor and franchisee perspectives reflect components of the exchange.

The theory that the franchise relation centers exclusively on protecting the integrity of the franchisor's system and the value of the trademark leads to what I call a rule of "business judgment" in resolving cases that go beyond the simple paradigm of a franchisee violating a standard operating requirement. Facing the more problematic cases in which a franchisor terminates a franchisee, the courts routinely adopt the view that since the franchise relationship is a "business" relation the franchisor needs only to articulate a plausible business justification. . . .

Essentially, this view frees the franchisor to behave opportunistically in making those decisions. The business judgment approach rejects any scrutiny of whether the decision results in the extraction of sunk costs from the franchisee, or of whether the franchisor's decision is one that is profitable to the franchisor only at the expense of the franchisee's investments. The problem with this approach should now be clear: By enforcing the franchise contract in this way, courts fail to protect fully one half of the interests necessary to support the franchise relationship.

What are the consequences of this failure by the courts? There are a number of ways to approach this question. It seems clear that franchisees generally enter the franchising relation expecting to exchange conscientiousness, investment, and hard work for a successful management package. Thus, from a relational perspective, courts undermine the exchange by routinely ignoring this expectation in resolving contract disputes. Courts fail to appreciate that franchisees did not enter into the exchange as if they were purchasing stock in the franchisor's company, that is, putting up their money and taking all risks short of fraud, self-dealing, or other failures of corporate responsibility. Franchisees make large, undiversified, long-term investments; normally there is no "meeting of the minds" granting the franchisor power to issue directives that profit the franchisor only at the direct expense of the franchisee despite covenants which require the franchisee to obey all directives. Treating franchisees as if they made deals they did not in fact make offends the traditional rationale for contract enforcement.

Under this rationale, contract law should provide a contracting instrument that enables parties to devise their own arrangements privately and enforces the commitments underlying those arrangements. In the world of classical complete contracting, courts read the contract and compare it with what has occurred. In franchising's world of incomplete contracting the courts must strive to identify the entirety of the commitment structure that underlies the franchise arrangement and then to enforce those commitments.

The incompleteness of franchise contracts is neither an accident nor a failure of adequate draftsmanship. Given this, parties and courts cannot merely pursue rules of interpretation intended to spur better drafting and competition among franchisors in designing a standard form contract. They must instead grapple with the difficult issues of relational interpretation.

A Relational Approach to Interpreting Franchise Contracts

. . . Courts should stop conceiving of the franchise relation as one solely dedicated to protecting the franchisor's trademark and goodwill. The franchise relation is a mutual exchange. Franchisees make large, sunk investments in operating outlets—investments the franchisor would otherwise have to make in order to retail its product or service in the franchisee's area—to reduce the risks of operating their own small business. In turn, franchisors obligate themselves to promote the profitability of franchisees. Franchisors risk the value of their trademarks in exchange for shifting sunk investments to franchisees. Franchisees commit themselves to complying with franchisor directives in anticipation that they will receive sound products, sound business advice, and support. The relation makes little sense unless the "contract" between franchisor and franchisee balances these mutual arrangements, establishing commitments on the part of both franchisee and franchisor.

The relational approach to contract interpretation requires sensitivity to the particularities of any one relation. The features of the abstract franchise relation should only be used as a guide to identify the features of a concrete franchise relation. In some cases, a "franchise" relation may differ in important respects from an abstract franchising relation. Using the abstract model as a guide, however, a court can distinguish those arrangements in which the typical franchising problems of control and opportunism are potential problems from those in which they are not.

Courts facing a dispute between a "franchisor" and a "franchisee" should first determine whether the arrangement includes sunk investments, a necessary prerequisite for opportunism. Many distributorship arrangements, for example, lack sunk investments; distributors may take on a manufacturer's product but make very few investments in distribution facilities and equipment specific to the product. In other cases, the franchisor agrees to purchase all of the franchisee's assets at market prices—that is, prices that include a component for the franchisee's future profits as a franchisee—in the event of termination. In these two types of cases, the franchisee needs little protection against opportunistic behavior which reduces the value of sunk investments. Therefore, there is no reason to believe that the relationship includes limits on manufacturer behavior other than those explicitly set out in the contract. If the court finds sunk investments, however, it must determine whether the challenged action falls within the scope of the relationship.

There are no hard-and-fast rules for accomplishing this end. Relational interpretation is a fact-specific exercise of attention, insight and judgment. Consider a recent case in which Volkswagen terminated a franchisee after it distributed Fords in the same dealership. The written contract contained a provision giving Volkswagen the right to prescribe standards for the showroom layout, but it did not prohibit a dual dealership. Manifesting a business judgment approach, the court upheld the termination, suggesting that the Ford line affected the showroom layout and finding that Volkswagen had made a valid business judgment that it was not in its interest to have dual dealerships. From a relational perspective, this resolution is troublesome. *Volkswagen* involves the typical scenario in which the franchisee has made large sunk investments in capital with the attendant danger that the franchisor will make "business judgments" that fail to take these investments into account. The court needs to ask: What prompted the franchisee to take on the Ford line? It is possible that sales of Volkswagens had declined to such a level that the scale of business no longer generated a reasonable return on the franchisee's investment in showroom space. Why was there no explicit clause prohibiting a dual dealership? It is possible that Volkswagen was unable to get franchisees to take on dealerships with such a prohibition at the time the contract was formulated, because of uncertainty about the popularity of the automaker's models and thus the wisdom of making sunk investments. Did the additional Ford line in fact substantially affect the showroom layout? Did the franchisee have an opportunity to make changes to the showroom in order to maintain the amount of space previously devoted to Volkswagens? Did the prohibition on a dual dealership benefit only the franchisor's interests, such as by promoting tangential trademark concerns at substantial cost to the franchisee, without promoting the franchisee's interest in selling more cars or creating a more profitable dealership? The court should have attempted to determine, in light of the particular relationship at hand, whether the franchisor's judgment took advantage of the franchisee's commitment of sunk resources.

Finding that the franchisor does not have a unilateral contract right to prevent a dual dealership does not forever doom the franchisor to this unhappy course. The point is that, like any other business, it may have to pay for achieving this result. The franchisor could have renegotiated with the franchisee, compensating the franchisee for giving up the Ford line, or it could terminate the franchise and pay damages calculated to compensate the franchisee for its loss, including repurchase of inventory and showroom and damages for lost future profits.

Looking more closely into the franchising relationship, courts should be sensitive to the ways in which franchisor and franchisee interests diverge.

Only then can they assess whether the franchisor's interests should trump the franchisee's. Franchisor policies or promotional programs that produce a profit for the franchisor, but only because franchisees must bear the cost, are one example. There are numerous others. A franchisor may prefer a particular location because its high visibility promotes consumer recognition of the trademark, but a franchisee might find that the same location's inaccessibility fails to generate a profitable volume of customers. Franchisor image advertising of the trademark may generate revenue from secondary licensing (such as T-shirts and movie releases) but produce very little in the way of business for the franchisees. Franchisors generally seek to increase gross volume per outlet because they receive royalties based on volume and because volume generates consumer recognition. On the other hand, franchisees, who must bear the cost of increasing volume, seek to increase profits, which requires operating at a volume somewhere below the maximum possible. When deciding whether to sell their franchise systems, franchisors have an interest in maintaining high sales and trademark value by keeping such plans quiet, whereas potential and existing franchisees can make better investment decisions and evaluate the risks if they have access to the franchisor's plans. In a declining franchise system, large corporate franchisors with alternative uses for resources may have an interest in withdrawing those resources at a time when franchisees with sunk investments are still able to earn a reasonable return from the system. A franchisor has an interest in transferring the assets of an existing franchise to a new franchisee at a low price or to a particular candidate, but the outgoing franchisee has an interest in recouping as much of its sunk investment as possible.

Accompanying these divergent interests is the franchisor's control power. This power allows the franchisor to pursue its interest at the expense of the franchisee's. When a franchisor decides to terminate and buy out a successful franchisee, is it taking advantage of a gamble that turned out well but not compensating the franchisee for the loss of future profits? When a franchisor withholds approval of the transfer of a franchise, does it do so in order to grant the franchise to a specific transferee it prefers, who will pay a lower price for the franchise? If a franchisor disapproves of the transfer price, does it do so because it reasonably fears that the proposed transfer price will overburden the new franchisee? By keeping future plans for the franchise system a secret, has the franchisor induced franchisees to make decisions, such as putting up investments or foregoing alternative arrangements, that a reasonable businessperson with the information would not have made? Are plans to alter the system in an effort to bolster a declining organization overly risky in light of the cost to franchisees? Is the decision to terminate old franchisees or to seek voluntary termination by changing the terms of the franchise an effort to take over the franchises at low cost or an effort to eliminate obsolete inventory?

Proponents of a business judgment approach to resolving franchise contract disputes are likely to protest that the courts are not competent to ask, much less to answer, these types of questions. First, I believe courts are capable of answering these types of questions; for the most part the questions simply involve widening the judicial lens to include the full landscape of events. Courts can guide their inquiry with a relatively simple analytical structure, one developed from an understanding of the nature of commitment, sunk costs, and opportunism. Second, even conceding that the courts will make mistakes in carrying out this inquiry, courts should be at least as competent in understanding franchise disputes as they are in analyzing medical malpractice cases, product liability issues, and antitrust suits, or in reviewing the decisions of modern administrative agencies, assessing the employment decisions of public and private organizations, or making determinations about the deals struck between large, equally sophisticated firms. Although the courts may be more or less flawed in performing many of their functions, the relative competence of the courts is a fixed feature of the modern judicial system and one of the costs of administering that system.

Third, while courts may err and see opportunism where it does not exist or fail to see it where it does, these errors should be largely random. Any bias resulting from the relational approach is likely to be far smaller than the bias we would expect to emerge if the franchisor were handed the judge's gavel. But a business judgment approach, by deferring to the franchisor's assessment of the decision to terminate a franchisee, does essentially this. Given the inherent difficulty of writing incomplete contracts and the central difficulty of balancing franchisors' interests in quality control against franchisees' interests in protection against opportunism, the franchise relationship needs an unbiased dispute resolution mechanism. The business judgment approach, because it will always be more biased than even a flawed relational approach, fails to provide this mechanism.

Finally, unless the courts wish to abandon their traditional role in enforcing the exchanges actually reached by contracting parties, there is no escaping relational interpretation in a world of incomplete contracts. The business judgment approach does not respect the basic franchising exchange and thus undermines it. Complete contracting is simply not feasible. Consequently, despite the difficulties of administering incomplete contracts, the judicial system must either grapple with this task or else deny accurate contract enforcement to arrangements such as franchising.

Notes on Termination Decisions

1. Both the Hadfield and Klein excerpts stress the potential conflict of interests between franchisors and franchisees. For example, if a local franchisee cuts corners on maintenance or quality control, some consumers may treat the lower-quality operation as evidence that the entire chain is of low quality, thus reducing the profits of the franchisor (and of other franchisees). As the local franchisees will reap all the benefits of the reduced expenditures, without having to suffer all the costs, the local franchisee may have a strong

incentive to engage in just this sort of cost cutting. Of course, similar incentives can distort the franchisor's incentives as well. For example, if the franchisor reduces expenditures on national advertising, all of the savings from the reduced expenditures will accrue to the franchisor, while some of the costs will be felt by individual franchisees who see their volume of business decline.

This conflict of interest is similar to that analyzed by Goetz and Scott in their article on marketing and supply contracts (supra at page 186). As long as each party must share any increase or decrease in profits with the other party, neither will have the full incentive to take care to maximize the overall profits.

2. If courts do not simply give all discretion to the franchisor, and instead try to police terminations and other franchisor misconduct, by what standard should the court try to evaluate the franchisor's actions? Should they adopt an efficiency-based standard? That is, should they hold the franchisor to a requirement of never doing anything that would reduce the total profits accruing to the franchisors and the franchisees as suggested in the Goetz and Scott excerpt on marketing contracts (supra at pages 184–85)? If so, how should courts decide what decision would have maximized total profits? Moreover, what if the franchisor failed to maximize total profits as a result of an innocent mistake? For example, suppose the franchisor made a decision that, at the time, seemed likely to maximize total profits, but suppose the decision turned out in hindsight to be wrong. Should franchise contracts be interpreted to give franchisees insurance against this kind of franchisor error? As the Hadfield excerpt discusses in its final three paragraphs (supra at page 209), standards of this sort raise obvious questions of judicial competence, of the sort that were raised earlier in the Schwartz excerpt on marketing and supply contracts (supra at pages 187–93).

3. In discussing mechanisms to deter cheating by the franchisor, Klein suggests that market forces will often provide an adequate check because franchisors who regularly cheat their franchisees will have a hard time finding new franchisees. The circumstances under which market incentives are likely to work well (or are likely not to work) are discussed at more length in Chapter 5, in connection with the doctrine of unconscionability. See, in particular, the excerpts by Duncan Kennedy (infra at pages 315–19) and Alan Schwartz (infra at pages 319–21).

4. Similar issues can arise when employers try to terminate employment relations by firing employees. Traditionally, employment contracts were "contracts at will," meaning that the default rule (unless the contract specified otherwise) was that either side could terminate the relationship for any reason whatsoever. In recent years, many courts have shifted to a different rule, under which employers are not allowed to fire employees for reasons that would not qualify as "good faith." For discussions of some of the issues raised by this doctrine, see Jeffrey L. Harrison, The "New" Terminable-at-Will Employment Contract: An Interest and Cost-Incidence Analysis, 69 Iowa Law Review 327 (1984); Richard A. Epstein, In Defense of the Contract at Will, 5

University of Chicago Law Review 947 (1985); Frances Raday, Costs of Dismissal: An Analysis in Community Justice and Efficiency, 9 International Review of Law and Economics 181 (1989).

§3.4 Modification by the Parties

The Law of Contract Modification

VAROUJ A. AIVAZIAN, MICHAEL TREBILCOCK, AND MICHAEL PENNY

The principles of law governing the enforceability of promises which modify pre-existing contractual obligations are currently undergoing what appears to be a dramatic transformation. Contractual modifications (mutually agreed changes in contractual terms made subsequent to the formation of the primary contract) have traditionally encountered problems of enforceability because of the requirement that enforceable promises must be given for consideration. Doctrine stipulates that a promise in exchange for a promise to perform an existing contractual obligation is, without more, unenforceable. This is often referred to as the pre-existing duty rule. In place of this rule, a growing body of case law, statute law and legal commentary would enforce promises, particularly modification promises, without consideration, where they are "fair and equitable," made "voluntarily," not made in "bad faith" or as a result of "extortion" or "coercion," or are supported by "legitimate commercial reasons."

The nature of the apparent analytical paradox presented by contract modifications can be stated briefly: on the one hand, why would any party to a contract agree, by way of modification, to pay more or accept less than originally contracted for, without an appropriate quid pro quo (consideration), unless the other party had obtained bargaining power in the course of the relationship that he did not possess at the time of the contract formation and that he now seeks to exploit? If this explains most modification situations, then it might be argued that the law should attempt to discourage extortionary, coercive, opportunistic or monopolistic behaviour by refusing to enforce most modifications, perhaps by means of a presumption of invalidity. The traditional legal doctrine might be close to what is socially optimal. On

Varouj A. Aivazian, Michael J. Trebilcock & Michael Penny, The Law of Contract Modification: The Uncertain Quest for a Bench Mark of Enforceability, 22 Osgoode Hall Law Journal 173, 173–74, 187–97, 200–202, 204–5 (1984). Copyright © 1984 by V. A. Aivazian, M. J. Trebilcock and M. Penny. Reprinted by permission.

the other hand, especially in commercial contexts where most litigated modification cases seem to arise, it might be argued that parties would typically not enter into modifications unless they both felt better off as a result relative to the position that would or might have been obtained without a modification. Hence, the law should respect the parties' assessment of what course of action best advances their joint welfare and enforce modifications, that is, apply a presumption of validity. This would reflect much contemporary legislative, judicial and academic thinking. Both propositions, despite being contradictory, have strong axiomatic appeal and hence, the apparent paradox that we attempt to deal with in this paper. . . .

An Economic Framework of Analysis

In all contractual settings, parties' choices as to possible trading partners and/or terms of trade are subject to constraint. Concepts of "coercion" or "voluntariness," standing alone, are not helpful in identifying the nature, source, and impact of that subset of constraints that render a resulting contract objectionable. In an attempt to break out of this vicious circle of inventing new language to restate old problems, we try, in the next two sections of this paper, to escape the sterility of metaphysical concepts, such as the "overborne will," and the functional indeterminacy, in the present context, of recent, highly abstract, philosophical theorizing about the concept of "coercion." We assume a rather prosaic objective for the law on modifications: minimizing transaction costs, and proceed to develop an economic framework that attempts to identify the variables that must be taken into account in formulating legal rules that advance this objective.

We shall follow Posner and initially distinguish two alternative sets of cases in which contract modifications might be sought. In one set of cases there are no changes in the underlying conditions governing the initial concept except that the promisee has acquired some monopoly power ex post and exploits this power by forcing higher returns than provided for in the initial contract. Posner argues for the non-enforcement of contract modifications in such cases. The second set of cases is characterized by changes in the underlying economic conditions, or the emergence of new information about the underlying economic conditions governing the contract which prevent or inhibit the promisee from completing the promised performance without a modification of the contract. Posner argues that modification is justified in such cases because without that ability mutually advantageous exchanges may be precluded.

First, in relation to both sets of cases, it must be emphasized that the potential for opportunism in the course of contractual performance is likely to be constrained in various ways. The party demanding a modification on threat of breach will need to take account of: the impact of this on future dealings with the other party if repeat transactions are envisaged; the reputation effects on other potential trading partners in the market; ease of substitution by the party from whom the modification is demanded; initial contractual terms that may make the latter party unreceptive to a modification (for example, liquidated damage or penalty clauses, if enforceable, performance bonds, or back-end loading in payment schedules); the possibility of the latter party obtaining specific relief in the form of an injunction or specific performance; exposure to a damages claim in the event that modification is refused and breach occurs. However, despite these constraints, there will be situations where there may be gains from engaging in opportunism—repeat transactions are not envisaged, market networks may imperfectly disseminate information about contractual performance, substitution may be difficult or costly, initial contractual provisions may not fully penalize or constrain opportunism and may be costly to negotiate in great detail, specific relief may be unavailable, and damages for breach may not fully compensate the non-breaching party for the costs associated with procuring a substitute and other consequential damages induced by the breach, or lack of exigible assets, limited liability or bankruptcy may preclude effective enforcement of a damages judgment.

Consider now the first set of cases. For these cases ex post contract modification is a zero sum game. What one party gains the other loses—in other words there is, by assumption, no room for co-operative recontracting. Assume an economic environment characterized by zero transaction costs, rational expectations and complete information about all contingencies. Assume also that the contract is initially drawn up under perfectly competitive conditions. We will argue that in such an environment whether the law enforces contract modifications is irrelevant from a resource allocation perspectives and has no bearing on the economic welfare of the contracting parties. All that is required is that the law be *unambiguous*.

Suppose the law does enforce contract modifications. Then the terms of the initial contract will reflect the optimal future (ex post) strategy of the promisee. Suppose the promisor initially enters into the contract recognizing that the promisee's optimal future strategy will be to force higher returns through contract modifications. Since the law will enforce such modifications, the initial contract terms will be adjusted to reflect these future payoffs to the promisee. Potential extortionary monopoly rents will thus be fully impounded—given perfect competition and rational expectations—into ex ante contract terms. As long as the promisee's optimal ex post strategy (whether opportunistic or not) is unique, it will be fully reflected in the initial contract in this rational expectations environment. . . .

However, if there are positive transaction costs, then the choice of legal regime may have a significant influence on resource allocation. Initial analysis suggests that a law which unambiguously disallows contract modifications will generally be more efficient. Such a law economizes on the transaction costs that would otherwise be incurred in a regime that allowed contract modification. These costs not only include the direct costs associated with contract modifications (contract renegotiation takes time and absorb other resources), but also those associated with various contractual and institutional arrangements designed to forestall opportunistic behaviour. These latter costs include those of writing, monitoring and enforcing detailed contractual provisions to penalize or constrain opportunism, as well as the costs of the resources expended by the promisee (for example bonding costs) to convince the promisor that he will not behave opportunistically. It is in the promisee's interest to incur such costs (which limit his potential opportunism) in a regime that allows contract modifications to induce an optimal initial level of mutually advantageous contracts. However, such costs engender a reduction in the initial exchange opportunity since the exchange process is more costly . . . . The parties ex ante are therefore worse off than in the case where no modifications are allowed.

With unobservable "performance quality" asymmetries across promisees, an adverse selection problem also arises in that promisees with non-opportunistic ex post strategies will not be adequately compensated for their above average performance "quality" and will withdraw from the exchange. Equilibrium contract terms will reward only those with opportunistic ex post strategies. A law which disallows contract modifications will be efficient since it will reduce the transactions costs associated with signalling or sorting activities designed to provide information on performance quality differences among promisees.

In short, in cases where contract modifications occur purely and simply as a result of changes in the strategic circumstances of the contracting parties, the enforcement or non-enforcement of modified contracts in a zero transaction costs environment with complete information about future contingencies and rational expectations will have no bearing on resource allocation or economic welfare. In an environment with positive transaction costs or incomplete information, a law which disallows contract modifications will economize on transaction costs and maximize the gains from contractual agreements. Hence, efficiency considerations dictate that contract modifications in this context be non-enforceable.

The second set of cases in which contract modifications may be sought (supervening changes in the economic environment of the contract) are those in which modifications can represent mutually advantageous positive-sum games. . . . If recontracting between the parties in a particular ex post state of nature is mutually advantageous, then it will occur, leading to an optimal restructuring of contractual terms. These considerations suggest that contract modifications are necessary for the attainment of Pareto efficiency and should be allowed by law in this second set of cases. However, as we will see, such a conclusion is premature since there are additional considerations that bear on the problem.

Suppose that the law does *not* allow contract modifications. The impact of such a law on resource allocation and economic welfare depends very much on the economic environment postulated. Consider an economy characterized by complete contingent markets. In such an economy contracts can be written costlessly pre-specifying allocations among the contracting parties in every foreseeable contingency or state of nature, and contracts can be costlessly monitored and enforced by the parties. If modifications of contractual terms are disallowed, initially complete contingent contracts will provide for Pareto efficient allocations in every possible state of the world, precluding any need for ex post modifications of contractual terms. Alternatively, in the absence of complete contingent contracts, legal rules, such as implied terms, damage remedies for breach, specific performance, or frustration, may be designed to induce Pareto optimal allocations in every ex post state of nature, again precluding any need for ex post modification of contractual terms.

Once one abandons the assumption of complete contingent contracts, or of legal remedies which fully substitute for contingent terms in contracts, the issue of whether the law should allow contract modification becomes more complicated. Consider an economy with incomplete contracts where ex ante contractual provisions do not exist to provide unambiguous Pareto optimal allocations in some states of nature. . . .

A strong economic argument can be made that when risk assignments are not explicit (and third-party insurance possibilities are missing), as in incomplete markets, the law should assign the risk of an adverse outcome to the party who is the superior risk bearer. Superior risk bearer simply means the party who can minimize the risk at least cost or, where risk is the product of exogenous events beyond the control of either party, can insure at least cost. Relative insurance efficiencies turn on relative ability (a) to appraise both the probability of a risk materializing and the magnitude of the costs that will follow from it, and (b) to spread the risk through market insurance or diversification.

Where the contract has specifically assigned certain risks, the presumption should be that they have been assigned to the superior (that is, most efficient) risk bearer and, moreover, that he has been adequately compensated for bearing them, thus removing any distributional objections to leaving him to bear the costs. Where the contract is not explicit about the allocation of a given risk, it seems a reasonable presumptive rule of interpretation to assume that the parties intended for it to be borne by the superior risk bearer and that it has simply been impounded in the categoric terms of the contract. In either case, the party bearing such risks should not be permitted to subsequently reallocate them to the other party by contract modification, taking advantage of limitations in the relief available to the party on breach in order to induce the modification.

If the law allows contract modifications, it imposes at least part of the risks of adverse outcomes on the promisor. To the extent that the promisee can modify risks by his activities, and there is imperfect monitoring of his activities by the promisor, a moral hazard problem will arise. The promisee will have incentives to increase his risk exposure (relative to the case where he fully bears the risks) over time by devoting fewer resources to risk prevention than is optimal and increasing the probability of occurrence of adverse states (for example, bankruptcy) in which contract modifications may be necessary. Thus, the enforcement of contract modifications may generate inappropriate incentives over time that affect the probabilities of alternative outcomes or states of nature. Inappropriate incentives are created not only for the immediate contracting parties but, even more importantly, for future contracting parties in similar circumstances. These dynamic moral hazard inefficiencies, as well as the transaction costs associated with monitoring, bonding, and enforcement activity through a variety of contractual or institutional arrangements designed to overcome these inefficiencies, have to be weighed against the static efficiencies resulting from contract modifications discussed earlier. . . .

In developing our analysis of the category two modification cases, we are, perhaps ironically, adapting in large part the framework of analysis developed by Posner and Rosenfield [supra at pages 141–45] in the context of the doctrine of frustration/impracticability. In effect, we view enforceable modifications as a substitute for the doctrine of frustration: contracting parties, facing the occurrence of some intervening event that substantially affects the cost of performance may, under some circumstances, rearrange their contractual rights and obligations either through invocation of the assistance of the courts pursuant to the doctrine of frustration or through private recontracting. In other cases, while the underlying factual circumstances of the contract may not have changed, new information about those circumstances may have been uncovered. The parties may have contracted on the basis of incomplete or inaccurate information about the underlying factual environment of the contract. This situation is the domain of the doctrine of mistake. The economic considerations bearing on permissible rearrangements by virtue of contract modifications, the doctrine of frustration, or the doctrine of mistake would seem to be similar. . . .

However, in applying this approach both to frustration and modification cases, several problems must be acknowledged. First, the contract may not clearly assign given risks and an objective inquiry into who is the superior risk bearer may sometimes be indeterminate (for example, on party may have superior ability to appraise the probability of a particular event occurring, but the other party may have superior ability to appraise the magnitude of the costs entailed if it does). Such a case might arise, for example, where there is a contract to produce a machine to the buyer's specifications and a strike prevents timely completion. The supplier can probably best judge the likelihood of a strike in his industry and may to some extent be able to control or provide for this risk, while the buyer may best be able to judge the costs to him of delayed delivery. Here, even though the risk may not be an uncommon one, in the absence of a fully specified contract it may be difficult to know where the presumption leads that risks should be treated as initially assigned to the superior risk bearer and that subsequent reassignments through, for example, frustration or modification should be foreclosed. However, one desirable effect of a no-modification rule in the particular example given may be to lead the machine manufacturer to negotiate a limited liquidated damages clause governing breach, thus leading to an efficient initial allocation of risks.

Another problem may arise in cases where it may be feasible to identify clearly the superior risk bearer but the risk in question is very remote. This is referred to in American law as the "unforeseen circumstances" exception to the traditional consideration requirement for valid modifications. The law in many contexts rather facilely tends to classify risks as foreseeable or unforeseeable, while economists would tend simply to view all risks as carrying some degree of probability with them. The fact that a risk is one of low probability does not necessarily mean that it has not been foreseen and appropriately discounted in the initial contract terms, or that there are not efficient risk reduction or risk insurance strategies available to the parties, or that in this latter respect one party is not a superior risk bearer to the other. Indeed, most insurance contracts involve low probability risks. However, it remains true that if the risk is remote enough, that is, carries a very low probability, the effect of assigning it to one party or the other may have little effect on contractual behaviour because the *expected* (that is, ex ante) costs to whoever bears the risk are so small as to warrant very little, if anything, in the way of efficient precautionary responses.

Taking these two problematic cases together—(a) indeterminacy in the identification of the superior risk bearer, and (b) highly remote risks—it can be argued that these are the transactional domains where contractual modifications should be permitted. In both cases, permitting modifications enables the party to whom a modification is proposed to capture some of the static gains from recontracting by avoiding losses he may well sustain in the event of a breach. Losses in dynamic efficiency associated with a rule that permits subsequent reassignment of initially efficiently assigned risks are likely to be small, in the first case because we cannot determine with confidence what is an efficient allocation of risks, and in the second case because while this can be determined, disturbing the efficient allocation of risks will have little impact on the long-run behaviour of contracting parties at large while enabling the static gains for the immediate parties from recontracting, which may be significant, to be realized.

Permitting modifications in these two classes of cases, in effect, permits a flexible form of risk sharing by the contracting partners. Assuming the parties to be risk averse and the risk in question one that neither can readily control or insure against, contract modifications may reduce the variance in possible contractual outcomes for the parties and thus be Pareto efficient. A countervailing consideration is, of course, that the promisor now is exposed to the risk of strategic behavior by the promisee designed to exploit or manipulate these two exceptions and to the adjudication costs attendant on courts attempting to apply rules that do not supply knife-edged sharpness in the characterization of situations as falling within or outside given rules or exceptions thereto.

Applications of the Economic Approach

. . . One situation that has arisen quite frequently in the cases involves construction contracts. One type of case we have already dealt with—where the builder "holds up" the site owner part way through performance, demanding extra payments by trading on the fact that the site owner will not be able to find ready substitutes and may possess incomplete remedies for breach. This is a category one case (pure strategic modification) and . . . should not be enforced.

Another type of construction case involves situations where the builder, in the course of contractual performance, encounters, for example, difficult and costly excavation problems because of unusual soil conditions. In the case of a large builder erecting many commercial buildings, it seems obvious that he will typically be the superior risk bearer relative to the site owner. He can appraise the risks ex ante more efficiently and can often diversify them away across a number of similar projects. Some excavation projects will be more difficult, some less difficult, than the norm but these are familiar risk to him and he can be expected to adjust his contractual terms accordingly. Modifications that reassign the risks to the site owner attenuate the incentive for builders in this situation to act efficiently and should not be enforceable. To enforce modifications in this context is tantamount to turning a fixed-price contract into a cost-plus contact, an arrangement the parties could have negotiated but manifestly did not, presumably for good reason given the desired allocation of risks.

On the other hand, in the case where a small contractor agrees, for a modest sum, to dig a cellar under the other party's house and encounters a hard crust of earth three feet deep under which is a quagmire of wet mud, it is arguable that both qualifications to the superior risk bearer approach may apply. Determining who is the superior risk bearer may be difficult; first, these conditions may fall entirely outside the realm of experience of such contractors (making it difficult to know whether he or the homeowner could best appraise the risks or take risk reduction precautions), and second, the builder's ability to diversify across similar projects may be highly circumscribed by virtue of the size or nature of the operations. Even if he is clearly the superior risk bearer, to deny the possibility of a modification here imposes costs not only on him but on the homeowner who must search out a substitute and attempt to recover costs for breach in a damage claim which, even if successful, may yield less than full compensation (for example, in the event of the contractor's bankruptcy). By allowing the mutual gains from modification relative to breach to be realized, the question arises whether this significantly attenuates the incentives of contractors in this type of situation to take efficient risk reduction or risk insurance precautions. If the risk in question was highly remote and the *expected* costs associated with bearing the risk so small that no significant changes in behaviour are likely to be induced by shifting it, then permitting a modification seems to enhance the welfare of contracting parties and is efficient. . . .

Various other types of long-term supply contracts for goods and services involving modifications induced by supervening events raise similar issues. Some leading examples of long-term contracts for the supply of goods which raise these issues are briefly reviewed.

In the American case of *Goebel v. Linn,* the defendant brewers contracted for the supply of ice at a fixed rate. The ice was a crucial component of the defendant's business, being used to preserve the large beer stocks kept on hand. Due to a mild winter, the plaintiff could not accumulate all required stocks of ice and informed the defendant that it would not perform its supply contact unless the brewery agreed to pay a higher rate. The defendants agreed and took delivery of its ice but later repudiated the contract as modified. The Court held the contract modification enforceable, stating that the brewing company "chose for reasons which they must have deemed sufficient at the time to submit to the company's demand and pay the increased price rather than rely upon their strict rights under the existing contract." The Court went on to observe:

> Suppose, for example, the defendants had satisfied themselves that the ice company under the very extraordinary circumstances of the entire failure of the local crop of ice must be ruined if their existing contracts were to be insisted upon, and must be utterly unable to respond in damages, it is plain that then, whether they chose to rely upon their contract or not, it could have been of little or no value to them. Unexpected and extraordinary circumstances had rendered the contract worthless, and they must either make a new arrangement, or, in insisting on holding the ice company to the existing contract, they would ruin the ice company and thereby at the same time ruin themselves.

Posner endorses the result in this case on the basis of his distinction between modifications entered into where there has been a change in underlying circumstances and those where there has not. However, our emphasis on efficient risk allocation suggests a more cautious approach to evaluating the correctness of this decision. If mild winters were one of the occupational hazards of running an ice business during the era in question, the ice company would seem clearly to be the superior risk bearer, both in terms of risk reduction (for example, making different inventory or stand-by sub-contractual arrangements) and in terms of risk insurance (that is, being better able to appraise the impact of climatic variations on the supply of ice and adjusting the initial contractual fixed price accordingly). Only where the winter in question was quite out of the ordinary so that the gains from recontracting were likely to outweigh the long-run costs of moral hazard problems associated with permitting recontracting in these circumstances, could one support the decision. Otherwise, permitting recontracting is inefficient.

Notes on Modification by the Parties

1. The Posner article referred to at the outset of this excerpt is Richard A. Posner, Gratuitous Promises in Economics and Law, 6 Journal of Legal Studies 411 (1977). For other analyses of contract modifications from an economic perspective, see Timothy J. Muris, Opportunistic Behavior and the Law of Contracts, 65 Minnesota Law Review 521 (1981); Daniel A. Graham and Ellen R. Peirce, Contract Modification: An Economic Analysis of the Hold-

Up Game, 52 Law and Contemporary Problems 9 (1989); Alan Schwartz, Relational Contracts in the Courts: An Analysis of Incomplete Agreements, 21 Journal of Legal Studies 271, 308–13 (1992); and Jason Scott Johnston, Default Rules/Mandatory Principles: A Game Theoretic Analysis of Good Faith and the Contract Modification Problem, 3 Southern California Interdisciplinary Law Journal 335 (1993). Recent analyses taking a less exclusively economic perspective include Robert A. Hillman, Policing Contract Modification Rules Under the UCC, 64 Iowa Law Review 849 (1979); and Subha Narasimhan, Modification: The Self-Help Specific Performance Remedy, 97 Yale Law Journal 61 (1987).

2. As this excerpt indicates, the modification of a contract following discovery of an unexpected problem raises issues that are in some ways similar to those raised by the doctrines of impracticability, frustration, and mistake. One difference is that in the typical case of impracticability, frustration, or mistake, the parties have not been able to agree to some ex post revision of their contract, so one party is asking the court to imply such a revision. Because modification cases involve an ex post agreement by the parties, these cases can also raise issues of duress and unconscionability, doctrines that will be examined at more length in the readings in Chapter 5. Modification cases may also raise issues relevant to the consideration doctrine, which is discussed in the following chapter.

3. As the excerpt by Aivazian, Trebilcock, and Penny points out (supra at page 213), the potential for a "hold up," or a forced reallocation of risks, is greatest in situations where the victim's normal remedy of damages for breach would not be fully compensatory. For example, if the victim's losses from nonperformance cannot be easily measured, the victim may lose less by agreeing to an unfavorable modification than he or she would lose by refusing to agree the modification, in which case the other party might refuse to perform, and the victim's only remedy would be to sue for less-than-compensatory damages. (Other possible limits on the remedies for nonperformance were discussed in Chapter 2, supra at pages 56–92.)

On the other hand, in cases where the victim's remedy for nonperformance would be fully compensatory, it will be much harder for the other party to force the victim to agree to an unfavorable modification. In these cases, should courts presume that any modification the victim agrees to is actually an efficient response to a change in circumstances? For an argument for this proposition, see Alan Schwartz, Relational Contracts in the Courts: An Analysis of Incomplete Agreements, 21 Journal of Legal Studies 271, 308–13 (1992).

4. Under § 2–209 of the U.C.C., an agreement to modify a contract is enforceable without consideration. However, the Comments to § 2–609 limit enforceability to those modifications arrived at in *good faith*. A standard of good faith may also be used to judge a company's performance under a long-term marketing contract (see the excerpt by Goetz and Scott, supra at pages 181–87), or to judge a franchisor's decision to terminate a franchisee (see the

excerpts by Hadfield and Klein, supra at pages 201–09), or to judge an employer's decision to fire a worker. Is this the same legal standard, applied to four different situations? Or is it four different legal standards, which (confusingly) happen to share the same name?

Consider the following argument for the proposition that "good faith" does not represent a single legal standard:

> In contract law, taken as a whole, good faith is an "excluder." It is a phrase without general meaning (or meanings) of its own and serves to exclude a wide range of heterogeneous forms of bad faith. In a particular context the phrase takes on specific meaning, but usually this is only by way of contrast with the specific form of bad faith actually or hypothetically being ruled out.

Robert S. Summers, "Good Faith" in General Contract Law and the Sales Provisions of the Uniform Commercial Code, 54 Virginia Law Review 195, 201 (1968). See also Robert S. Summers, The General Duty of Good Faith—Its Recognition and Conceptualization, 67 Cornell Law Review 810 (1982).

On the other hand, Steven Burton has argued that good faith has a core of common meaning in that it is violated whenever a party to a contract attempts to recapture an opportunity that that party gave up (perhaps only implicitly) by signing the contract. Steven J. Burton, Breach of Contract and the Common Law Duty to Perform in Good Faith, 94 Harvard Law Review 369 (1980); see also Steven J. Burton, More on Good Faith Performance of a Contract: A Reply to Professor Summers, 69 Iowa Law Review 497 (1984). Does this formulation succeed in unifying these various uses of "good faith"?

In some respects, the Burton formulation is similar to the method of selecting default rules advocated in an earlier excerpt by Posner and Rosenfield (supra at pages 28–29). Posner and Rosenfield argued that default rules should reflect the rules the parties would have agreed to if they had discussed the matter; they further argued that most parties would normally agree to whatever rules were most efficient, in the sense of maximizing the total value created by the proposed contract. If the obligation to perform in good faith is viewed as this sort of default rule—and if one realizes that determining what rule will be most efficient, or what opportunities the parties gave up on entering the contract, will require different inquiries in different situations— does it matter whether the good faith standard represents one rule or several different rules?

4

Contract Formation

Anglo-American law has never enforced all promises. The readings in the first section of this chapter discuss some of the doctrines that determine which promises will be enforced. The first two sets of readings—by Melvin Eisenberg, Jay Feinman, and Edward Yorio and Steve Thel—discuss consideration and reliance as alternative bases for enforcement, thus continuing the debate begun in Chapter 1 (in the readings by Patrick Atiyah and Charles Fried). The Eisenberg essay also discusses the use of seals and other formalities to make a promise enforceable. The role and significance of formalities is discussed at more length in the final two readings in the first section by Duncan Kennedy and Patricia Williams.

The readings in the second section of this chapter analyze the various issues involved in the formation of contracts, with particular attention to the rules of offer and acceptance. The rules of contract formation have not been as extensively analyzed by contracts scholars, but many of the issues raised by these readings are similar to those discussed in the first section of this chapter. Questions such as "When would a promisor *want* his or her words to be interpreted as making a legally binding commitment?" and "When should the law protect a party who relies to his or her detriment on another's words?" are common to many of the readings in both sections.

Several of the readings in this chapter also continue the discussion of default rules begun in Chapters 1 through 3. Some of the rules of contract formation are clearly default rules, in the sense that they can always be changed by language to the contrary. For example, the traditional "mailbox rule"—acceptance of an offer binds the offeror when the acceptance is dispatched—can be altered by a contrary specification in the offer. The status of other formation doctrines is more controversial—for example, should parties be allowed to "opt out" of the rule that promises unsupported by consideration will not be enforced? The Eisenberg article on consideration and the Kennedy article on formalities both bear on this question.

§ 4.1 Bases for Enforcement

4.1.1 Consideration

The Principles of Consideration

MELVIN ARON EISENBERG

A promise, as such, is not legally enforceable. The first great question of contract law, therefore, is what kinds of promises should be enforced. The answer to that question traditionally has been subsumed under the heading "consideration." Properly understood, that term merely stands for the set of general principles defining the conditions that make promisees enforceable, and it might profitably be replaced by the more descriptive term "enforceability." Over the last hundred years or so, however, a more confined approach developed, under which consideration doctrine was made to turn on a bargain pivot, and was articulated through a set of highly particularized rules. This approach was part of a school that was characterized by an attempt to derive contract law through logical deduction from received axioms. The purpose of this Article is to reconstruct applicable doctrine along modern lines. To this end, I shall develop an analysis in which bargain is only one of several alternative conditions of enforceability, and the law concerning the enforceability of promises is expressed in principles that are sufficiently open-textured to account for human reality, and to permit growth of doctrine as principles unfold and social facts change over time. . . .

Melvin Aron Eisenberg, The Principles of Consideration, 67 Cornell Law Review 640, 640, 643, 649–51, 653–54, 659–62 (1982). Copyright © 1982 by Melvin Aron Eisenberg. Reprinted by permission.

The Element of Bargain

The determination of whether any given type of promise is legally enforceable should turn on both substantive and administrative considerations. As a substantive matter, the state (speaking through the courts) may justifiably take the position that its compulsory processes will not be made available to redress the hurt caused by every broken promise, but only to remedy substantial injuries, prevent unjust enrichment, or further some independent social policy, such as promotion of the economy. As an administrative matter, the state may fairly take into account the extent to which enforcement of a certain type of promise would involve difficult problems of proof. Cutting across both substantive and administrative categories is the question of whether the type of promise at issue is normally made in a deliberative manner, so as to accurately reflect the promisor's wants and resources.

By these standards, bargain promises clearly should be enforceable. The injury to the promisee is typically substantial. Usually he will have relied upon the promise, and often he will have seen the promisor enriched as a result of that reliance. The state has an independent interest in the enforcement of such promises. Exchange creates surplus, because each party presumably values what he gets more highly than what he gives. A modern free-enterprise system depends heavily on private planning and on credit transactions that involve exchanges over time. The extent to which private actors will be ready to engage in exchange, and are able to make reliable plans, rests partly on the probability that bargain promises will be kept. Legal enforcement of such promises increases that probability.

Other criteria for enforceability point in the same direction. For example, if the bargain has been half-completed, that a valuable performance has been rendered to an unrelated party helps satisfy the administrative concern for evidentiary security. Even if the transaction is wholly executory, bargains are not easy to fabricate from whole cloth. And because bargain promises are typically rooted in self-interest rather than altruism, they are likely to be finely calculated and deliberatively made.

Ironically, however, the axiomatic school, having adopted the bargain theory of consideration, stopped short of giving the bargain element its full scope, and instead wrongly adopted rules that denied enforcement to several important classes of bargain promises. . . .

Illusory Promises

The fallacy of the illusory-promise doctrine is that it treats transactions involving illusory promises as if they were failed bilateral contracts, intended to involve a promise for a promise. In fact, however, such transactions are often successful unilateral contracts, intended to involve a promise for an act—the act of giving the promisor a chance. The party who makes the real promise in these cases does not do so for altruistic reasons. Rather, he seeks to advance his own interests by inducing the promisee to give him a chance to show that his performance is attractive, so as to convince the promisee to transact. Giving a real promise in an illusory-promise transaction achieves that objective in two closely related ways. First, such a promise conveys information— the information that the promisor is so confident his performance will be found attractive that he is willing to limit his freedom of action to get a chance to demonstrate that attractiveness. In this sense the promise resembles a money-back guarantee or an extended warranty, which transmits information concerning a seller's confidence in his product. Second, such a promise is designed to alter the promisee's incentives. Giving a chance is not cost-free, and presumably the promisor believes that without the promise the promisee's incentives to give the promisor a chance would be insufficient. In effect, there is a disparity of information and incentives between promisor and promisee. The promisor has a degree of confidence in the attractiveness of his performance which he believes the promisee does not share. To increase the likelihood of exchange, the promisor makes a promise that is intended to change the promisee's incentives sufficiently to induce him to give the promisor a chance. If the promisee gives the chance, the inducing promise should be enforceable under standard unilateral contract analysis.

This proposition can be illustrated by the following hypothetical, involving an illusory promise comparable to that in the second Illustration to section 77 [of the *Restatement (Second) of Contracts*]:

> *Confident Student.* A, a third-year law student whose grades are only fair but whose confidence is great, interviews the well-known Washington litigation firm, F, G & H. After the interview, which A feels went very well, he writes to G as follows:
>
> Dear G:
>
> I very much enjoyed meeting you at my recent interview. I know my grades are below the level F, G & H usually requires. However, I also know they are not a fair indicator of my skills, particularly in litigation. (As you may recall, I did exceptionally well in several moot court settings.) I am sure that if you gave me a chance you would be more than pleased with my work. In order to induce you to give me a chance I make you the following offer: I will work for you for one year at $12,000 (one-third of F, G & H's normal starting salary), beginning September 1. If you concur in my proposal, you can nevertheless change your mind at any time before that date, and in addition you may discharge me at any time thereafter, without notice, and with no questions asked.
>
> Sincerely yours,
A
>
> G concurs, and A begins work. After three months, however, A leaves for another job, over G's objection. F, G & H bring suit for breach of contract.

Under the illusory-promise doctrine, as exemplified in Illustrations 1 and 2 to section 77, A would not be bound to his promise. Clearly, however, A has made a bargain. If, in *Confident Student*, F, G & H had paid $500 for A'spromise, the promise would clearly be enforceable. The act of giving A a chance to prove himself may be worth much more to A (and may cost F, G &

H considerably more) than that amount. A received exactly what he bargained for and should be bound by his promise.

Of course, it is possible in a bargain involving an illusory promise that the party who made the real promise was not bargaining for a chance, and was unconscionably fast-talked into believing that a real promise was made to him. Indeed, as with the legal-duty rule, the illusory-promise doctrine may have been a crude technique for covertly introducing issues of fairness into contract law at a time when courts believed it improper to deal openly with such issues. Many if not most of the illusory-promise cases, however, involve transactions between merchants who are unlikely to have misperceived what each was giving and receiving.

Even today, the illusory-promise doctrine has only a precarious toe-hold in the law. It is often avoided by some modern equivalent of the peppercorn, and its application is frequently rejected on flimsy if not specious grounds. Now that the doctrine of unconscionability as been explicitly recognized, the illusory-promise doctrine should be abandoned. Any potential for unconscionability in bargains involving illusory promises should be treated directly, by scrutinizing the transaction to ensure that it did not involve unconscionable fast-talking. . . .

Firm Offers

A firm offer is a legal molecule consisting of two atoms: an offer, and a promise to hold the offer open for a fixed or reasonable time. The axiomatic school, applying the bargain theory of consideration, adopted the rule that the offeror was not obliged to keep the offer open, because the promise to do so was revocable unless paid for. But just as enforcing bargain promises is socially desirable because surplus is created through exchange, so too the law should enforce promises that facilitate or augment the likelihood of exchange. Firm offers obviously fall within that category. As stated in the *Restatement Second* itself, "[t]he fact that the option is an appropriate preliminary step in the conclusion of a socially useful transaction provides a sufficient substantive basis for enforcement . . . ."

Furthermore, the logic of the rule that firm offers are not enforceable is flawed in exactly the same way as the illusory-promise doctrine. A firm offer is made, not for altruistic motives, but to advance the offeror's interests by inducing the offeree to deliberate. In deciding whether to accept an offer, an offeree must often make an investment of time, trouble, and even money. The offeree is more likely to make such an investment if he is sure the offer will be held open while the investment is being made than he is if the offer may be revoked during that period. Like the real promise in a bargain involving an illusory promise, a promise to hold an offer open normally has two closely related purposes, both of which are advanced by making the commitment legally enforceable. The first purpose of such a promise is to convey information—the information that the offeror is so confident his offer is attractive that he is willing to limit his freedom of action to demonstrate that attractiveness. The second purpose is to alter the offeree's incentives. Presumably, the offeror believes that without the promise, the offeree's incentives to deliberate on the offer would be insufficient. In effect, there is a disparity of information and incentives between offeror and offeree. The offeror has a degree of confidence in the attractiveness of his offer which he believes the offeror does not share. To increase the possibility of creating a surplus through exchange, the offeror makes a commitment that is intended to provide the offeree with additional information and change the offeree's incentives to deliberate. Given the offeror's intent to induce such an investment, the likelihood that the investment will be made, the difficulty of proving the investment by direct means, and the probability that more exchanges will take place if firm offers are enforceable than if they are not, the law should respond by assuming that the offeror has received the investment he wanted to induce. . . .

The Element of Form

Given that unrelied-upon donative promises are normally unenforceable, the question arises whether the law should recognize some special form through which a promisor with the specific intent to be legally bound could achieve that objective. "It is something," said Williston, "that a person ought to be able . . . if he wishes to do it . . . to create a legal obligation to make a gift? Why not? . . . I don't see why a man should not be able to make himself liable if he wishes to do so."

At early common law the seal served this purpose. In modern times, most state legislatures have either abolished the distinction between sealed and unsealed promises, abolished the use of a seal in contracts, or otherwise limited the seal's effect. The axiomatic school, however, never rejected the rule that a seal makes a promise enforceable, and that rule is now embodied in section 95(1)(a) of the *Restatement Second,* which provides that "[i]n the absence of statute a promise is binding without consideration if . . . it is in writing and sealed . . . ."

The *Restatement Second* makes no attempt to justify this rule. That is not surprising, because justification would be hard to find. Originally, the seal was a natural formality—that is, a promissory form popularly understood to carry legal significance—which ensured both deliberation and proof by involving a writing, a ritual of hot wax, and a physical object that personified its owner. Later, however, the elements of ritual and personification eroded away, so that in most states by statute or decision a seal may now take the form of a printed device, word, or scrawl, the printed initials "L.S.," or a printed recital of sealing. Few promisors today have even the vaguest idea of the significance of such words, letters, or signs, if they notice them at all. The *Restatement Second* itself admits that "the seal has come to seem archaic." Considering this drastic change in circumstances, the rule that a seal renders a promise enforceable has ceased to be tenable under modern conditions. The rule has been changed by statute in about two-thirds of the states, and at least one modern case held even without the benefit of statute that the rule should no longer be strictly applied. Other courts can and should follow suit.

Should the law then recognize some new formality to play the role once played by the seal? An obvious candidate is nominal consideration—that is, the form of a bargain—because it can be safely assumed that parties who falsely cast a nonbargain promise as a bargain do so for the express purpose of making the promise legally enforceable. A rule that promises in this form were enforceable would have obvious substantive advantages, but would also involve serious difficulties of administration. As a practical matter, such a form would be primarily employed to render donative promises enforceable. Both morally and legally, however, an obligation created by a donative promise should normally be excused either by acts of the promisee amounting to ingratitude, or by personal circumstances of the promisor that render it improvident to keep the promise. If Uncle promises to give Nephew $20,000 in two years, and Nephew later wrecks Uncle's living room in an angry rage, Uncle should not remain obliged. The same result should ordinarily follow if Uncle suffers a serious financial setback and is barely able to take care of the needs of his immediate family, or if Uncle's wealth remains constant but his personal obligations significantly increase in an unexpected manner, as through illness or the birth of children.

Form alone cannot meet these problems. Thus the French and German Civil Codes, while providing special forms that enable a donative promise to be rendered legally enforceable, also provide extensive treatment of improvidence and ingratitude as defenses. For example, under article 519(1) of the German Civil Code, a promisor may refuse to keep a donative promise "insofar as, having regard to his other obligations, he is not in a position to fulfill the promise without endangering his own reasonable maintenance or the fulfillment of obligations imposed upon him by law to furnish maintenance to others." Under article 530 (1), a donative promise may be revoked "if the donee, by any serious misconduct towards the donor or a close relative of the donor shows himself guilty of gross ingratitude." Similarly, under articles 960–966 of the French Civil Code, a donative promise made by a person with no living descendants is normally revoked by operation of law upon the birth of a child. Under articles 953 and 955, a donative promise can be revoked on the ground of ingratitude that involves serious cruelty, wrongs, or injuries.

As these rules suggest, the common law could not appropriately make donative promises enforceable solely on the basis of a form unless our courts were also prepared to develop and administer a body of rules dealing with the problems of improvidence and ingratitude. Certainly such an enterprise is possible. It may be questioned, however, whether the game would be worth the candle. An inquiry into improvidence involves the measurement of wealth, lifestyle, dependents' needs, and even personal utilities. An inquiry into ingratitude involves the measurement of a maelstrom, because many or most donative promises arise in an intimate context in which emotions, motives, and cues are invariably complex and highly interrelated. Perhaps the civil-law style of adjudication is suited to wrestling with these kinds of inquiries, but they have held little appeal for common-law courts, which traditionally have been oriented toward inquiry into acts rather than into personal characteristics. The question is whether the social and economic benefits of a facility for making donative promises enforceable would be worth its social and economic costs. The answer is that benefits and costs are in rough balance, so that nonrecognition of such a facility is at least as supportable as recognition would be.

Notes on Consideration

1. Professor Eisenberg states that "an obligation created by a donative promise should normally be excused either by acts of the promisee amounting to ingratitude, or by personal circumstances of the promisor that render it improvident to keep the promise (supra at page 229). Does this mean (a) that no promisor should be allowed to make a stronger commitment than this— that is, no promisor should be allowed to make a promise that would not be subject to these excuses—even if the promisor truly wants to do so? Or (b) that no promisor would ever want to make so strong a commitment? Or (c) that the number of promisors who might want to make the stronger commitment is so small that it wouldn't be worthwhile for the law to provide a mechanism (such as a seal) by which those few would be able to do so— especially given the risk that the same mechanism might also be used by promisors who didn't understand that it was locking them into a promise from which there would truly be no escape? Under interpretation a, the law's refusal to enforce some donative promises would clearly represent a restriction on the promisor's freedom of contract. Under interpretations b or c, the conflict with freedom of contract is less direct.

2. The Eisenberg excerpt also states that "exchange create surplus, because each party presumably values what he gets more highly than what he gives" (supra at page 225). The converse of this view is that donative promises, or promises that do not involve exchanges, represent "sterile transmissions," which do not "increase wealth" in the way that exchange promises do. This sentiment is often attributed to Claude Bufnoir, Propriété et Contrat (Paris: Rousseau et Cie., 2d ed. 1924), p. 487.

Most economists today would deny that gifts are "sterile transmissions," or that they are any more or less sterile than exchanges. Compare the following two cases:

a. A promises to give B A's car because B has promised in return to give A B's boat, which A has always admired and from which A expects to derive great pleasure.
b. A promises to give B A's car without receiving anything in return because B is a niece whom A has always admired and making a gift to this niece is something from which A expects to derive great pleasure.

In each case, B ends up with a car, and A ends up (or expects to end up) with great pleasure. In each of these cases, then, each party "values what he gets more highly than what he gives" even though only one of these cases involves an exchange. To economists, any promise that increases (or is expected to increase) the pleasure of both parties is not "sterile."

3. In discussing two categories of commercial promises—illusory promises and firm offers—Eisenberg argues that the promisors in such cases will often have a good reason for wanting to bind themselves in order to induce the other party to rely on the promise. The maker of a donative promise could have a similar reason to want to bind himself or herself—that is, to induce the other party to rely on the promise. For example, if A promises to pay for B's trip to Europe, but B knows that such a promise is not legally binding, B may be reluctant to embark on the trip. For discussions of this motive for making a donative promise, see Charles J. Goetz and Robert E. Scott, Enforcing Promises: An Examination of the Basis of Contract, 89 Yale Law Journal 1261, 1273 (1981); Richard A. Posner, Gratuitous Promises in Law and Economics, 6 Journal of Legal Studies 411 (1977); Steven Shavell, An Economic Analysis of Altruism and Deferred Gifts, 20 Journal of Legal Studies 401 (1991); Andrew Kull, Reconsidering Gratuitous Promises, 21 Journal of Legal Studies 39 (1992). (This motive is also relevant to the readings in the following subsection, discussing reliance as a basis for enforcing promises.)

4. On the other hand, a binding promise is not cost free to the promisor. In either a donative or a commercial setting, subsequent events may lead the promisor to regret having made the promise. For example, the confident student A might later decide that he doesn't want to work for F, G & H for an entire year; or an offeror who promises to keep her offer open for two weeks might get a better offer from someone else during that two week period and might, therefore, want to withdraw the offer she made to the original offeree.

Viewed from this perspective, someone considering whether to make a binding promise will have to balance three factors: (a) the risk that events will lead the promisor to change his or her mind and regret being bound; (b) the benefit that the promisor expects to get from the promise—for example, inducing the promisee to change his or her behavior in some way, or (in a purely altruistic promise) simply making the promisee happy; and (c) the extent to which making a binding promise increases the promisor's likelihood of receiving this benefit. For example, if there was a good chance that the firm of F, G & H would hire A if A merely told the firm that he would probably work for them for a full year, without making a binding commitment to do so, A would have less reason to make a binding commitment. (In Eisenberg's hypothetical, this might not have been the case, as the firm might have been unlikely to give A a chance unless they could be sure that A would continue to work for them for at least a year.) Similarly, some nephews might be willing to embark on a trip to Europe even if they were merely told that their uncle would *probably* compensate them for their expenses—that is, even if the uncle had not made a binding commitment to that effect. If so, the uncle might have less reason to make a binding commitment.

The strength of each of these three factors is likely to vary from case to case. That is, in some cases there may be little risk that the promisor will come to regret the promise, while in other cases that risk might be quite high. Similarly, in some cases the promisee may not lose very much if the promise is later withdrawn, and thus might be willing to engage in the desired conduct even without an absolutely binding commitment. In other cases, the promisee's conduct may leave him or her at great risk if the promise is withdrawn: in these cases, the promisee is less likely to act without a binding commitment. As Eisenberg notes (supra at page 228), someone who makes what is clearly a binding promise can normally be presumed to have balanced these factors and decided that "without the promise[,] the promisee's incentives . . . would be insufficient." But what if it isn't clear whether the so-called promise was really meant to be binding? Can a court make its own estimate of these factors to decide whether it is likely that the alleged promise was meant to be binding, or to decide whether most people in this position would have wanted to make a binding promise? Would such a balance tell us—as Eisenberg seems to conclude—that most donative promisors would not want their promises to be interpreted as binding?

The difficulty of interpreting promises that may or may not have been meant as binding is discussed in the excerpt by Jay Feinman (infra at pages 232–38). Similar issues arise in interpreting ambiguous offers or acceptances (as discussed infra at pages 252–66).

5. Additional information on the French and German law of donative promises, referred to in the Eisenberg excerpt, can be found in John Dawson, Gifts and Promises: Continental and American Law Compared (New Haven: Yale University Press, 1980).

6. Consideration issues can also arise in cases of extorted modification, as discussed in the earlier excerpt by Aivazian, Trebilcock, and Penny (supra at pages 211–19. When the consideration doctrine is used to deny enforcement of contracts whose content or whose process of formation seems unfair in some way, it also has much in common with the doctrine of unconscionability discussed in Chapter 5.

§ 4.1.2 **Reliance**

Promissory Estoppel and Judicial Method

JAY M. FEINMAN

. . . Consistent with the abstract approach characteristic of traditional law, the typical doctrinal formulation of promissory estoppel holds out as its paradig-

Jay M. Feinman, Promissory Estoppel and Judicial Method, 97 Harvard Law Review 678, 689–94, 712–16 (1984). Copyright © 1984 by the Harvard Law Review Association. Reprinted by permission.

matic case a clear promise manifesting a commitment of future action, to which the promisee responds, as the promisor should have foreseen, by undertaking a specific act of substantial reliance sufficient to ensure that nonenforcement of the promise would be a manifest injustice. The key concepts in this vision are difficult to apply, and among the variety of fact situations presented by the cases are few that even approach the paradigm. Often a promise is not explicit, definite, or complete, or the reliance is not substantial, neatly linked to the promise, or clear. The problem for courts is to explicate the doctrine or to find a method for determining in which of the many possible cases promissory estoppel should actually apply. . . .

The Promise

Perhaps the principal issue in applying promissory estoppel is whether the promisor has made a promise that justifies the promisee's reliance. The manner in which courts address this issue exemplifies their general ambivalence toward the doctrine. Section 2 of the *Restatement (Second) of Contracts* defines a promise as "a manifestation of intention to act or refrain from acting in a specified way, so made as to justify a promisee in understanding that a commitment has been made." When the promisor makes a specific, definite statement regarding her future conduct, the courts need only adhere to a narrow definition of promissory estoppel to find the promise actionable. In such cases, promissory estoppel performs its traditional role as a consideration substitute that validates a gratuitous promise. (The promise is "gratuitous," of course, only in the sense that it lacks consideration; in most recent cases, clear promises are not donative, but are instead the products of exchange situations.)

In circumstances that depart from the promissory norm, courts are variously strict and flexible in determining whether a manifestation of intention may furnish a basis for actionable reliance. The strict view holds that a statement that is not specifically demonstrative of an intention respecting future conduct or that is indefinite or limited cannot be the basis for promissory estoppel. Under a more flexible view, courts have held such statements to suffice.

The strict view of promise carefully distinguishes promises, which are future oriented, from statements of belief, which concern only the present. For example, in cases involving franchisors who have made representations about the business potential that franchisees may expect or about their own policies of support for franchisees, some courts have held that such representations may not reasonably be relied upon as indications of future conduct. Courts usually view these representations, unlike promises to grant franchises, as statements of "belief" (that is, predictions or statements of current company policy) rather than intention. The statements are, therefore, not promises.

The strict view also requires that the promise be definite and unequivocal.

The court may determine that the promisor's expression concerning its future conduct is insufficiently certain and specific to give rise to promissory estoppel. Similarly, if the expression is made in the course of preliminary negotiations when material terms of the agreement are lacking, the degree of certainty necessary in a promise is absent. Finally, when the intention manifested is conditional, the promissory ideal is not met unless the event constituting the condition occurs. This rule holds good even when the event is within the control of the promisor, as is a condition of approval by a home office or higher official or a condition of execution of a final written agreement. The conclusion courts have drawn in all of these instances is that the facts of the case are not sufficiently compelling to invoke the limited reliance doctrine.

The alternative, more flexible approach to promise allows reliance recovery in a wider variety of settings. Courts adopting this perspective hold that a promise need not be explicitly expressed but may be inferred from, for example, statements about future conduct or factual representations about a present state of affairs. The standard, consistent with the definition in section 90, is not whether the promisor clearly made a promise, but whether, given the context in which the statement at issue was made, the promisor should reasonably have expected that the promisee would infer a promise. This standard may be met not only by a particular promise or representation, but also by general statements of policy or practice, such as published plans for employee compensation or benefits. In appropriate cases of this type, the court may infer a promise even in the face of inconsistent expressions.

A recurrent example of the flexible approach to promise is found in the courts' treatment of construction bidding cases, which have repeatedly generated important promissory estoppel decisions. In the typical case, a general contractor preparing to bid on a construction project receives bids on parts of the job from subcontractors and suppliers. The general then prepares its own bid on the basis of the lowest reliable subcontract bids. Subcontractors occasionally miscalculate, in part because they often compute their bids and telephone them to the general only hours before the general's bid is due. A subcontractor may also intentionally submit a low bid in the hope of receiving the contract and renegotiating the price. Conflict typically arises when, after the general has calculated and submitted its own bid and won the contract, a subcontractor notifies the general that the subcontractor has made an error or an intentionally low bid and refuses to perform.

Under traditional contract analysis, the subcontractor could withdraw with impunity, because its bid was regarded as an offer, revocable until accepted, to enter into a bilateral contract. In the leading case of *Drennan v. Star Paving Co.*, however, Justice Traynor held that the business context of the bid required that promissory estoppel apply to make the subcontractor's offer irrevocable until the general contractor had an opportunity to accept after being awarded the prime contract. The general's acceptance of the subcontractor's bid then created a traditional bilateral contract, for breach of which the subcontractor was required to pay as damages the difference between its bid and the higher price the general had to pay another subcontractor to perform the work. Cases since *Drennan* have held that promissory estoppel normally binds a subcontractor to the terms of its bid. Although the subcontractor does not make an explicit promise to keep its bid open, the court infers such a promise.

Some courts have taken a similar approach to promises that, under a strict view, might be considered preliminary or conditional. In *Hoffman v. Red Owl Stores, Inc.,* one of the seminal cases exemplifying this approach, the court held that manifestations of intention during negotiations constituted enforceable promises. Emphasizing the importance of good faith in negotiations and of protecting the reliance interest even in the face of limited commitment, *Hoffman* and subsequent cases have enforced what are actually conditional promises. . . .

Mediating Concepts: "Reason to Know"

Situation-sense leads a court to focus its search for an appropriate result on the facts of the individual case. The method falters because the process of identifying and generalizing from those facts ensnares the court in value choices not determined by the facts. Normative analysis, on the other hand, can only direct the court to statements of principle or policy; as the court attempts to apply the principles to a particular case, the subjectivity of the application cripples the integrity of the method. A third approach attempts to mediate between the abstraction of standard-based doctrine and the particularity of factual context in individual cases. Unlike situation-sense, a mediating concept overtly recognizes the inevitability of normative analysis in the decision of cases. But unlike the method of normative analysis, a mediating concept is neither limited, abstract, nor subjective. It incorporates a variety of norms and combines them in a simple statement that can be applied to cases much more easily than can general norms such as protection of the reliance interest.

The mediating concept I focus on here is foreseeability or, as formulated in a recent important article by Richard Speidel, "reason to know." The concept of foreseeability is ubiquitous in modern law, not only in its obvious applications in contract and tort, but also in areas as diverse as criminal law and civil procedure. In contract law it has manifested itself most notably in the limitation of recoverable damages under the rule of *Hadley v. Baxendale,* but its use extends well beyond that context. Speidel discusses "reason to know" as an element of the test set forth in section 204 of the *Restatement (Second) of Contracts* for the treatment of omitted terms in a contract, but in the process he points out its broader use in the second Restatement as the "connecting link between the alleged promise in a particular bargain and the surrounding context." The same use occurs in section 90's formulation of promissory estoppel: the availability of a section 90 remedy depends on whether, in the context in which the promise was made, the promisor had reason to know that the promise might induce reliance. As the "connecting link," "reason to know" guides courts in analyzing facts that form the basis of legal determinations. Although policy judgments may "color the edges" of the "reason to know" determination, the concept offers guidance "to connect a number of standards . . . to the relevant commercial context and to guide the court in determining which party's understanding should be preferred." "Reason to know" represents an effort to collapse such norms as the protection of reasonable reliance, the preservation of freedom of action, and the promotion of transactional security into a readily administrable formulation. It thus attempts to mediate the classical methodological opposition between abstraction and particularity, as well as the contradiction between freedom and coercion, by providing a more determinate approach to resolving the conflicting values of contract law in particular cases.

Speidel suggests that the "reason to know" test can be broken into five questions:

> 1. Did A understand in fact that B had made a promise? If not, there is no manifestation of assent.
>
> 2. If so, did B know (that is, have actual knowledge) at the time of his conduct of A's understanding? If so, a promise is made.
>
> 3. If B did not know, of what facts in the "total situation" did B have actual knowledge?
>
> 4. Given this knowledge, and taking B's level of intelligence into account, would B infer that if he acted or spoke in a certain way, A would understand that a commitment was made? If so, B has "reason to know" of that understanding.
>
> 5. If not, would B infer that there was a substantial chance that A would understand that a commitment was made? If so, B does have a duty to act with reasonable care to avoid misunderstanding; a failure to proceed with reasonable care is, apparently, tantamount to "reason to know."

Employment cases illustrate the way the "reason to know" test operates in promissory estoppel situations. Two kinds of manifestations of intention recur in the cases. The first is a relatively clear, simple, and direct promise by an employer. For example, in *Grouse v. Group Health Plan, Inc.,* Grouse, relying on a promise of employment as a pharmacist, gave notice to his then-current employer, refused another job offer, and was therefore injured when the promise was withdrawn. The second kind of manifestation is inferred from a variety of expressions, written and oral, which are often complex and even inconsistent; a case involving such a "promise" is *Perlin v. Board of Education,* in which the manifestation was contained in a school district's administrative compensation plan and other documents.

In the case of a simple promise, the first two of Speidel's questions regarding "reason to know" might seem to conclude the issue of promissory estoppel's application. In *Grouse,* the promisee understood that Group Health, the defendant, had promised him a job. If Group Health knew of Grouse's understanding, the promise was complete and Grouse's reliance rendered Group Health liable. If Group Health did not have actual knowledge of Grouse's belief, the judgment proceeds to the next of Speidel's questions, but the analysis required is still not too complex. Given what Group Health did know, it should have inferred either that Grouse would understand or that there was "a substantial chance" that Grouse would understand that he had been promised a job. The inference is justifiable not only because the particular facts of the case point to it, but also because, absent contrary factors, a direct promise in an employment setting—or, for that matter, in any other commercial setting—raises such an understanding.

In fact, few cases involve promises so simple that analysis can stop at the second of Speidel's questions. A typical defense in a promissory estoppel case is that facts known to the plaintiff at the time of the manifestation of intention negate any promissory effect. In *Grouse,* for example, Group Health could argue that it did not foresee reliance on the promise, because under the circumstances Grouse should have understood that a reference check by the employer was a precondition to employment. A case like *Perlin* requires still more complex analysis: the *Perlin* court inferred a promise from several documents, including the school board budget, the administrative compensation plan, and the school board's personnel rules. It is from such an array of facts that the answer to Speidel's fifth question—whether the promisor should have inferred that the promisee would probably regard the manifestation as a committed intention upon which it was safe to rely—must be drawn. Thus, the determination of obligation depends on complex factual scrutiny.

If "reason to know" succeeded in providing a method for applying vague standards of promissory reliance to particular fact situations in a disciplined manner, it would strike a balance between indeterminate particularism and arbitrary value judgment and between protection of freedom and undue coercion. But it fails to provide such a method. As applied in the case, "reason to know" conceals policy judgments in the guise of simple factual determinations and commonsense normative findings.

Commentators have long recognized that foreseeability in tort law is a vehicle for policy decisions, not a reflection of empirical reality. The same observation is not as frequently made about the applications of foreseeability in contract law, though of course it is as true of that area of the law as it is of tort law. In promissory estoppel cases, just as in negligence cases, "reason to know" analysis is based on policy, not empiricism. Speidel's fifth question, in fact, formulates the promisor's duty—a duty to exercise reasonable care to avoid misunderstanding—in terms identical to those commonly used to state a liability rule in tort.

The "reason to know" approach to promissory estoppel shifts to the promisor the risk of the promisee's reliance in cases in which the promisor has failed to act with reasonable care in manifesting her intentions. As a first step, then, "reason to know" makes a covert choice: it reflects the notion that, when the promisor had reason to know of the possibility of reliance, she ought to be liable. Furthermore, the determination of whether a given promisor had "reason to know" of the possibility of reliance is not based on the subjective expectation of the parties or on a survey-type analysis of beliefs in the relevant commercial community. Rather, the determination is made by the court as an application of the court's own views of the proper norms of commercial conduct.

"Reason to know" thus provides the court with no simple link between fact and norm. The determination of whether the promisor has violated a duty of care involves the weighing of principles of freedom of the promisor, security of the promisee, paternalism, the desirability of a precontractual cause of action in modern commercial society, and so forth. "Reason to know" makes an assumption and poses a question. In *Grouse,* for example, the assumption is that the court should hold Group Health liable for Grouse's reliance if the company had reason to know that there was a substantial chance of reliance, and the question is whether the court should hold that the company had reason to know. Because both the assumption and the question have normative elements, casting the policy judgment in quasi-empirical terms obscures rather than illuminates the underlying problem: Is it desirable social policy to hold the defendant liable?

The policy considerations underlying even this simple case of promissory reliance reflect the conflict between the classical image of contract and more modern views. The paradigm of the discrete promise is grounded on a classical notion of individualism that attempts to protect the promisor's freedom of action and the promisee's security simultaneously. But the reliance paradigm of promissory estoppel rejects the classical bargain concept in favor of a tort-like notion of loss-shifting. The underlying economic image is different as well: the reliance paradigm embraces a new vision of an economy characterized not by discrete transactions among autonomous actors, but rather by integrated exchanges that require the imposition of precontractual liability. In adopting and implementing "reason to know," a court makes choices that reflect these different visions.

The Promissory Basis of Section 90

EDWARD YORIO and STEVE THEL

. . . The framework set out in the *First Restatement* is best understood by bearing in mind that Section 90 has the effect of eliminating the requirement of bargained-for consideration. But Section 90 ensures that some of the functions of the bargain requirement are served in other ways. Among other

Edward Yorio & Steve Thel, The Promissory Basis of Section 90. Copyright © 1991 by The Yale Law Journal Company. Reprinted by permission of the Yale Law Journal Company and Fred B. Rothman & Company from The Yale Law Journal, volume 101 pages 116–22, 124–27, 161–66.

things, the presence of a bargain helps to identify promises on which the promisor has seriously reflected. If she made the promise to get something in return, her promise was sufficiently well considered to justify legal enforcement. By contrast, donative promises may be made impulsively. Section 90 identifies well-considered donative promises for enforcement by requiring that reliance be within the reasonable expectation of the promisor, thereby distinguishing between relatively casual promises and promises that, precisely because reliance by the promisee is expected, are likely to be more serious; a promise-based law of contracts would enforce only the latter.

For Williston and the *First Restatement,* it was not enough that the reliance be reasonably foreseeable; the expected action or forbearance also had to be "of a definite and substantial character." Subsequent critics find the language "definite and substantial" problematic, primarily because in their view the nature and degree of the promisee's reliance matters only in computing damages. Both requirements have been eliminated from the *Second Restatement.* Professor Braucher, the reporter for the *Second Restatement,* justified this change on the ground (among others) that "the requirement of reliance of a definite and substantial character becomes doubtful" with the addition of a second sentence in Section 90 allowing courts to limit the remedy awarded as justice requires.

. . . [T]he critics proceed from a different perspective than Williston's and, not surprisingly, reach a different conclusion. If Section 90 is designed to protect reliance, they argue, there is no reason to require that the reliance be definite and substantial. Any reliance should be protected; if indefinite or insubstantial, it will be compensated accordingly. But Williston's discussion of the language "definite and substantial" focused on the promisor, not on the promisee. For Williston, the requirements of definiteness and substantiality served not to mark the level of reliance that deserved protection, but to establish the level of foreseeable consequences that would justify holding the promisor to her promise:

> [Definite and substantial] actions should reasonably have been expected by the promisor . . . . [I]t will not be enough that some action of the promisee, even if [of] substantial character[,] has been induced by the promise.

> [T]his section covers a case where there is a promise to give and the promisor knows that the promisee will rely upon the proposed gift in certain definite ways.

> . . .

> . . . The uncle should certainly expect that the next thing that would happen would be that Johnny would run around the corner to the Buick agency and get a car and perhaps sign some promissory notes for it.

> . . .

> We have confined the Section to the case where a reasonable person would say that the promisor expected the man to do just what he did or that he ought to have expected it.

Because his objective was to enforce certain types of promises, not to protect reliance, Williston rejected the view that "*any* reliance on a gratuitous promise will render the promise enforceable . . ." For Williston, the problem was to identify those promises that ought to be enforced. His solution was to require that the promisor reasonably expect her promise to induce action or forbearance of a definite and substantial character.

Of the two limitations, the more revealing is the requirement of definiteness. If solicitude for the promise were the primary concern of Section 90, it would be hard to justify a requirement that the promisor foresee the particular act in reliance, as the word "definite" requires. Any reliance—or at least any nontrivial reliance—would be sufficient to justify requiring the promisor to compensate the promisee for the harm suffered. But requiring that the promisor foresee the particular act in reliance helps to ensure that Section 90 comes into play only when the seriousness of her promise was apparent to the promisor.

The requirement of substantiality may be read in the same vein; rather than justifying enforcement on the degree of a promisee's reliance, it helps instead to identify promises that the promisor understood might have serious consequences. A promisor is less likely to make a promise casually if she expects it to induce substantial action in reliance by the promisee.

Taken together, the words "definite and substantial" make clear what the language of the *First Restatement* accomplishes: If the promisor made a promise expecting it to induce action or forbearance of a definite and substantial character, the promise was likely well considered and deserving of legal enforcement. Thus, the "definite and substantial" clause in Section 90 of the *First Restatement* serves the same cautionary function that reciprocity serves in the context of bargains. . . .

The Importance of Promise

The routine award of expectancy relief, the absence of a requirement of inducement or detriment, and the refusal in some cases to grant relief despite detrimental reliance all show that the reported cases cannot be explained on the basis of reliance. In fact, reliance theory does not explain why in Section 90 cases courts insist that there be a promise. If the basis of recovery were harm caused by the defendant's conduct, it should not matter whether the conduct constituted a promise.

The critical and difficult question about Section 90 in the courts is not whether to protect reliance, but whether to enforce the promise at issue. It is neither sufficient nor necessary that the promise induce the promisee to rely to her detriment. Every promise may influence the promisee's behavior, and yet not every relied-upon promise is enforceable. What distinguishes enforceable from unenforceable promises is the quality of the commitment made by the promisor.

Separating serious from frivolous promises is one of the functions of the doctrine of consideration. This function was served in Section 90 of the First Restatement by the requirement that the promisor reasonably expect the promise to induce definite and substantial action (or forbearance). Although the Second Restatement dropped this requirement, in *every* case it cites under subsection 1 of Section 90 that grants relief, the promise was expected to induce particular and substantial action (or forbearance).

The promisor's contemplation of particular and substantial reliance is important not in and of itself, but because it signals the quality of her commitment. Other circumstances may establish that a promise was sufficiently well considered for legal intervention. Thus, a promisor who expects a benefit in return for her promise is fairly charged with the seriousness of the promise. The promisee's situation may also indicate that the promisor's commitment was well considered. In cases of marriage settlements, for example, promises may be enforced even in the absence of actual inducement because of the possibility, foreseen by the promisor, that the promisee might change her behavior in reliance on the promise. Alternatively, promises made in the marriage context may be regarded as inherently serious because they involve what is traditionally viewed as a very important part of personhood.

The central importance of promise under Section 90 is well illustrated by charitable pledges and by promises to procure insurance. Pledges are often enforced when there is no reliance. Indeed, even the normal Section 90 foreseeability requirement may not be satisfied by a pledge to a general fund campaign of a charity because the promisor does not contemplate any particular act by the charity in reliance on the pledge. Yet many courts enforce this type of charitable subscription. Other courts pointedly reject the *Restatement*'s invitation to enforce pledges outside of the normal requirements of Section 90. These courts emphasize the lack of consideration and the fact that the pledge was not directed to a specific project expected to be undertaken by the charity.

Although courts differ somewhat on the enforcement of pledges, the likelihood of enforcement generally varies with the quality of the promisor's commitment. Perhaps the easiest case is a pledge made in exchange for consideration furnished by the charity, such as naming a building for the donor. Somewhat more problematic—but still routinely enforced—are pledges made not in return for consideration, but with the promisor's knowledge that the charity might use the pledged funds for some particular and substantial project. If the promisor does not receive consideration or contemplate reliance by the charity, the risk is greater that the pledge was not seriously considered. Not surprisingly, therefore, some courts refuse to enforce such a pledge. Other courts apparently regard every promise made to a charitable institution engaged in good work as a promise that ought to be enforced. While it may seem wrong to break any promise, it may seem more wrong for someone to break a promise to support the work of an organization that she has recognized (by her promise) selflessly does good work.

Like charitable pledges, promises to procure insurance may be enforced in the absence of reliance by the promisee. Here, too, enforcement depends on the quality of the promisor's commitment. The *Second Restatement* states that courts should be cautious about enforcing insurance promises because of the severe consequences to the promisor: for failing to pay a small premium, she may be liable for the amount of a very large policy. Despite its warning, the *Restatement* proceeds to catalogue various factors that may justify enforcement even absent actual reliance, including the formality of the promise, part performance by the promisor, a commercial setting, or potential benefit to the promisor.

If the objective of Section 90 were to protect reliance, the formality of a promise could not substitute for reliance. On the contrary, formality matters only as evidence of the promisor's seriousness. Similarly, partial performance indicates that the promise was made with the intention of performing. Promises made in a commercial setting or with the expectation of obtaining a potential benefit are more likely to be serious than promises made in a purely donative context. Thus, each of the factors adduced by the *Restatement* indicates that the promise was likely to have been well considered; none relates to reliance by the promisee.

In addition, the promisor's commitment is generally likely to be serious because of the possibility, even if remote, that the promise might keep the promisee from obtaining insurance. This possibility should alert the promisor to the potentially severe consequences of making and breaching her promise. Thus, the *Restatement*'s argument for caution in enforcing these promises actually cuts the other way: a promise to procure insurance is almost certainly serious because performance involves relatively little cost and the consequence of not performing may be substantial loss to the insured.

Cases that fall within subsection 1 of Section 90, cases of charitable pledges and cases of insurance promises, all involve nonbargain promises. Enforcement depends not on the promisee's reliance, but on the seriousness of the promisor's commitment. The central importance of commitment is also shown by the many cases in which a fault with the promise leads the court to deny the plaintiff any relief. In nine of the ten cases cited by the *Second Restatement* in which the plaintiff does not obtain any relief, the outcome turns largely on a problem with the promise. When the plaintiff obtains partial relief, virtually the only substantive explanation is a defect in the promise or in the defendant's proof of the promise.

When courts award reliance damages in cases of defective promise or misrepresentation, they are vindicating the plaintiff's interest in not being harmed by the defendant's conduct. The wrong resembles that done by driving an automobile negligently and injuring a pedestrian. By putting the plaintiff in the same position as before the deficient statements were made, reliance damages serve the same objective served by damages awarded to compensate for injury caused by negligence.

The theory of liability in these cases differs radically from that in the typical Section 90 case. Courts enforce promises under Section 90 when they view the promises as serious and deserving of enforcement *qua* promise; they do not enforce them out of solicitude for promisees. The promisor's commitment may be shown to be sufficiently serious by her contemplation of particular and substantial reliance, by the formality of the promise, by the situation of the promisee, or by a chance of benefit to the promisor. The importance to courts of promise explains why the remedy for breach of a Section 90 promise is invariably expectancy relief (if measurable); why the absence of inducement and detriment is irrelevant; why some promises are not enforced despite detrimental reliance; and why the outcome (in terms of both liability and remedy) generally turns on some aspect of promise.

Notes on Reliance

1. There is a long tradition of associating reliance and Section 90 with concepts and policies deriving from tort rather than from contract. The classic statement of this view is Grant Gilmore, The Death of Contract (Columbus: Ohio State University Press, 1974). Another example can be found in the Atiyah excerpt (supra at pages 4–9), which argued that redressing the harm from reliance on a promise had more in common with tort than with contract. An earlier and perhaps even more influential writing in this tradition was an article by Lon L. Fuller and William R. Perdue, Jr., The Reliance Interest in Contract Damages, 46 Yale Law Journal 52 (Part I), 373 (Part II) (1936–37). Fuller and Perdue pointed out a large number of cases in which courts awarded reliance damages as a remedy, thus protecting the promisee's reliance interest rather than the promisee's expectation interest. (A portion of this article, dealing with judicial manipulation of the "foreseeability" test under *Hadley v. Baxendale,* was quoted in the notes supra at page 78.)

2. Is the measure of damages awarded in promissory estoppel cases dispositive on the question of whether Section 90 represents a "tort" or a "contract" notion? Suppose that courts limited promisees to their reliance damages, but did so on the ground that most promisors would be willing to commit to making good the promisees' reliance expenses if they failed to perform but unwilling to commit to performing or paying full expectation damages. If such a result did in fact carry out the wishes of most promisors, it could easily be described as a typical "contract" notion, in spite of its calling for reliance damages rather than for expectation damages.

For other discussions of the relation between reliance damages, promissory estoppel, and tort law, see the excerpt by Alan Farnsworth on precontractual reliance (infra at pages 278–84); and Randy E. Barnett and Mary E. Becker, Beyond Reliance: Promissory Estoppel, Contract Formalities, and Misrepresentations, 15 Hofstra Law Review 443 (1987). The tort/contract distinction is also discussed in the excerpts by Mark Pettit (infra at pages 266–70) and Clare Dalton (infra at pages 273–76). The Speidel article quoted in the Feinman excerpt is Richard E. Speidel, *Restatement Second*: Omitted Terms and Contract Method, 67 Cornell Law Review 785 (1982).

3. A different perspective on liability under Section 90 would view it as a kind of default rule or interpretation doctrine. Under current law, any promisor who clearly specifies that his or her promise is *not* legally binding ("Mind you, I intend to perform, but you should not rely unless you're willing to do so at your own risk") is almost sure not to be held liable under Section 90. Similarly, a promisor who clearly specifies his or her intent to be liable will generally be bound. Promisors who leave their intentions unclear will have to accept the courts' interpretation of their status, as discussed by Feinman and by Yorio and Thel.

Does the fact that the promisor is free to specify his or her intent more clearly mean that Section 90 is therefore more "contractual" than "tort-like"? Or are some default rules more contractual than others, depending on whether the content of the default rule is or is not designed to reflect the preferences of most contracting parties?

4. How should courts interpret the statements of a party who has not clearly indicated whether he or she is making a legally binding promise? Consider the analysis in the Eisenberg excerpt (supra at pages 225-28) explaining why a promisor such as the Confident Student might want to make his or her commitment legally binding in order to give the other party the security that party might need to rely on the promise in some way. As discussed in the notes following that excerpt (supra at page 231), such a promisor would have to balance (a) the benefits of any extra reliance that might be induced by making a binding promise against (b) the risk that a change of events might lead the promisor to subsequently regret being bound. If, in any given case, this balance suggests that the promisor would have benefited by making a binding commitment, that might justify a holding that the commitment was binding, thus putting the burden on the promisor to specify otherwise if he or she was not willing to be bound. On the other hand, if this balance suggests that the promisor probably would *not* have benefited from being bound, that might justify a holding that the promisee should have relied only at his or her own risk, thus putting the burden on the promisee to secure a more definite commitment from the promisor if he or she wants to be legally protected. Similar concerns may enter into the interpretation of offers and acceptances, as discussed in the excerpt by Peter Klik (infra at pages 257–62).

Conceivably, the courts could instead adopt a "penalty default" approach to deal with this problem. (See the excerpt by Ayres and Gertner, supra at pages 22-27.) For example, courts could threaten to interpret *any* ambiguous remark as a binding commitment in order to force every defendant who didn't want to be bound to state his or her intention more clearly from the outset. However, this approach could significantly increase the cost of communicating by requiring that even the most innocuous remarks ("The weather looks nice today") be accompanied by a disclaimer clearly disavowing any intent to make a binding promise.

§ 4.1.3 **Formalities**

Form and Substance in Private Law Adjudication

DUNCAN KENNEDY

The first dimension of rules is that of formal realizability. I will use this term, borrowed from Rudolph von Ihering's classic *Spirit of Roman Law,* to describe the degree to which a legal directive has the quality of "ruleness." The extreme of formal realizability is a directive to an official that requires him to respond to the presence together of each of a list of easily distinguishable factual aspects of a situation by intervening in a determinate way. Ihering used the determination of legal capacity by sole reference to age as a prime example of a formally realizable definition of liability; on the remedial side, he used the fixing of money fines of definite amounts as a tariff of damages for particular offenses.

At the opposite pole from a formally realizable rule is a standard or principle or policy. A standard refers directly to one of the substantive objectives of the legal order. Some examples are good faith, due care, fairness, unconscionability, unjust enrichment, and reasonableness. The application of a standard requires the judge both to discover the facts of a particular situation and to assess them in terms of the purposes or social values embodied in the standard.

It has been common ground, at least since Ihering, that the two great social virtues of formally realizable rules, as opposed to standards or principles, are the restraint of official arbitrariness and certainty. The two are distinct but overlapping. Official arbitrariness means the sub rosa use of criteria of decision that are inappropriate in view of the underlying purposes of the rule. These range from corruption to political bias. Their use is seen as an evil in itself, quite apart from their impact on private activity.

Certainty, on the other hand, is valued for its effect on the citizenry: if private actors can know in advance the incidence of official intervention, they will adjust their activities in advance to take account of them. From the point of view of the state, this increases the likelihood that private activity will follow a desired pattern. From the point of view of the citizenry, it removes the inhibiting effect on action that occurs when one's gains are subject to sporadic legal catastrophe.

It has also been common ground, at least since Ihering, that the virtues of

Duncan Kennedy, Form and Substance in Private Law Adjudication, 89 Harvard Law Review 1685, 1687–89, 1697–1701 (1976). Copyright © 1976 by Duncan Kennedy. Reprinted by permission.

formal realizability have a cost. The choice of rules as the mode of intervention involves the sacrifice of precision in the achievement of the objectives
lying behind the rules. Suppose that the reason for creating a class of persons
who lack capacity is the belief that immature people lack the faculty of free
will. Setting the age of majority at twenty-one years will incapacitate many
but *not all* of those who lack this faculty. And it will incapacitate some who
actually possess it. From the point of view of the purpose of the rules, this
combined over- and underinclusiveness amounts not just to licensing but to
requiring official arbitrariness. If we adopt this rule, it is because of a judgment that this kind of arbitrariness is less serious than the arbitrariness and
uncertainty that would result from empowering the official to apply the standard of "free will" directly to the facts of each case. . . .

Relationship of the Formal Dimensions to One Another

. . . [T]he main disadvantage of general rules is their over- and underinclusiveness from the point of view of the lawmaker's purposes. In the context of formalities the problem is that general rules will lead to many instances
in which the judge is obliged to disregard the real intent of the parties choosing between alternative legal relationships. For example, he will refuse to
enforce contracts intended to be binding (underinclusion), and he will enforce
terms in agreements contrary to the intent of one or even both parties
(overinclusion). Since we are dealing with formalities, this is an evil: the
lawmaker has no substantive preferences about the parties' choice, and he
would like to follow their wishes.

The Argument for Casting Formalities as Rules

The response is that the problem of over- and underinclusiveness has a special
aspect in the case of formalities because the lawmaker can enlist the energies
of the parties in reducing the seriousness of the imprecision of rules. The
parties have an interest in communicating their exact intentions to the judge,
an interest that is absent when they are engaged in activity the legal system
condemns as immoral or antisocial. But this communication has a cost and
involves risk of miscarriage. The lower the cost, and the greater the probability that the judge will respond as expected, the more the parties will invest in
getting the message across.

   The lawmaker can take this private calculus into account in designing the
formalities. He can reduce the cost of learning the language of form by making his directives as general as possible. A "technical" system composed of
many different rules or standards applying to closely related situations will be
difficult to master and confusing in practice. For example, Williston's formulation of the parol evidence rule involves a rule of "plain meaning of the writing
on its face" to determine whether a given integration embodies the total

agreement of the parties. But this is subject to exceptions for fraud and duress. Another rule applies in determining whether the integration was intended to be "final," and yet another to the problem of agreements whose enforceability was meant to be conditional on the occurrence of events not mentioned in the document. It is hard to imagine a layperson setting out to master this doctrinal tangle.

If generality can reduce the cost of formal proficiency, formal realizability should reduce the risk that the exercise of judicial discretion will bring formal proficiency to naught. Standards discourage investment in two ways. The uncertainty of the outcome if the judge is at large in finding intent, rather than bound to respond mechanically to ritual acts like sealing, will reduce the payoff that can be expected from being careful. Second, the dangers of imprecision are reduced because the judge may bail you out if you blunder. The result *may* be a slippery slope of increasing informality that ends with the legal system treating disputes about wills as though they were automobile accidents litigated under a fault standard.

If general rules lead people to invest in formal proficiency, at least as compared to standards, the result should be the reduction of their over- and underinclusiveness. In other words, the application of the rule should only very rarely lead to the nullification of the intent of the parties. The rare cases that do occur can then be written off as a small cost to pay for the reinforcement of the sanction of nullity. People will miss fewer trains, the argument goes, if they know the engineer will leave without them rather than delay even a few seconds. Standards, by contrast, are dynamically unstable. Rather than evoking private action that compensates their inadequacies, they stimulate responses that aggregate their defects.

Finally, rules encourage transaction in general. If an actor knows that the use of a formality guarantees the execution of his intentions, he will do things that he would not do if there were a risk that the intention would be defeated. In particular, actors will rely on enforcement of contracts, trusts, and so forth, in making investments. Since we are dealing with formalities, it is a matter of definition that the legal system is anxious to encourage this kind of activity so long as private parties desire to engage in it. . . .

The Critique of the Argument for Rules

The argument for casting formalities as rules rests on two sets of assumptions, each of which is often challenged in discussions of actual legal institutions. The first set of assumptions concerns the impact on real participants in a legal system of the demand for formal proficiency. If the argument for rules is to work, we must anticipate that private parties will in fact respond to the threat of the sanction of nullity by learning to operate the system. But real as opposed to hypothetical legal actors may be unwilling or unable to do this.

The contracts of dealers on produce exchanges are likely to use the most exquisite and most precisely manipulable formal language. Poor consumers, by contrast, are likely to be formally illiterate. Somewhere in between lie the businessmen who have a highly developed understanding of the mechanics of their deals, yet persistently—and perfectly rationally, given the money cost of lawyers and the social and business cost of legalism—fail to master legal technicalities that return to plague them when things go wrong. We must take all the particular variations into account. In the end, we may decide that a particular formal system works so smoothly that a refusal to fill the gaps with general rules would be a wanton sacrifice of the parties to a judicial prima donna. But others work so badly that little is lost by riddling them with loopholes. . . .

In those situations in which some parties *are* responsive to the legal system, a regime of formally realizable general rules may intensify the disparity in bargaining power in transactions between legally skilled actors who use the legal system constantly, and unskilled actors without lawyers or prior experience. At one extreme there is a kind of fraud that is extremely difficult to police effectively: one party knows that the other party does *not* know that the contract must be in writing if it is to be legally binding. At the other is the bargaining confrontation in which the party with the greater skills legitimately relies on them to obtain a result more favorable than would have occurred if everyone knew that the issue *had* to be left to the judge's discretion.

The second set of assumptions underlying the argument for rules concerns the practical possibility of maintaining a highly formal regime. A great deal of legal scholarship between the First and Second World Wars went into showing that legal directives that looked general and formally realizable were in fact indeterminate. Take, for example, the "rule" that a contract will be rescinded for mutual mistake going to the "substance" or "essence" of the transaction, but not for mistakes as to a "mere quality or accident," even though the quality or accident in question was the whole reason for the transaction. We have come to see legal directives of this kind as invitations to sub rosa balancing of the equities. Such covert standards may generate more uncertainty than would a frank avowal that the judge is allocating a loss by reference to an open textured notion of good faith and fair dealing.

In other situations, a "rule" that appears to dispose cleanly of a fact situation is nullified by a counterrule whose scope of application seems to be almost identical. Agreements that gratuitously increase the obligations of one contractual partner are unenforceable for want of consideration. *But,* such agreements may be binding if the judge can find an implied rescission of the old contract and the formation of a new one incorporating the unilaterally onerous terms. The realists taught us to see this arrangement as a smokescreen hiding the skillful judge's decision as to duress in the process of renegotiation, and as a source of confusion and bad law when skill was lacking.

The critic of the argument for rules can often use this sort of analysis to show that what looks like a rule is really a covert standard. It is also often possible to make a plausible claim that the reason for the "corruption" of what was supposed to be a formal regime was that the judges were simply unwilling to bite the bullet, shoot the hostages, break the eggs to make the omelette and leave the passengers on the platform. The more general and the more formally realizable the rule, the greater the equitable pull of extreme cases of over- or underinclusion. The result may be a dynamic instability as pernicious as that of standards. There will be exceptions that are only initially innocuous, playing with the facts, the invention of counterrules (for example, waiver and estoppel), the manipulation of manifestations of intent, and so forth. Each successful evasion makes it seem more unjust to apply the rule rigidly in the next case; what was once clear comes to be surrounded by a technical and uncertain penumbra that is more demoralizing to investment in form than an outright standard would be.

Alchemical Notes: Reconstructing Ideals from Deconstructed Rights

PATRICIA J. WILLIAMS

Some time ago, Peter Gabel and I taught a contracts class together. Both recent transplants from California to New York, each of us hunted for apartments in between preparing for class and ultimately found places within one week of each other. Inevitably, I suppose, we got into a discussion of trust and distrust as factors in bargain relations. It turned out that Peter had handed over a $900 deposit, in cash, with no lease, no exchange of keys and no receipt, to strangers with whom he had no ties other than a few moments of pleasant conversation. Peter said that he didn't need to sign a lease because it imposed too much formality. The handshake and the good vibes were for him indicators of trust more binding than a distancing form contract. At the time, I told Peter I thought he was stark raving mad, but his faith paid off. His sublessors showed up at the appointed time, keys in hand, to welcome him in. Needless to say, there was absolutely nothing in my experience to prepare me for such a happy ending.

I, meanwhile, had friends who found me an apartment in a building they owned. In *my* rush to show good faith and trustworthiness, I signed a detailed, lengthily-negotiated, finely-printed lease firmly establishing me as the ideal arm's length transactor.

As Peter and I discussed our experiences, I was struck by the similarity of what each of us was seeking, yet in such different terms, and with such polar

Patricia J. Williams, Ideals from Deconstructed Rights, 22 Harvard Civil Rights–Civil Liberties Law Review 401 (1987). Permission granted by the President and Fellows of Harvard College. Copyright © 1987 Harvard Civil Rights–Civil Liberties Law Review.

approaches. We both wanted to establish enduring relationships with the people in whose houses we would be living; we both wanted to enhance trust of ourselves and to allow whatever closeness, whatever friendship, was possible. This similarity of desire, however, could not reconcile our very different relations to the word of law. Peter, for example, appeared to be extremely self-conscious of his power potential (either real or imagistic) as a white or male or lawyer authority figure. He therefore seemed to go to some lengths to overcome the wall which that image might impose. The logical ways of establishing some measure of trust between strangers were for him an avoidance of conventional expressions of power and a preference for informal processes generally.

I, on the other hand, was raised to be acutely conscious of the likelihood that, no matter what degree of professional or professor I became, people would greet and dismiss my black femaleness as unreliable, untrustworthy, hostile, angry, powerless, irrational and probably destitute. Futility and despair are very real parts of my response. Therefore it is helpful for me, even essential to me, to clarify boundary; to show that I can speak the language of lease is my way of enhancing trust of me in my business affairs. As a black, I have been given by this society a strong sense of myself as already too familiar, too personal, too subordinate to white people. I have only recently evolved from being treated as three-fifths of a human, a sub-part of the white estate. I grew up in a neighborhood where landlords would not sign leases with their poor, black tenants, and *demanded* that rent be paid in cash; although superficially resembling Peter's transaction, such "informality" in most white-on-black situations signals distrust, not trust. Unlike Peter, I am still engaged in a struggle to set up transactions at arms' length, as legitimately commercial, and to portray myself as a bargainer of separate worth, distinct power, sufficient *rights* to manipulate commerce, rather than to be manipulated as the object of commerce.

Peter, I speculate, would say that a lease or any other formal mechanism would introduce distrust into his relationships and that he would suffer alienation, leading to the commodification of his being and the degradation of his person to property. In contrast, the lack of a formal relation to the other would leave me estranged. It would risk a figurative isolation from that creative commerce by which I may be recognized as whole, with which I may feed and clothe and shelter myself, by which I may be seen as equal—even if I am stranger. For me, stranger–stranger relations are better than stranger–chattel.

Notes on Formalities

1. The concept of a "formality" (or a "formally realizable rule") discussed in the Kennedy excerpt is very similar to the "penalty default rule" discussed

in the excerpt by Ayres and Gertner in Chapter 1 (supra at pages 22–27). Ayres and Gertner used the term "penalty default" to refer to any legal doctrine that threatened a contracting party with an outcome that would not be very much to that party's liking, in order to induce that party to take the time and trouble to specify explicitly the outcome it really desired. In Kennedy's terms, a penalty default encourages at least one of the contracting parties to "invest in formal proficiency."

The rules of consideration were characterized as formalities in a famous article by Lon L. Fuller, Consideration and Form, 41 Columbia Law Review 799 (1941). Fuller suggested that the legal requirement of consideration sometimes served a "cautionary function" (the transfer of consideration would alert the parties to the fact that a legally binding agreement was being created) and a "channeling function" (by exchanging consideration, the parties could signal to the courts that they wanted their agreement to be legally enforceable). Of course, some parties who intended their agreements to be enforceable did not exchange any consideration, and thus failed to achieve their goal of a legally binding agreement. On this view, courts' refusal to enforce such agreements would make other parties all the more careful to make sure they included consideration if they wanted their agreements to be enforceable— thereby strengthening the law's cautionary and channeling functions.

2. Both the Kennedy excerpt, and the earlier excerpt by Ayres and Gertner, discuss the cost of complying with a formality as a relevant policy consideration, though neither excerpt described those costs in any detail. Some costs are obvious: the cost of learning the legal requirements, the cost of drafting or otherwise conforming one's contract to whatever the law requires, and the cost (or risk) of the bad consequences that follow if one forgets or fails to comply with the applicable requirements.

The Williams excerpt identifies some other factors that might also be considered costs of formal requirements. For example, some requirements might impose psychic costs on the parties by placing them in attitudes of what seem to be distrust or inferiority. Some requirements might also be costly because of the signal they send to the other contracting party—for example, "this is a person who is going to be troublesome enough to insist on his every legal right," or "this is a person who isn't very comfortable with formal contracting procedures."

Interestingly, economists and game theorists have also analyzed situations in which the adoption of a particular clause (or a particular kind of contract) reveals to one party information about the party who insists on that kind of clause. For recent applications of these models to the legal literature, see, for example, Ian Ayres and Robert Gertner, Strategic Contractual Inefficiency and the Optimal Choice of Legal Rules, 101 Yale Law Journal 729 (1992); Jason Scott Johnston, Strategic Bargaining and the Economic Theory of Contract Default Rules, 100 Yale Law Journal 615 (1990); Steven Shavell, An Economic Analysis of Altruism and Deferred Gifts, 20 Journal of Legal Studies 401 (1991).

§ 4.2 **Offer and Acceptance**

§ 4.2.1 **Interpreting Offers**

The Language of Offer and Acceptance

PETER MEIJES TIERSMA

In John Searle's philosophy of language, each attempt to communicate is some type of speech act. Speech acts consist, inter alia, of two other types of acts: propositional acts and illocutionary acts. Propositional acts refer and predicate. A proposition such as "Sam smokes habitually" refers to Sam, and it predicates that he smokes habitually. The proposition "Donna smokes habitually" has the same predicate but a different reference: Donna.

Notice that the same proposition may occur in utterances of different types:

1. Sam smokes habitually.
2. Does Sam smoke habitually?
3. Sam, smoke habitually!
4. Would that Sam smoked habitually.

Because the proposition in these sentences is the same, each sentence performs the same propositional act. Nonetheless, the speaker's intent with respect to the proposition is quite different. In Searle's terminology, each has a different "force." The first sentence *asserts* the proposition, the second *asks* a question, the third makes a *command,* and the fourth expresses a *desire.* Searle calls these acts—asserting, asking, etc.—*illocutionary acts,* which account for the *illocutionary force* of utterances.

Often a specific verb expresses the illocutionary force of an utterance. In the examples above, adding such a verb makes the illocutionary force explicit, as in "I *assert* that Sam smokes habitually" or "I *ask* you whether Sam smokes habitually" or "I *command* Sam to smoke habitually." Other illocutionary verbs include greet, state, describe, warn, command, promise, object, demand, and argue.

More often than not, the illocutionary force of an utterance is not made explicit by one of these verbs. If someone tells her neighbor, "I'll fix your fence tomorrow," there is no expressed illocutionary verb to indicate whether

Peter Meijes Tiersma, The Language of Offer and Acceptance: Speech Acts and the Question of Intent. Copyright © 1986 by the California Law Review, Inc. Reprinted from California Law Review, volume 74, number 1 (January, 1986), pages 194–95, 206–12, by permission.

this is a mere statement of intent or an actual promise with some moral commitment. This ambiguity is evident in that the neighbor can respond, "Is that a promise?" If the illocutionary force were made more explicit, as in "I promise to fix your fence tomorrow," it would no longer be appropriate to ask the speaker if she had indeed promised to do so. An illocutionary verb can almost always be added to an otherwise vague or ambiguous utterance, thereby rendering explicit the illocutionary force. . . .

Interpretation

In normal conversation the content of an utterance, including its illocutionary force, is often expressed indirectly. The circumstances and context of the utterance assume great importance in deciphering the meaning of such speech acts. The sentence "Do you think that you could open the door?" might under some circumstances be a question equivalent to "I ask you whether you think that you are able to open the door." This would be the correct interpretation if the parties are standing before the five meter high portals of a medieval castle. On the other hand, when the speaker is returning home with large bags of groceries, the same sentence is a request or even a command. The rest of this Part suggests some techniques which participants in the speech setting use to interpret an utterance in context and to determine whether it has the requisite elements of an offer or acceptance.

Ambiguity and Meaning

Often utterances seem hopelessly ambiguous. Even words which have only one dictionary meaning can develop other meanings through metaphorical extension. At the other extreme, a word such as "right" has a large variety of dictionary meanings. Context will usually allow the hearer to extrapolate the intended meaning. For example, "right" is clearly a directional term in the sentence "The house is on the right."

Ambiguity results when the intended meaning does not emerge from the context. If a driver asks her passenger, "Should I turn left here?" the passenger might reply, "Right." Most meanings of "right," such as "conservative" or "privilege," are eliminated by the linguistic context. Still, two competing meanings remain: the opposite of "left" and the opposite of "wrong." Unless a difference in intonation or something in the extralinguistic circumstances indicates the intended meaning (for example, if a right turn were physically impossible), the utterance "right" is ambiguous.

A formula such as "I hereby accept your offer" is the most explicit way to make an offer or acceptance. It is probably more common, however, for assent to be made indirectly, where the words do not literally indicate the force of assent. An example is an offer that states "Will you take ten dollars for that watch?" Obviously, this question creates greater possibilities for ambiguity than does the use of a formula. Nonetheless, where such indirect utterances are equivalent to explicit performatives, the utterance will count as putting the speaker under a commitment to abide by its terms. The rest of this Section will deal with how words can indirectly commit the speaker to the terms of an agreement, and will further demonstrate how ambiguities, at times, can be reduced or eliminated.

Idiomatic Meaning

A sentence such as "Can you open the door?" has at least two meanings. The first is a fairly literal meaning, one in which the whole of the sentence is roughly equal to the sum of its parts. In its most direct sense, "can" refers to the *ability* to carry out a task. Yet "can you" is often used idiomatically to give an order. This accounts for the other meaning of this sentence. . . .

Other idiomatic expressions, such as "would you" and "will you," may be used to issue directions in addition to simply asking questions. Examples are "Will you open that door for me?" "Would you swat that fly?" or "Would you like to wash the dishes?" These and other expressions, therefore, may have a conventional illocutionary force different from or in addition to their literal meaning.

There are a number of expressions that do not literally indicate the illocutionary force of offer but that do so idiomatically. For example, auxiliary verbs that indicate possibility or ability can occur in sentences that have the illocutionary force of an offer. "We *can* offer you," "We *are able* to offer you," or "It *is possible* for us to offer you" may all, in the proper circumstances, have the force of offer despite the superficial waffling.

Phrases that literally refer to a positive state of mind also may have the illocutionary force of offer. Examples are "We *would like* to offer you p" or "We *are pleased* to offer you p." Normally it is not enough that one "would like" to offer or accept, but in ordinary language these expressions usually count as committing the speaker to an offer. The sentence "We would like to offer you a job at Smith and Smith" has the illocutionary force of an offer despite the presence of "would like." On the other hand, in the sentence "We would like to offer you a job at Smith and Smith, but have insufficient funds this year," the context militates against the nonliteral interpretation.

Verbs referring to authority may co-occur with the word "offer" without robbing the expression of the illocutionary force of offer: "We have been *authorized* to offer you p," or "We have been *permitted* to offer you p." This is true even though literally being authorized or permitted is simply a preliminary step to offering.

Finally, offers may be made without using the word "offer" at all. For example, sentences such as "Will you take five dollars for that hat?" or "Will you sell me your bicycle for fifty dollars?" may count as committing the speaker to pay fifty dollars for the bicycle. Although these utterances are literally requests for information, in a bargaining context they usually function as the equivalent of "I (hereby) offer that I will promise to pay you fifty dollars and you will promise to give me your bicycle."

Acceptance also can be expressed idiomatically. In response to an offer, a person could reply "You're on, "It's a deal," or "You've got yourself a bargain. If asked the question "Will you sell me your bicycle for fifty dollars?", the offeree could simply respond with "yes." The crucial issue is whether the speaker has uttered the equivalent of an explicit performative, and so has committed herself to the terms of the bargain.

Grice's Maxims of Conversation

The use of idiomatic expressions often leads to superficial ambiguity, since the language generally will have both a literal (word-for-word) meaning as well as the more idiomatic meaning. The hearer must determine which of these meanings is intended. Native speakers of the language continually make such judgments. Courts would frustrate the will of the parties if they looked only to the most literal meaning of the words used.

Judges have become increasingly amenable to interpreting less explicit speech acts as native speakers of the language would. In *Embry v. Hargadine, McKittrick Dry Goods Co.,* an employer responded to a worker's question about renewal of his employment with the words: "Go ahead, you're all right; get your men out and don't let that worry you." The words "don't let that worry you," among others, were found to constitute acceptance of the offer. In the terms of this analysis, they were found the equivalent of "I hereby accept your offer." Had the court relied only on the literal meaning of the words, it could never have reached this conclusion. But the court recognized that "don't worry about it" or similar phraseology often is used idiomatically as a promise to remedy the source of the worry. If two people enjoy an expensive dinner and one finds himself short of cash, his dining partner may say, "Don't worry about it," thereby implying that he will pay for the meal. Notice that this implication only applies to phrases closely resembling "Don't worry about it." "Don't be so paranoid" and "Worrying will get you nowhere" do not normally constitute a promise to pay.

A useful method for determining the meaning of these kinds of ambiguous utterances is a notion advanced by the philosopher Grice, which he calls *conversational implicature*. Conversational implicature is based on a set of maxims that operate as rules of conversation. While the maxims may seem self-evident or even simplistic at first, they are useful both in eliminating unintended meanings and in inferring meanings that are not immediately evident. In other words, these maxims help make explicit what it means to say that utterances can only be understood in context or under particular circumstances.

What Grice calls the "cooperative principle"—an overarching principle governing most conversation—is central to conversational implicature. This principle rests on the observation that participants in a conversation normally cooperate with one another. For example, a person will usually try to make each utterance on a particular topic relevant to the preceding utterance. This being so, parties to a conversation will interpret an utterance as being cooperative—in this case, as being relevant to what went on before.

Various maxims derive from the cooperative principle, which is divided into a number of categories. The category of *quantity* includes two maxims: make your contribution to a conversation as informative as necessary and, conversely, do not make your contribution more informative than is required. The category of *quality* also contains two maxims: a contribution should be true and the speaker should have adequate evidence for it. The category of *relation* contains a maxim mandating that a contribution be relevant. Finally, the category of *manner* includes maxims promoting brevity, clarity, and the like.

These maxims clarify the result in *Embry v. Hargadine*. The primary reason that "Don't worry about it" can constitute acceptance of the employee's implicit offer has to do with Grice's conversational maxim of relevance. A literal interpretation of the employer's statement would not be a relevant answer to the employee's concern. Therefore, this possible interpretation can be eliminated. All that remains is the more idiomatic meaning of the phrase as a promise to dispose of the employee's concern. The employer's response was relevant only as an acceptance.

The maxims of relevance and quantity also account for the force of many of the less direct offers discussed above. Consider the question, "Will you take five dollars for that bowl?" If a friend says this to a potter, after the potter has said that she hopes to sell it for six dollars, it would be relevant as a question and would be interpreted as such. But if the utterance is made by a potential customer, in response to the potter's offer to sell it for six dollars, the utterance is most relevant not as a question, but as a counter-offer. Similarly, suppose that a job applicant receives a letter stating, "We are authorized to offer you a position." Under these circumstances, the letter in its literal sense communicates both too much information and too little—it announces the employer's authorization while failing to disclose whether an offer will be forthcoming. Only interpreting the letter as an actual offer is responsive to the applicant's implicit question as to whether he will in fact obtain an offer.

Assent by Implication

Grice's maxims also help interpret speech acts where the meaning is not idiomatic but must be *inferred* from the circumstances. In other words, the maxims also apply to utterances that *imply* an offer or acceptance. In such cases the maxims do not eliminate inappropriate meanings, as when an idiomatic meaning competes with a literal meaning, but rather give words an interpretation that they would not have otherwise. If a duke tells his servant, "There's a draft in here," there is little doubt that this is an order to close the window. The order, though not one of the conventional meanings of this utterance, is implicit in the circumstances, especially the relationship of the duke to his servant.

A similar example is where a job applicant receives a letter thanking him for applying and telling him that the firm has completed its hiring process. The applicant can easily infer that he has been rejected. Grice's maxim of quantity, which states that one should say neither too much nor too little for the purposes of the conversation, provides an explanation. If the applicant had gotten the job, one would expect the letter to say so explicitly; otherwise the letter would not have been adequately informative for the purposes of the "conversation." Lack of an offer where one would be expected implies that no offer is forthcoming.

Much the same principle operates when a professor writes a letter of recommendation for a student. Suppose that the letter, after the customary greeting, states "I have had Jan in several intimate classes and can assure you that her attendance record is exemplary. Yours very truly, etc." Here the professor has flouted the maxim of quantity, for he knows that more information is desired. In the context of a letter of recommendation, it is expected that a professor will communicate as much positive information as possible. The omission of certain expected sorts of information can only lead to the inference that in those areas the professor is unimpressed.

Mass Media and Offers to the Public

PETER KLIK

Mass media can be used by someone who contemplates engaging in contracts. Examples of the use of mass media that come to mind are advertising on radio or television, in newspapers or more or less specialized magazines; sending out handbills, price lists, circulars, mail-order catalogs; displaying goods in a store window; putting up goods for sale at an auction; employing vending machines; or providing information through computer terminals. No one single individual or entity is approached by use of one of these mass media; the information is instead made available to a larger group, or to the public in general. . . .

American and Dutch legal doctrine both recognize that the question whether a general offer has been made ultimately is a question of fact. The two systems differ, however, in that in American doctrine the presumption is against general offers, whereas in Dutch literature the presumption favors general offers.

Neither system provides specific guidelines as to how concrete cases should be resolved. Both systems fail to specify in detail the situations in which an exception to the "presumption" rule is warranted. Therefore, the same results can be reached in both systems, despite the different abstract rules. There is not sufficient case law on the subject to state that American and Dutch case law are the same; the results, however, tend to be similar, and the discussion of the case law will at the least illustrate that in both systems the concerns are the same.

Peter Klik, Mass Media and Offers to the Public: An Economic Analysis of Dutch Civil Law and American Common Law, 36 American Journal of Comparative Law 235, 235, 245–51, 254–55, 262–67, 269–71, 277–78 (1988). Copyright © 1987 by The American Association for the Comparative Study of Law, Inc. Reprinted by permission.

For purposes of the analysis in the remainder of this article, we will ignore
the fact-specific results of both American and Dutch law, and instead use the
different presumptions adhered to in the systems to construe opposing models
as tools of analysis. In what hereinafter will be referred to as the American
model, the "offeror" generally is not bound by the communication to the
public; a legally binding obligation can only come into being through subsequent negotiations by the parties. In what will be called the Dutch model,
offerees generally can conclude a contract through an unilateral act, the acceptance, but certain defenses are available to the offeror. By juxtaposing a
Dutch and American model, we can analyze the benefits and costs relating to
both models, and thus to the different "presumption" rules adhered to in
American and Dutch law. . . .

General Considerations in Formulating an Optimal Model

Reliance

Advertisements not only convey information, but are also calculated to induce behavior. The communicator hopes the offerees' interest will be aroused
to the extent of establishing contact between them. If, however, after inducing
a contract the "offeror" refuses to conclude a contract, the "offeree" will have
wasted resources such as time and cost of communicating, and possibly have
foregone other opportunities.

Parties generally accept some costs of shopping around. For one who
seeks to deal, the search costs are justified by the increase in value of the
deals located by the search. During the search, decisions on expenditure of
additional search costs will hinge on the expected benefits of increased value
to be derived from the additionally located prospects. Once the search is
over, however, and a deal has been selected, additional resources will have
to be expended to conclude the chosen bargain. For example, consider the
customer who reads several newspaper advertisements, selects one "super
special," drives to the store and learns the product is sold out. Reading the
newspaper is a search cost. The expenses connected with the drive are
reliance costs, expenses induced by information about the prospect of a
favorable deal. When the deal turns out to be unavailable, those reliance
costs are wasted. Reliance on the unavailable prospect may also involve the
loss of other opportunities; if the person had not gone after the "super
special" there might have been time to profit from a "daily special" at
another store. . . .

Reasons for Making "Real" General Offers

Communications to the public can serve several purposes: they spread general
information, help potential customers select, and persuade them to enter into
negotiations. They also can be used to make "real" offers. We noted that a
cost of favoring general offers is the increased risk of being bound to un-

anted contracts. Under certain circumstances, however, the advertiser might ally want to be bound upon acceptance by members of the public.

There is one very good reason why one might wish to make a "real" general offer. A real general offer not only informs about potential opportunities, but also assures the members of the public that the opportunity is actually available. General offers give the offeree control over whether or not a contract will be concluded. As a general rule, the acceptor will value the fact that he is put in control. Where a widget is offered for the same price, a buyer will generally choose to deal with the seller who puts the buyer in control rather than running the risk that the offer will be refused. When a "real" offer has been made, the offeree is given the assurance that expended reliance costs will not be wasted. The existence of option contracts, the very object of which is to put the acceptor in control, demonstrates that to be put in control is of value to offerees. Offerees not only want to avoid refusals, but also periods of uncertainty.

How important it is for the offeror to give offerees control over the situation will depend on several factors, such as the general competitiveness of the market and the array of possibilities available to distinguish oneself from one's competitors. It is clear, however, that making real offers can be of value to the offeror because it increases the value the communication has to the offeree. If offerees are more likely to consider seriously proposals which take the form of general offers, and if making a "real" general offer increases the likelihood of favorable responses, offerors will find it in their interests to communicate a willingness to be bound. Thus, a benefit of a system which favors general offers is that it meets the offeror's need to make general offers.

An Optimal Model

From the previous subsections we can derive the elements of an optimal solution to the problem of the legal effect of communications to the public through mass media. An optimal solution allows offerors to bind themselves through the use of general offers when they think this will be in their best interest, while at the same time it does not expose them to the risk of being bound against their will. The optimal solution would also protect the "offerees" from incurring wasted reliance costs. This optimal result can only be reached if, in his communication, the "offeror" indicates specifically the conditions under which he is willing to be legally bound. Typically, those who communicate to the public do not specify the legal effects of the communication, and for a good reason: it entails high transaction costs.

Specifying exactly under which circumstances and to what extent one is willing to be legally bound involves enormous costs. It seems clear that the benefits to be derived from forcing the "offeror" to be this explicit are outweighed by the costs. Thus, the law has to provide a rule. The legal rule is used in the adjudication of cases, and, more importantly, serves as guidance to parties in their behavior. A "pure" American rule creates wasteful audience reliance, and disables offerors from obtaining the benefits of making real offers. A "pure" Dutch rule deters the publishing of valuable market information. Optimal results can be reached by sometimes applying an American rule, and sometimes employing a Dutch rule. We have seen that both rules allow for this flexibility in that they are phrased as presumptions. In light of the variety of situations covered by these rules adopting either of the "pure" rules would be clearly undesirable. The question remains which "presumption" rule should be adopted. Before this question can be answered, we need to examine additional potential costs of both models. . . .

Specific Issues

Duration and Revocability

After a communication to the public has been sent out, circumstances can change in such a fashion that the "offeror" regrets the terms of the offer. This in itself is not disquieting. In fact, it is inevitable and part of the system. People are constantly assessing their needs, and trying to predict future prices. The seller offers to sell because of an assumption that the price will go down, and the buyer buys because of an assumption that the price will go up, and one of the two will be mistaken and take a loss. When a general offer has been made, a miscalculation can lead to increased potential liability, because there are many offerees. Are offerors willing to take that risk? Certainly such willingness will depend on the particular circumstances. As a general rule it can be stated that the willingness to take the risk will be influenced by the legal rules concerning the duration of the offer and concerning the offer's revocability. . . .

Quantity

When someone communicates to the general public without stating a quantity that person can be confronted with a situation where he or she has to handle more business than contemplated, due to a large number of respondents, the quantity of orders, or both. If the offer is for a limited quantity the potential acceptors find themselves in a competitive market. They will have to calculate the risk that other acceptors will have exhausted the limited quantity.

We can try to solve these "quantity" problems by considering the options available to a person wanting to communicate with the public. We will investigate the different variations of the message, "I am selling widgets," changing each time the information provided as to quantity. For sake of simplicity we assume elements such as price, size, quality, etc., are stated. . . . We will look at how specific the information is, the apparent intent of the offeror and the reliance induced, and whether the offeror's liability is (implicitly) limited. Several possibilities are set out below:

1. I am selling widgets to everyone, as many as they order;
2. I am selling widgets to everyone, limit 12 per person;

3. I am selling widgets to the first 10 persons, limit 12 per person;
4. I am selling widgets, limit 12 per person;
5. I am selling widgets while stocks last;
6. I am selling 120 widgets;
7. I am selling widgets;
8. Maybe I will be willing to sell widgets.

The last message is clearly not an offer. The other messages are either clearly offers or they can be interpreted as offers.

The first two messages clearly indicate that the offeror wants to make an offer. Such unambiguous statements will rarely be found. Even in the American model, these messages will be considered binding: They are examples of a party purposely avoiding the rule. The offeror wants to make a real offer, and induces reliance by giving the acceptor complete control over the situation. The seller might end up in the situation where he or she has to do more business than desired, but is willing to take this risk. . . .

In the third message, "I am selling widgets to the first 10 persons, limit 12 per person," a potential acceptor is not given complete control over the situation because of a competitive market for the goods. The acceptor must weigh the chances and calculate the risk that other acceptors have exhausted the limited quantity. The phrasing of the offer, however, a promise to sell to the first 10 persons, clearly indicates that the offeror makes a real offer. The offeror has exposed himself to a limited liability only. The information conveyed is that of a real offer. In *Lefkowitz v. Great Minneapolis Surplus Store, Inc.,* the store included "First come, first served" in a newspaper advertisement for a very limited number of items; the court held an offer had been made.

Message #4, "I am selling widgets, limit 12 per person" resembles message #2; only the unambiguous "to everyone" is deleted. The offeror's potential liability, although limited because every acceptor can only take 12 items, is undetermined. Considering this to be an offer needs a strong justification in as much as it is unclear what quantity risk the offeror is willing to undertake. If either the intent of the offeror to undertake unlimited exposure or significant reliance costs induced can be proven there is reason to classify it as a true offer. The per person limit stated tends to make the expenditure of significant reliance costs by any responder less probable or reasonable. Nevertheless, it seems that the offeror has estimated the number of acceptors, and calculated how many goods can be sold to each. The message could be read as merely providing the information that no one will be able to order more than 12 items, so that everyone who wants more items need not respond. The sophistication of the message, however, strongly suggests an intent to make an offer, and it will probably be relied on as such.

In message #5, "I am selling widgets while stocks last," the offeror knows the extent of his liability. But, like example #3, the acceptors are not put in control of the situation. Again, we can reason that the additional phrase shows an intent to make an offer. "While stocks last" could be interpreted as intended to provide additional non-binding information, but when dealing with "limit 12 per person" in the previous message we could not justify that conclusion, and the clause "while stocks last" seems even less amenable to such an interpretation. A "while stocks last" clause conveys little concrete information, it merely suggests an intent to sell everything in stock. Message #5 should be interpreted as a real offer. This is in conformance with both Dutch and American law.

There are strong arguments in favor of considering message #6, "I am selling 120 widgets," as an offer. The offeror provides specific information to potential acceptors, which will induce reliance. On the other hand, we want to encourage those who want to communicate non-committedly to the public to be as specific as possible. The message provides the useful information that only 120 widgets are available. Including this information in the message reduces reliance costs because some potential acceptors will discard the message immediately on the basis of this extra information. We find ourselves in the contradictory situation that providing specific information both decreases and increases reliance. The addition of "120 widgets" to "I am selling" does not seem to reveal a clear intent to make a real offer; the additional terms in messages #3, #4, and #5 did. The message does not subject the offeror to unlimited liability; this "quantity" aspect is therefore not troubling. A decision concerning message #6 ("I am selling 120 widgets") will depend on the general considerations developed earlier, a weighing of the information function, the intent of the offeror, and reliance induced.

Message #7 is the basic message, "I am selling widgets." In this message no quantity is stated at all. We can treat this message as an offer for unlimited quantity, an offer for a limited quantity, or a non-offer.

When we treat this message as an offer for unlimited quantity, we do not give the offeror any defense, and he is bound by every acceptance. This does not seem a very practical solution. In its worst consequences it will drive the offeror out of business and into bankruptcy, which means the acceptors will not derive much benefit from their contracts. An even more serious disadvantage is that such a rule puts a stiff penalty on communications with the public, which will put a high cost on communicating, because everyone will take great pain to avoid getting into this situation. Neither in Dutch law nor American law is this rule advocated. . . .

The problem of the offeror's unlimited liability can be solved by implying a limit. As stated above, American law generally treats the message "I am selling . . ." as a non-offer, and therefore refuses to imply an offer for a limited quantity. Dutch authors state that generally an offer of goods is considered to have been made under an "as long as stocks last" clause, but they do not cite cases in support of this statement.

In the Dutch model, offerors who want to communicate without committing themselves must resort to messages such as #8. In the American model, real offerors are forced to include clauses such as "while stocks last" in their offers. It seems easier for real offerors to include a quantity statement than for non-offerors to clearly express that no real offer is intended. The American model also encourages real offerors to articulate more precisely conditions as to quantity, thereby improving the quality of the information provided to the market. . . .

Notes on Interpreting Offers

1. The analysis of advertiser's motives in the Klik excerpt is similar to the analysis of promisor's motives presented in the Eisenberg excerpt on consideration (supra at pages 225–28). This similarity stems from the fact that both a legally binding promise and a legally binding offer (that is, an offer that becomes enforceable as soon as an offeree accepts) limit the promisor's or offeror's future options. Both promisors and offerors may be reluctant to make this sort of commitment if there is any danger that a change of circumstances (or a change of mind) will lead them to regret this commitment. On the other hand, both promisors and offerors may want to induce some form of reliance by the promisee/offeree, and in some situations the promisee/offeree may be unwilling to rely unless he or she receives a binding commitment. In these situations, the promisor/offeror will have to balance (a) the potential benefits of inducing the other party to rely in the desired way against (b) the potential risk of being committed in a way that he or she may later come to regret. (This trade-off was discussed at more length in the notes on consideration, supra at pages 231–32).

2. As the Klik excerpt points out, a potential offeror's willingness to strike this balance in one direction or the other will depend in part on the ancillary legal rules determining the legal effect of an offer. For example, if offers are deemed to have unlimited duration, or if there is no quantity limit implied in an offer, or if any offeree can accept simply by posting an acceptance in the mailbox, the potential offeror may be less willing (all else equal) to make the commitment represented by an offer. However, all else will rarely be equal, for the increased risk that a broad commitment poses to a potential offeror will generally be at least partly offset by the increased benefit that a broad commitment would represent to a potential offeree. For example, reliance on an offer that may be revoked at any time before the acceptance reaches the offeror is somewhat risky to a potential offeree, so offerees may be more willing to rely on offers that are deemed to have unlimited duration or which can be accepted simply by posting a letter in a mailbox.

In short, the net effect of a rule extending an offer's duration (or extending the offeror's potential commitment in any other respect) depends on the same balance between the offeror's desire to induce reliance and the offeror's desire to remain free of potentially risky commitments. This suggests that the balance discussed in the Klik excerpt (and in the earlier excerpt by Eisenberg) may be useful in analyzing the ancillary rules governing offer and acceptance, as well as in analyzing the specific question of whether a mass communication should be interpreted as an offer.

3. A potential offeror's stance toward this balance may also depend on a number of facts about the offeror's (and offeree's) respective situations. For example, some stores stock inventory in such large quantities that they can estimate their expected demand with relatively little risk of serious error. These stores may be quite willing to have their advertisements interpreted as

unlimited offers, especially if they are trying to induce significant reliance on the part of their potential customers. By contrast, a private individual advertising a single automobile for sale has much less flexibility, and therefore would be less willing to have his or her advertisement interpreted as an unlimited offer. This is why Klik suggests that any rules governing the interpretation of offers should be presumptions at most, and that these presumptions should be overcome if the facts indicate that in any particular case the balance ought to be struck differently.

However, it will usually be important to both the offeror and the offeree that they *know* whether an offer has been made—and, more generally, that they know the exact extent of the offeror's commitment. This suggests, at a minimum, that the legal presumptions should be overturned more freely in cases where both parties know the facts that would justify overcoming the presumption (for example, the fact that the offeror had only a single car to sell). It also suggests that the legal presumptions, and the conditions under which those presumptions can be overcome, should be as predictable as possible. Is it possible to bring predictability to the law of offer and acceptance if each case must turn on an assessment of the balance described above?

4. A very different approach to the law of offer and acceptance would make greater use of "formalities," as discussed in the Kennedy excerpt (supra at pages 245–49). For example, the law could adopt an irrebuttable presumption that any communication using the word "offer" would be an offer, while any communication not using that word would be a mere inquiry inviting further negotiations. (The presumption would also have to specify the ancillary details of the offer—e.g., "any communication using the word 'offer' represents an offer of *unlimited* duration, which can be accepted simply by posting the acceptance.") Faced with such a rule, sellers who wanted to give their customers the assurance of a definite commitment could do so by using the word "offer," while sellers who were unwilling to make such a commitment could avoid using that word. If sellers could be induced to signal their intent in such a way, neither customers nor courts would be faced with any uncertainty about how any particular seller preferred to strike the balance between encouraging customer reliance and remaining free from potentially risky commitments. In other words, such a formality would have all the virtues of the "penalty default rules" discussed by Ayres and Gertner (supra at pages 22–27).

However, such a formality would also have all the defects discussed by Kennedy and by Ayres and Gertner. Buyers and sellers who didn't know the rule, or who forgot to use the word "offer" in the precise way dictated by the rule, would not benefit from the legal certainty and might actually be misled. A mechanical application of such a rule might lead to results that would not be expected even by those who knew the rule—for example, is a letter saying "please contact me if you wish to make me an offer" a "communication using the word 'offer' " (and thus itself an offer) within the meaning of the rule? Moreover, there are many different kinds of offers that sellers might wish to make: revocable and irrevocable offers, offers of varying duration, and offers inviting various means of acceptance. It seems unlikely that a separate formality could be devised for every possible kind of offer. It seems even less likely that buyers and sellers could be expected to remember all of the different formalities such a system would entail.

5. The proper interpretation of offers (and of acceptances) has received relatively little theoretical analysis in recent years. Some of the reasons for this gap are discussed in Avery Katz, The Strategic Structure of Offer and Acceptance: Game Theory and the Law of Contract Formation, 89 Michigan Law Review 215, 218–23 (1990). Katz suggests that there may be a scholarly consensus that the proper method of communicating an offer or an acceptance is whatever method is defined by the prevailing linguistic conventions. On this view—a view with which Katz disagrees—the only role for scholarship is to identify what those conventions are, and the only role for law is to protect people who adhere to those conventions.

The normal meaning of many sentences can be called "conventional" because they depend on a prior understanding or *convention* about how certain words will be used. Just as there is no logical reason why a green stoplight signals "go" while a red one signals "stop," there is no logical reason why the legal relations created by an offer are represented by the English word "offer" (and not by some other word, such as "cat"). There is, however, a widely shared understanding about the literal meaning of the word "offer," just as there is a widely shared understanding about the significance of red and green stoplights. There are also widely shared understandings about the practical significance of some sentences—often called *idioms,* or idiomatic usages—whose significance is commonly understood to extend beyond their literal meaning (e.g., "I am authorized to make you this offer").

These purely conventional meanings (whether literal or idiomatic) can be thought of as a sort of preestablished code. For example, the prevailing code assigns the expression "I hereby make you this offer," the meaning that corresponds to the legal concept of an offer. Such sentences cannot be interpreted (or decoded) by anybody who does not know the preestablished code—for example, the English language and its modern idioms. More important, anybody who *does* know the code can decode the normal meaning of those sentences without knowing anything else—for example, without knowing any of the factual circumstances that might bear on the speaker's willingness to make a potentially risky commitment. In this respect, "conventional meanings" such as these can be thought of as a kind of formality. That is, anybody who says "I hereby make you an offer" will normally be treated as having made a legal offer without any judicial inquiry into whether such a commitment would have been sensible for someone in that party's position.

6. One problem addressed in the Tiersma excerpt (and recognized by many other philosophers of language) is that many inferences and implications have not been defined in advance by a code. For example, there is probably no preexisting code telling the duke's servant that the statement "there's a draft in here" really means "please close the window." Depending on the context, the same statement by the duke could also mean "please close

the door," "please fetch the queen a shawl," or even "this is where I want you to hang the new wind chimes." Philosophers of language often refer to these as the *pragmatic* meanings (as opposed to the *semantic* meaning) of the duke's utterance. Rather than relying on a set of predefined conventions or codes, the duke's servant must rely on his knowledge of the duke's likely purposes, and of the possible ways in which those purposes might be carried out. In other words, this servant must know a good deal more than the code establishing the literal and idiomatic meanings of the duke's sentence. (The servant might adopt a penalty default rule and refuse to take any steps the duke had not spelled out with literal precision. However, such a servant probably would not remain employed very long.)

Courts face similar issues in their interpretation of offers and acceptances. If the courts are unwilling to adopt penalty default rules to force parties to spell out their intentions with literal precision, the courts' interpretations must then rely on the courts' ability to ascertain the parties' likely purposes. Viewed from this perspective, the Klik excerpt (and the earlier excerpt by Eisenberg) can be read as an attempt to spell out the parties' likely purposes in cases involving commercial advertisements.

7. It is generally accepted that contract law must make use of default rules to spell out the ancillary obligations governing parties who have entered into a contract but who have not spelled out every last detail of their respective obligations. It may be more controversial to suggest that contract law may use default rules to govern the process of contract formation. Applying default rules to the contract formation process could be seen as imposing an obligation—at the very least, an obligation to specify otherwise if they do not want to be so governed—on people who have not yet agreed to enter into any contract. Nevertheless, as the Klik and Tiersma excerpts both demonstrate, the law often does use default rules to govern the contract formation process.

§ 4.2.2 Unilateral Contracts

Modern Unilateral Contracts

MARK PETTIT, JR.

The most notable expansion of unilateral contract analysis has occurred in disputes between employers and employees. . . .

In *Sylvestre v. State,* six retired Minnesota state district judges brought an action seeking a declaration that the state could not reduce their retirement benefits by amending the judges' retirement statute. The Supreme Court of Minnesota characterized the retirement statute that was in effect when the plaintiffs began their judicial careers as an offer for a unilateral contract. The court stated the offer as follows: "If you will stay on the job for at least 15 years and then retire after having reached the specified retirement age, we will pay you a part of your salary for the remainder of your life." Two of the plaintiffs began work under this statute, but did not reach retirement age until after the legislature amended the statute to decrease benefits. The court held that as soon as the judges took office the state was "irrevocably bound" by the terms of the statute in effect at that time, and that the statutory amendment violated the contract clauses of the state and federal constitutions.

The contract analysis employed by the court in *Sylvestre* typifies the approach taken in many modern employment benefits cases. Courts continue to characterize an employer's benefit offer as an offer inviting a unilateral contract, despite the criticism that this concept has received. The primary reason for the popularity of unilateral contract analysis in the employment area is that the concept allows a finding of promissory liability of the employer without the necessity of finding a return promise by the employee. These cases almost never involve any explicit promise by the employee. Furthermore, courts are reluctant to infer promises from the employee's conduct. Few legal principles are more widely shared than the notion that, unless he explicitly agrees to work for a fixed term, an employee makes no promise of continued service to his employer. The judges in *Sylvestre* had no contractual obligation to continue to work for the state for fifteen years. Of course, their choice not to do so would forfeit their rights to receive pension benefits, but they would not incur any liability to the employer. . . .

Searching for Limits on the Use of Unilateral Contract

The crucial, difficult question facing the courts in the modern cases is not whether to choose unilateral or bilateral contract analysis, but whether to employ any contractual analysis at all. Judges and lawyers have been expanding contract analysis into new areas and situations. It is no coincidence that they have resorted to unilateral contract in this process. The need to find an enforceable return promise provides some limitation on the use of bilateral contract theory to impose liability. Plaintiffs pursuing a bilateral theory have to prove, and sometimes subject themselves to, their own promissory obligation. Plaintiffs using the unilateral contract device, like tort plaintiffs, have to prove only the defendant's obligation.

More often than not, the source of the defendant's obligation in modern unilateral contracts cases is an implied, rather than an express, promise. Moreover, the implied promise often takes the following form: "As long as you keep doing what you've been doing, or are about to do, I'll keep doing what I've been doing, or am about to do." In other words, the defendant's alleged promise is a promise to maintain the status quo, and the plaintiff's performance is simply continuing the status quo. The only limit on the use of unilateral contract theory is the court's willingness to find the alleged implied promise. Any assessment, therefore, of the appropriateness of modern judicial use of unilateral contract theory must focus on the promissory basis of the asserted obligation: Did the defendant really make a promise and, if so, should that promise serve as the basis for determining the existence and extent of liability?

The Promissory Basis of Unilateral Obligation

The *Second Restatement of Contracts* defines a promise as "a manifestation of intention to act or refrain from acting in a specified way, so made as to justify a promisee in understanding that a commitment has been made." As this definition suggests, whether it is appropriate to infer a promise depends on the facts of each case; general pronouncements cannot provide much assistance. Charles Fried begins his book, *Contract as Promise,* by articulating the principle underlying the promissory basis of obligation (he calls it the "promise principle") as "that principle by which persons may impose on themselves obligations where none existed before." The concept of the self-created obligation provides the key for evaluating judicial use of unilateral contract theory to impose liability. Are courts imposing liability because of words or actions of the defendant manifesting an intention to make a commitment, or are they using unilateral contract as a device to enforce obligations arising from some sense of community standards?

In most cases, of course, both the defendant's actions and community standards play a role in determining liability, and it is difficult to separate and weigh the contributions of each. If the defendant's commitment-making actions do not predominate, unilateral contract is an inappropriate rationalization that allows judges to avoid confronting their true motivations. . . .

Although judges usually are reluctant to find legally enforceable obligations of employees in employment benefit cases, they often are willing to recognize employer obligations. Sometimes employers make express promises (to maintain a certain level of pension contributions, for example) in written or oral employment agreements. Courts have little difficulty enforcing such promises, usually by saying that an employee accepted the employer's promissory offer by beginning to work or by continuing to work for the employer. In a majority of the reported decisions involving unilateral contract analysis, however, employers have not made express promises.

When courts do not find express promissory language, they are not reluctant to infer promises. If a personnel manual or statement of company policy contains a description of existing benefit plans, courts often infer a promise that the employer will maintain these plans as described, even though the manual or policy statement does not contain any language about the future existence of the plans. Sometimes the employee does not even offer evidence of "promise," beyond a past pattern of paying the benefits in question. Even in these situations, courts often are willing to find an implied promise.

Although whether it is reasonable to infer employer promises depends on the particular facts of each case, in most cases judicial recognition of an employer's implied promise to retain a benefit plan that he describes to a new or continuing employee seems justifiable. But what if the employer's description of the benefit plan (or the plan itself) expressly negates any promise? What if the plan states:

> This Plan shall not be deemed to constitute a contract between the Company and any employee or to be a consideration for, or a condition of, the employment of any employee.

Or what if the employer states that his action is "voluntary" and "is not to be deemed to create a contractual obligation"?

It is obviously much more difficult to find an implied promise by an employer when he expressly says he is making no such promise. In recent litigation, however, neither of the above clauses proved fatal to the employee's case. It seems stretching the point but still defensible to say that sometimes actions speak louder than words, and that if an employer's actions indicate that he is intentionally creating expectations or inducing reliance, a verbal disclaimer should be ineffective. For example, if every year for twenty years an employer provides a Christmas bonus equal to two weeks' wages, it might be reasonable to infer a promise to continue to pay that bonus even if there is a written bonus plan somewhere that says that the employer is making no such promise. On the other hand, if the judges are saying not only that the specific disclaimers in these particular cases were ineffective, but also that all attempts at disclaimer would necessarily be ineffective, then a promise by the employer cannot explain the decisions to impose liability. . . .

Public Policy Limitations

The previous section suggested that considerations of public policy sometimes impel judges to find implied promises despite the absence of words or actions that could be reasonably understood as manifesting an intention to induce expectations and reliance. Public policy concerns can push in the other direction as well, causing judges to deny the existence of implied promises (or to refuse enforcement of promises without denying their existence). Cases involving contractual claims by students against schools provide a good example.

In *Blatt v. University of Southern California,* a former law student employed a unilateral contract theory in an action seeking to compel his admission into a national honorary legal society. He alleged that the defendants represented to him that if he graduated in the top ten percent of his class he would be eligible for membership in the honor society. He further alleged that he ranked in the top ten percent of his class, but was refused membership because of his failure to work on the school's Law Review, a requirement which the plaintiff alleged was never mentioned to him. The court held that the complaint did not present a justiciable issue: "We hold that in the absence of allegations of sufficient facts of arbitrary or discriminatory action, membership in the Order is an honor best determined by those in the academic field without judicial interference." Although the court went on to discuss and reject the plaintiff's unilateral contract and promissory estoppel arguments on narrower grounds, the most significant aspect of the decision is the deference the court accorded the academic decisionmaking process.

Solicitude for the policies and decisions of academia continues to play an important role in the judicial disposition of claims brought by students against schools. The general trend in the school cases, however, is toward less deference—toward more public control of the actions of private schools. It seems safe to assume that the increased invocation of contract law in this area results primarily from changing views about the student-school relationship and not from increased promise-making activity by schools.

The citizen-government cases evidence a similar trend toward increased contractual analysis of relationships previously thought to be outside the realm of contract law. Although it certainly can be argued that in the last half-century government has been making more promises and creating more expectations, the primary explanation for increased judicial intervention lies in social, intellectual, and political developments and not in the "promise principle." As indicated earlier . . . judges continue to be cautious in their expansion of contract analysis in government-citizen cases because of concerns about the political and financial consequences of locking governments into long-term commitments, not because of the inability to create plausible implied-promise arguments.

These relational cases, then, seem to support those who argue that contract is becoming indistinguishable from tort because in modern contract cases, as in tort cases, the existence and extent of obligation is being determined primarily by social and political factors and not by the promise principle. It would be a mistake, however, to conclude that because the idea of the self-created obligation is insufficient to explain the results of the cases it is irrelevant. Most of the modern unilateral contract cases involve some combination of promissory and non-promissory sources of obligation.

Notes on Unilateral Contracts

1. As this excerpt points out, many of the employment cases hold (in effect) that the employer is bound to continue to offer the benefit plan, but the employees are not bound to continue working for the employer. Is this lack of mutuality of obligation undesirable? Or can it be justified on grounds similar to those offered in the excerpts by Eisenberg (supra at pages 224–40) and Klik (supra at pages 257–62)?

It might be argued that employers would be willing to commit themselves to continue the benefit plan in order to induce their employees to rely in some desired way (e.g., by continuing on the job). Such a commitment would be risky to the employers, of course, for conditions might change so as to make continuation of the benefits unprofitable. On the other hand, without such a commitment the employers might have a harder time getting employees to work for them, or they might have to offer a higher salary (to make up for asking the employees to take the risk that the benefit plan might be discontinued). Viewed in these terms, the question is whether the riskiness of the commitment is greater or less than what the employer would have to pay to get a similar level of employment without making such a commitment.

2. The Pettit excerpt also suggests that in suits against schools and universities, the courts have been deferential to the schools' need for flexibility and have therefore been unwilling to treat schools as having bound themselves by offering a unilateral contract. What is the distinction between the school cases and the employment cases? Is the risk or cost of a commitment to schools any greater than the risk or cost of a commitment to employers? Or are students more willing to rely even without a contractual commitment than employees are?

3. Asking whether an employer made a unilateral commitment is very similar to asking whether the employer's words and actions should be interpreted as a *promise*. As the Feinman excerpt pointed out (supra at pages 232–38), deciding whether one party's words or actions should be interpreted as a promise is also a key issue in many promissory estoppel cases. Indeed, many cases involving unilateral promises by employers are decided on promissory estoppel grounds rather than on a finding of a unilateral contract. The relationship between these two doctrines is discussed in a portion of the Pettit article not reprinted here. Pettit, Modern Unilateral Contracts, 63 Boston University Law Review 551, 591–93 (1983). See also Peter Meijes Tiersma, Reassessing Unilateral Contracts: The Role of Offer, Acceptance and Promise, 26 U.C. Davis Law Review 1 (1992).

4. The Pettit excerpt raises questions concerning the boundary between contract and tort. That is, are the obligations imposed in the cases Pettit discusses better regarded as "contractual" obligations that are assumed voluntarily by the offeror, or as "tort" or "social" obligations that are imposed by law in order to promote some other social policy? (See the earlier discussion of this question in the notes on reliance and promissory estoppel, supra at pages 243–44. A similar issue is raised in the excerpt by Clare Dalton, infra at pages 273–76.)

The answer to this question could depend on whether the imposition of unilateral contract liability is merely a default rule, which employers (or other promisors) could avoid by appropriate language. For example, if most employers would have been willing to make a binding commitment to their employees in the circumstances described by Pettit, liability would then be reflected in the way that most employers would have wanted their behavior interpreted. It could then be argued that the resulting obligation was entirely contractual.

Indeed, even if most employers would not have wanted their behavior interpreted as a unilateral commitment, such a commitment could still be called "contractual" as long as employers were free to escape liability by clearly disclaiming any commitment, using language of the sort quoted by Pettit (supra at page 269). In that event, the decisions imposing liability on employers who failed to include any such disclaimer would represent a penalty default rule, designed to encourage the use of explicit disclaimers in order to alert employees to their lack of legal protection.

On the other hand, the Pettit excerpt also notes that courts often refuse to give effect to such disclaimers. Such holdings are particularly likely when the disclaimers were contradicted by other statements by the employer's representatives, or when the disclaimers appeared in fine print contracts that most employees were unlikely to ever read. Possible arguments for overriding explicit disclaimers such as these will be discussed in Chapter 5 in connection with the reasonable expectations and unconscionability doctrines.

§ 4.2.3 **Implied Contracts**

Economic Analysis of Law

RICHARD A. POSNER

A doctor chances on a stranger lying unconscious on the street, treats him, and later demands a fee. Has he a legal claim? The law's answer is yes. The older legal terminology spoke of an implied contract between the physician and the stranger for medical assistance. This idea has been attacked as a fiction, and modern writers prefer to base the physician's legal right on the principles of unjust enrichment. This term smacks of morality, but the cases are better explained in economic terms, and the concept of an implied contract is a useful shorthand for an economic approach; for it underscores the continuity between issues in express contracts and the issues nowadays treated under the rubric of unjust enrichment.

In the case of the doctor the costs of a voluntary transaction would be prohibitive. The cause of high transaction costs in that case is incapacity; in other cases it might be time (e.g., the stranger is conscious but bleeding profusely and there is no time to discuss terms). In such cases we must consider whether, had transaction costs not been prohibitive, the parties would have come to terms and if so what (approximately) the terms would have been. If a court can be reasonably confident both that there would have been a transaction and what its essential terms would have been (that the doctor use his best efforts and that the patient pay the doctor's normal fee for treatment of the sort rendered), it does not hesitate to write a contract between the parties after the fact.

Should it make a difference whether the defendant can prove that he was a Christian Scientist and would not, if conscious, have contracted for the doctor's services? It should not—unless in the other cases of treating unconscious people doctors are given premiums to compensate them for the risk that the unconscious person doesn't really want (and hence won't be made to pay for) their services.

But now suppose that a man stands under my window, playing the violin beautifully, and when he has finished knocks on my door and demands a fee for his efforts. Though I enjoyed his playing I nonetheless refuse to pay anything for it. The court would deny the violinist's claim for a fee—however reasonable the fee might appear to be—on the ground that, although the violinist conferred a benefit on me (and not with the intent that it be gratuitous), he did so officiously. Translated from legal into economic terminology, this means he conferred an unbargained-for benefit in circumstances where the costs of a voluntary bargain would have been low. In such cases the law insists that the voluntary route be followed—and is on firm economic grounds in doing so.

An Essay in the Deconstruction of Contract Doctrine

CLARE DALTON

In the earlier part of the nineteenth century, a will theory of contract dominated the commentary and influenced judicial discussion. Contractual obligation was seen to arise from the will of the individual. This conception of contract was compatible with (and early cases appear sympathetic to) an emphasis on subjective intent: Judges were to examine the circumstances of a case to determine whether individuals had voluntarily willed themselves into positions of obligation. In the absence of a "meeting of the minds," there was no contract. This theory paid no particular attention to the potential conflict between a subjective intention and an objective expression of that intention.

Clare Dalton, An Essay in the Deconstruction of Contract Doctrine. Copyright © 1985 by The Yale Law Journal Company. Reprinted by permission of The Yale Law Journal Company and Fred B. Rothman & Company from The Yale Law Journal, volume 94, pages 1012–18.

The idea that contractual obligation has its source in the individual will persisted into the latter part of the nineteenth century, consistent with the pervasive individualism of that time and the general incorporation into law of notions of liberal political theory. Late nineteenth-century theorists like Holmes and Williston, however, began to make clear that the proper *measure* of contractual obligation was the formal expression of the will, the will objectified. Obligation should attach, they reasoned, not according to the subjective intention of the parties, but according to a reasonable interpretation of the parties' language and conduct. Enforcement of obligation could still be viewed as a neutral facilitation of intent, despite this shift, if the parties are imagined as selecting their language and conduct as accurate and appropriate signals of their intent. Thus, even in this objectified form, the will theory of contract was equated with the absence of state regulation: The parties governed themselves; better yet, each party governed himself.

The Realists made it impossible to believe any longer that contract is private in the sense suggested by this caricature. By insisting that the starting point of contract doctrine is the state's decision to intervene in a dispute, the Realists exposed the fiction of state neutrality. As Morris Cohen argued:

> [I]n enforcing contracts, the government does not merely allow two individuals to do what they have found pleasant in their eyes. Enforcement, in fact, puts the machinery of the law in the service of one party against the other. When that is worthwhile and how that should be done are important questions of public policy.

From this vantage point, the objectivist reliance on intent as the source of contractual obligation was a blatant abdication of responsibility, a failure to address and debate the substantive public policy issues involved in decisions about when and how courts should intervene in disputes between contracting parties. . . .

The Implied Contract Story: Wrestling with the Problem of Knowledge

The implied-in-law or quasi-contract plays a crucial role in sustaining the notion that contract law is essentially private. The implied-in-law contract is portrayed as essentially non-contractual and public, in contrast to the implied-in-fact contract in which the private is dominant. In this account, the implied-in-fact contract is presented as kin to the express contract, the only difference being that the former is constituted by conduct and circumstance rather than words. An examination of how and when courts choose to impose quasi-contractual obligations, however, reveals the essential similarity between the decision and the supposedly dissimilar decision that a given situation evidences implied-in-fact contractual obligations. Thus, although the distinction between the two types of implied contracts accords with our experience—we intuitively know that being bound by one's word is different from being bound by an externally imposed obligation—the methods of legal argument used for over one hundred years to distinguish the two situations do not and cannot hold. . . .

Hertzog v. Hertzog, decided by the Pennsylvania Supreme Court in 1857, is reputedly the first American case to distinguish the quasi-contract from the implied-in-fact contract. The themes and method of analysis present in *Hertzog* still reverberate in the treatment of implied contract found in the *Second Restatement* and in modern case law.

In *Hertzog,* an adult son lived and worked with his father until his father's death, at which point the son sued the estate for compensation for services rendered. The trial judge instructed the jury that John Hertzog could recover only if an employment contract existed between father and son. Two witnesses gave testimony that could be interpreted as evidence of such an agreement: One Stamm testified that he "heard the old man say he would pay John for the labour he had done," while one Roderick swore that the father "said he intended to make John safe." The jury found for John, and the defendant appealed, successfully.

Pennsylvania Supreme Court Justice Lowrie begins the opinion by distinguishing express, implied-in-fact, and implied-in-law contracts. In advancing this categorization, Lowrie particularly criticized Blackstone for failing to distinguish the implied-in-fact from the implied-in-law contract.

Blackstone had suggested that "[i]mplied [contracts] are such as reason and justice dictate; and which, therefore, the law presumes that every man undertakes to perform." Lowrie, true to his advanced understanding of the implications of the will theory of contract, observes, "There is some looseness of thought in supposing that reason and justice ever dictate any contracts between parties, or impose such upon them. All true contracts grow out of the intentions of the parties to transactions, and are dictated only by their mutual and accordant wills." The only "contracts" that reason and justice dictate, according to Lowrie, are "*constructive* contracts" in which the contract is "mere fiction," a form adopted solely to enforce a duty independent of intention. "In one," says Lowrie, "the duty defines the contract; in the other, the contract defines the duty."

Lowrie offers this definition of quasi-contract:

> [W]henever, not our variant notions of reason and justice, but the common sense and common justice of the country, and therefore the common law or statute law, impose upon any one a duty, irrespective of contract, and allow it to be enforced by a contract remedy, [this is] a case of [quasi-]contract.

For Justice Lowrie, quasi-contract, unlike contract proper, reflects public norms. Public norms, however, require legitimation, and Lowrie offers two types—one positivist, the other dependent on natural law. The norms are "positively" binding because they are part of the body of common law or statute recognized as authoritative. They are "naturally" binding because they reflect "common sense and common justice." While Lowrie distinguishes these public obligations from obligations based on consent, he invokes consent to legitimize public norms: Consent underlies his distinction between *"variant* notions of reason and justice" and *"common* sense and common justice."

Lowrie avoids the need to devote more time and attention in *Hertzog* to quasi-contract by stating that "[i]n the present case there is no pretence of a constructive contract, but only of a proper one, either express or implied." The focus of the opinion, then, is on whether John Hertzog can demonstrate the existence of a contract by words spoken or by an account of the relationship and circumstances.

As to the express contract, Lowrie explicitly uses the parties' relationship and their circumstances to "frame" the words spoken in such a way that they become words of "non-contract" instead of contract:

> The court told the jury that a contract of hiring might be inferred from the evidence of Stamm and Roderick. Yet these witnesses add nothing to the facts already recited, except that the father told them, shortly before his death, that he intended to pay his son for his work. This is no making of a contract or admission of one; but rather the contrary. It admits that the son deserved some reward from his father, but not that he had a contract for any.

The father-son relationship clearly influences Lowrie's conclusion. *Hertzog* thus illustrates that words of intention are inconclusive until they are shaped by a judicial reading of the context in which they are uttered. Even the paradigmatically self-sufficient "express" contract, in which "the terms of the agreement are openly uttered and avowed at the time of the making," is invaded by "publicness" in its interpretation and enforcement.

In regard to the implied-in-fact contract, Lowrie says that "[t]he law ordinarily presumes or implies a contract whenever this is necessary to account for other relations found to have existed between the parties." In *Hertzog,* Lowrie's willingness to find an employment contract will therefore turn on whether the parties are related: He assumes that strangers assist one another only on the expectation of reward, whereas precisely the opposite is true of employment between intimates.

Lowrie thus bases his conclusion that no implied contract exists almost entirely upon "the customs of society" and commonly accepted notions about human nature in general and family relationships in particular. But his reliance on such customs and commonalities hopelessly undermines his distinction between contracts implied-in-fact and quasi-contracts. Lowrie's treatment of the absence of a contract proper could just as easily be read as an account of the absence of a quasi-contract. Plainly he has decided that common sense and common justice demand a finding that no contract exists here. The advantage of his contractual analysis is that it permits public considerations to be introduced as if they were private, without the elaborate scrutiny of their source and justification that a quasi-contractual analysis would require.

Notes on Implied Contracts

1. Can the distinction between the "private" aspects of contract doctrine and the "public" aspects of contract doctrine be defined precisely? The Dalton excerpt describes nineteenth-century doctrine as holding that private obligations are controlled by the parties, while public obligations are controlled by the state. But how do contract law's default rules fit into that distinction? Consider three possible interpretations:

a. As long as any party who wishes to contract around a default rule is free to do so, the default rule (and the resulting obligation) is "private." Thus, the only contract duties that are "public" are those that are implied when the parties are in no position to disclaim them (e.g., the unconscious patient discussed in the Posner excerpt) and those that are not disclaimable as a matter of law (e.g., the rule against excessively high liquidated damage clauses).

b. If the law lets parties contract around the default rule if they wish and adopts a default rule designed to reflect the obligation most contracting parties would want, the resulting obligation is still "private" because the law's only purpose is to carry out the parties' preferences. But if the law adopts a default rule chosen on some other basis—for example, to promote an equitable distribution of income, or some other goal *not held by the contracting parties*—the resulting obligation is "public," even if the parties were free to contract around the default rule had they disliked it sufficiently.

c. If the law adopts a penalty default rule, which is designed to *and succeeds in* inducing the parties to spell out the obligation they prefer, the resulting obligation is "private" because it was selected by the parties. But if the law adopts a majoritarian default rule—even a rule designed to reflect what the courts or legislature think most contracting parties would prefer—and if the parties leave that default rule in effect, the resulting obligation is "public" (even though the parties were left free to contract around the default rule) because it was selected by the courts or the legislature.

2. Under any of these definitions of the public/private distinction, what turns on whether an obligation is labeled "public" or "private"? Do we have to decide whether an obligation is properly labeled public or private in order to decide whether that obligation is a good one?

It may be more helpful to focus on the individual components of each definition without trying to apply a summary label such as "public" or "private." For example, we can ask whether the rule in question is a default rule which can be contracted around, or a mandatory rule which cannot, or whether the content of the rule was chosen to reflect what most contracting parties would want, or whether it was chosen for some other purpose, without ever having to ask whether the rule in question is "public" or "private." The Posner excerpt follows roughly this strategy and never even uses the term "public" or "private." If we were to rewrite the Posner excerpt to explicitly identify each obligation he discusses as either "public" or "private" (using any one of the definitions listed above), would that advance the excerpt's analysis?

3. Similar questions can be asked about the distinction between contracts "implied in fact" and contracts "implied in law," or about the distinction between contract and tort discussed in several of the preceding excerpts. (See, for example, the excerpt by Feinman, supra at pages 232–38, or the excerpt by Pettit, supra at pages 266–70). As the Dalton excerpt notes, "public" obligations and contracts "implied by law" sound more "tort-like," while "private" obligations and contracts "implied in fact" sound more "contractual." However, it is difficult to define precisely how these distinctions would apply to contract law's default rules. Any of the three distinctions suggested in note 1 could be applied here with "contractual" or "implied in fact" substituted for "private." Would these substitutions advance the analysis in any way?

4. Contracts implied in fact or implied by law are also closely related to the contracts that are created when one party's silence is deemed to count as acceptance of another party's offer. For recent analyses of the acceptance-by-silence doctrines, see Avery Katz, The Strategic Structure of Offer and Acceptance: Game Theory and the Law of Contract Formation, 89 Michigan Law Review 215, 249–72 (1990); and Avery Katz, Transaction Costs and the Mechanics of Legal Exchange: When Should Silence in the Face of an Offer Be Construed as Acceptance? 9 Journal of Law, Economics, and Organization 77 (1993).

§ 4.2.4 **Preliminary Negotiations**

Precontractual Liability and Preliminary Agreements

E. ALLAN FARNSWORTH

Courts have traditionally accorded parties the freedom to negotiate without risk of precontractual liability. Under the classic rules of offer and acceptance, there is no contractual liability until a contract is made by the acceptance of an offer; prior to acceptance, the offerer is free to back out by revoking the offer. This "freedom from contract" is enhanced by a judicial reluctance to read a proposal as an offer in the first place.

Furthermore, courts traditionally take a view of this precontractual period that relieves the parties of the risk of any liability, whether contractual or not, and that results in a broad "freedom of negotiation." As a general rule, a party to precontractual negotiations may break them off without liability at any time and for any reason—a change of heart, a change of circumstances, a better deal—or for no reason at all. The only cost of doing so is the loss of that party's own investment in the negotiations in terms of time, effort, and expense.

At the root of this view of the precontractual period is what I call the common law's "aleatory view of negotiations: a party that enters negotiations in the hope of the gain that will result from ultimate agreement bears the risk of whatever loss results if the other party breaks off the negotiations. That loss includes out-of-pocket costs the disappointed party has incurred, any worsening of its situation, and any opportunities that it has lost as a result of the negotiations. All is hazarded on a successful outcome of the negotiations; all is lost on failure. As an English judge expressed it, "he undertakes this work as a gamble, and its cost is part of the overhead expenses of his business which he hopes will be met out of the profits of such contracts as are made . . . ." This aleatory view of negotiations rests upon a concern that limiting the freedom of negotiation might discourage parties from entering negotiations.

If the negotiations succeed and result in ultimate agreement, a party that has behaved improperly can be deprived of the bargain on the ground of misrepresentation, duress, undue influence, or unconscionability. But if the negotiations fail because of similar behavior, courts traditionally have been reluctant to impose precontractual liability. Although a duty of fair dealing is now generally imposed on the parties to a contract, that duty is not formulated so as to extend to precontractual negotiations. The past several decades, however, have seen inroads into this traditional freedom of negotiation. . . .

Unjust Enrichment as a Basis of Liability

The duty to make restitution of benefits received during negotiations is perhaps the most fundamental ground for precontractual liability. A negotiating party may not with impunity unjustly appropriate such benefits to its own use. To prevent such unjust enrichment, the law imposes liability measured by the injured party's restitution interest. Claims to restitution commonly involve either ideas disclosed or services rendered during negotiations.

The clearest cases are those that involve misappropriation of ideas. Suppose, for example, that the owner of a business, in the course of negotiating for the sale of the business to a prospective buyer, discloses an idea in confidence in order to enable the buyer to appraise the value of the business, and when the negotiations fail the buyer makes use of the owner's idea. The buyer may be liable for misappropriation of the owner's idea on several grounds of which restitution is only one. . . .

A party may also get restitution for services rendered, as distinguished from ideas misappropriated, during unsuccessful negotiations. But the mere circumstance that the party's services benefitted the other party does not give a claim to restitution, for under the aleatory view of negotiations, a court may treat benefit as well as loss as being at risk in the negotiations. The time, efforts, and activities expended during negotiations are, as one court put it, "the common grist of negotiations aimed toward consummation of an agreement," and "the endeavors by either side, if they fail, do not warrant a claim that one party has been unjustly enriched at the expense of the other. Each side's efforts were for the purpose of advancing its own interests." This conclusion is sometimes buttressed by the statement that it is only "*unjust* enrichment which is to be avoided" by the law of restitution.

A particularly appealing kind of claim to restitution is that presented by an architect or builder who has rendered services to a developer during the planning stages of a development, only to have the developer award the contract to another. *Hill v. Waxberg,* decided by the Ninth Circuit in 1956, is such a case. Hill asked Waxberg, a contractor, to help him prepare for the construction of a building, on the understanding that if FHA financing could be arranged, Waxberg would get the building contract. Waxberg did so, expecting to be compensated out of his profits from that contract. When the FHA issued the commitment, Hill and Waxberg were unable to agree on the contract, each blaming the other for the failure. When Hill made a contract with another, Waxberg sued for his expenditures and for the reasonable value of his services, which had been of some value to Hill. The court held that Waxberg was entitled to restitution of "the value of the benefit which was acquired," because

> something in the nature of an implied contract results where one renders services at the request of another with the expectation of pay therefor, and in the process confers a benefit on the other . . . . It makes no difference whether the pay expected is in the form of an immediate cash payment, or in the form of profits to be derived from a contract, the consummation of which would or should be anticipated by reasonable men . . . .

The Restatement (Second) of Restitution supports recovery in such a case if the defendant's conduct "appears unconscionable," and if the loss or expense does not result from "a risk fairly chargeable" to the plaintiff rather than the defendant, a qualification consistent with the common law's aleatory view.

There is scant judicial gloss on the Restatement's requirements, for few courts have entertained claims for restitution of benefits conferred during failed negotiations. Even under *Hill v. Waxberg,* the claimant's services must have been furnished at the other party's request. But a claimant who can show a request for its services can often make out a claim based on a contract implied-in-fact for the reasonable value of those services and thus avoid having to prove that the services actually benefitted the other party. For this reason claimants have usually made claims based on contracts implied-in-fact in addition to or instead of demands for restitution, and courts that have allowed recovery have generally favored the former ground. . . .

Misrepresentation as a Basis of Liability

Misrepresentation, another fundamental basis of precontractual liability, has been no more popular with claimants than restitution has been. A negotiating party may not with impunity fraudulently misrepresent its intention to come to terms. Such an assertion is one of fact—of a state of mind—and if fraudulent, it may be actionable in tort. But few such actions have been brought.

Markov v. ABC Transfer & Storage Co. is a rare exception. There, a lessor of warehouse facilities intentionally misrepresented to the lessee that it intended to renew the existing lease for another three years, while "it was at the same time quietly negotiating for a sale of the premises." The lessor's motive was to have the premises occupied during the negotiations for their sale and to be assured that, should those negotiations fail, the lease could be renewed. When, only a few weeks before the lease expired, the purchaser gave the startled lessee a notice to vacate the premises, the lessee claimed damages from the lessor for fraud. The Supreme Court of Washington concluded that the lessor had fraudulently promised "to renew the lease . . . and to negotiate the amount of rentals in good faith" and explained that it is enough to show fraud "if the promise is made without care or concern whether it will be kept."

The court upheld an award of damages based on the lessee's reliance losses. The award included not only extra expense incurred by the lessee as a result of its precipitous move to another warehouse, but also profits lost to the lessee when its principal customer left because no preparations had been made for that move. The court thus counted in the lessee's reliance damages its lost opportunity of continuing to serve its principal customer. The lessee's problems of proof were minimized because the opportunity that was lost involved the continuation of an existing arrangement.

Implicit in the act of negotiating is a representation of a serious intent to reach agreement with the other party. The rationale of the *Markov* case therefore generally applies, even in the absence of any explicit representation, if a party enters into negotiations without serious intent to reach agreement. It also applies if a party, having lost that intent, continues in negotiations or fails to give prompt notice of its change of mind. It supports recovery in other situations as well. Thus, for example, a negotiating party that misrepresents its authority to act as agent for another could be held liable for reliance damages if negotiations failed when the other party discovered the misrepresentation. Indeed, it would seem that a party to failed negotiations might have a claim based on any misrepresentation, including one by nondisclosure, that upon being discovered caused the negotiations to fail. . . .

But courts have rarely applied the law of misrepresentation to failed negotiations. Perhaps this is because parties are rarely tempted to misrepresent their intent. While it is a common negotiating technique for a party to over-state its willingness to break off negotiations, a party would seldom have reason to overstate its eagerness to negotiate. One possible reason is suggested by *Markov*. In other less likely scenarios, a party might negotiate with a competitor in order to prevent it from taking advantage of another deal, or with an old customer in order to avoid telling it that its business was no longer wanted. Few other reasons suggest themselves. And even when such misrepresentations are made, the injured party may be discouraged by the difficulty of proving fraudulent intent or—unaware of the possibility of counting lost opportunities—of showing substantial loss. . . .

Specific Promise as a Basis of Liability

A more common basis for precontractual liability would seem to be the specific promises that one party makes to another in order to interest the other party in the negotiations. In the past two decades, courts have come to recognize that liability may be based on such a promise. A negotiating party may not with impunity break a promise made during negotiations if the other party has relied on it.

This liability has been developed largely in cases arising out of negotiations between franchisors and prospective franchisees in which prospective franchisees have recovered reliance damages. In 1965 the Supreme Court of Wisconsin, in *Hoffman v. Red Owl Stores*, went beyond the earlier franchise cases and held Red Owl Stores, a supermarket franchisor, liable to Hoffman, a prospective franchisee, as a result of a promise that the $18,000 cash that he could contribute would suffice to obtain a franchise. Extended negotiations followed, during which Hoffman sold his bakery, bought and sold a small grocery store, moved to another town, and made a down payment on a lot. Finally, the negotiations collapsed when Red Owl demanded a substantially larger contribution and Hoffman refused. The court concluded that "the promise necessary to sustain a cause of action for promissory estoppel" does not have to "embrace all essential details of a proposed transaction . . . so as to be the equivalent of an offer." . . .

The case affirms that liability based on a specific promise has a place in the law of precontractual liability, even though the promise falls short of being an offer. Indeed, it can be argued that the case for liability is stronger when there is no offer, for there is nothing for the promisee to accept as an alternative to acting in reliance on the promise. Nevertheless, during the two decades since *Red Owl* was handed down, its influence has been more marked in the law reviews than in the law reports. In part, the small number of cases in which claimants have sought recovery may be due to a fear that even courts that allow recovery will impose a heavy burden on the claimant to show that a promise was made and reliance on the promise was justified in the context of the negotiations. But it seems as likely that it is due to an unawareness by lawyers of the possibility of counting lost opportunities in reliance damages, so as to give a claim that is substantial enough to justify litigation. Whatever the reason, the impact of *Red Owl* has not lived up to its promise. . . .

General Obligation as a Basis of Liability

Some scholarly writers have generalized from the cases decided on the grounds of misrepresentation and specific promise to argue that a general obligation of fair dealing may arise out of the negotiations themselves, at least if the disappointed party has been led to believe that success is in prospect. Thus Summers wrote that if courts follow *Red Owl,* "it will no longer be possible for one party to scuttle contract negotiations with impunity when the other has been induced to rely to his detriment on the prospect that the negotiations will succeed." American courts, however, have been unreceptive to these arguments and have declined to find a general obligation that would preclude a party from breaking off negotiations, even when success was in prospect. Their reluctance to do so is supported by the formulation of a general duty of good faith and fair dealing in both the Uniform Commercial Code and the Restatement (Second) of Contracts that, at least by negative implication, does not extend to negotiations.

European courts have been more willing than American ones to accept scholarly proposals for precontractual liability based on a general obligation of fair dealing. But even in Europe it is difficult to find cases that actually impose precontractual liability where an American court would clearly not do so on other grounds. In 1861 the German jurist Rudolf von Jhering advanced the thesis that parties to precontractual negotiations are contractually bound to observe the "necessary diligentia." A party who commits a breach of this contractual obligation—a *culpa in contrahendo* (fault in contractual negotiation)—is therefore liable for reliance damages. Jhering himself was concerned primarily with such problems as the effect of *culpa* on contracts concluded by mistake and never applied his thesis to situations involving failed negotiations. It was a French scholar, Raymond Saleilles, who in 1907 advanced the view that after parties have entered into negotiations, both must act in good faith and neither can break off the negotiations "arbitrarily" without compensating the other for reliance damages. . . .

It is perhaps not surprising that American courts have rarely been asked to hold that a general obligation of fair dealing arises out of the negotiations themselves when they have reached a point where one of the parties has relied on a successful outcome. Often the reason may be that, as suggested earlier, the disappointed parties to negotiations are unaware of the possibility of a generous measure of precontractual liability. But it may also be that they are not greatly dissatisfied with the common law's aleatory view of negotiations and recognize that the few claims that arise are fairly treated under the existing grounds of restitution, misrepresentation, and specific promise. As long as these grounds are not often invoked and have not been pushed to their limits, there will be little pressure to add a general obligation of fair dealing.

There is ample justification for judicial reluctance to impose a general obligation of fair dealing on parties to precontractual negotiations. The common law's aleatory view of negotiations well suits a society that does not regard itself as having an interest in the outcome of the negotiations. The negotiation of an ordinary contract differs in this way from the negotiation of a collective bargaining agreement, in which society sees itself as having an interest in preventing labor strife. Although it is in society's interest to provide a regime under which the parties are free to negotiate ordinary contracts, the outcome of any particular negotiation is a matter of indifference.

There is no reason to believe that imposition of a general obligation of fair dealing would improve the regime under which such negotiations take place. The difficulty of determining a point in the negotiations at which the obligation of fair dealing arises would create uncertainty. An obligation of fair dealing might have an undesirable chilling effect, discouraging parties from entering into negotiations if chances of success were slight. The obligation might also have an undesirable accelerating effect, increasing the pressure on parties to bring negotiations to a final if hasty conclusion. With no clear advantages to counter these disadvantages there is little reason to abandon the present aleatory view.

Notes on Preliminary Negotiations

1. As the Farnsworth excerpt notes, the case of *Hoffman v. Red Owl* has been much discussed in the law reviews, even if it has not been much followed in the courts. The possibility that there might be a general obligation to negotiate in good faith has also received a good deal of scholarly attention. Recent analyses include Geoffrey Cauchi, The Protection of the Reliance Interest and Anticipated Contracts Which Fail to Materialize, 19 University of Western Ontario Law Review 237 (1981); Randy E. Barnett and Mary E. Becker, Beyond Reliance: Promissory Estoppel, Contract Formalities, and Misrepresentations, 15 Hofstra Law Review 443 (1987); Richard Craswell, Performance, Reliance, and One-Sided Information, 18 Journal of Legal Studies 365 (1989); and Juliet P. Kostritsky, Bargaining with Uncertainty, Moral Hazard, and Sunk Costs: A Default Rule for Precontractual Negotiations, 44 Hastings Law Journal 621 (1993). The classic article on this topic is Friedrich Kessler and Edith Fine, *Culpa in Contrahendo,* Bargaining in Good Faith, and Freedom of Contract: A Comparative Study, 77 Harvard Law Review 401 (1964).

2. The "aleatory view" discussed in the Farnsworth excerpt holds that no negotiating party will ever want to be bound to the other, unless and until one party's offer has been accepted (at which point both parties have been bound). Is there any reason to think this is an accurate description of *most* negotiating parties?

As a counterexample, consider the analysis in the excerpts by Eisenberg (supra at pages 224–30) and Klik (supra at pages 257–62), each of which suggested reasons why one party might want to be bound even though the other was not. Many of the considerations identified in those excerpts could also be applied to the situations discussed by Farnsworth. For example, the Red Owl company might want Mr. Hoffman to begin preparing for a franchise right away, so that the franchise (if and when one was awarded) would be profitable sooner. However, Mr. Hoffman might not be willing to begin such extensive preparations unless he had a binding commitment from Red Owl. Moreover, Red Owl might have believed it was very likely that the negotiations would end successfully, thus reducing the risk they would subsequently regret such a commitment. In such a situation, the Red Owl company might well have been willing to make some kind of commitment to Mr. Hoffman, especially if that was the only way they could get him to rely.

3. Should courts try to determine, on a case-by-case basis, which negotiating parties would be willing to make some sort of commitment and which would not? Or would this result in too unpredictable a legal rule?

In arguing against a generalized duty of fair dealing, the Farnsworth excerpt argues (in its concluding paragraph) that it would be hard for courts to determine the point in the negotiations at which one party would be obliged not to back out of the negotiations without a good reason. Is this any harder than it is for courts to decide when a contract ought to be implied in the cases discussed in the Dalton excerpt (supra at pages 273–76)? Or to decide when an employer has made a promise that an employee can accept without being bound in return in the cases discussed in the Pettit excerpt (supra at pages 266–70)? For that matter, is it any harder than it is for courts to decide when the retention of benefits provided by the other party is "unjust" in the restitution cases discussed in the Farnsworth excerpt?

4. In a section not reprinted here, the Farnsworth article discusses cases in which one party explicitly assumed some obligation to the other party during the negotiation process. Courts sometimes refuse to enforce such obligations on the grounds that they lack consideration (one party is bound but not the other) or that they are too indefinite to enforce. Farnsworth is critical of this judicial attitude, arguing that courts should be more willing to enforce explicit precontractual obligations. Farnsworth, Precontractual Liability and Preliminary Negotiations: Fair Dealing and Failed Negotiations, 87 Columbia Law Review at 249-69 (1987).

If parties can explicitly undertake precontractual obligations, does this mean that a party's failure to explicitly undertake such an obligation should be interpreted as evidence that no precontractual obligation was intended? That is, should the law make the absence of such an obligation the default rule, thus requiring any parties who do intend a precontractual obligation to clearly specify their intent? The law could also adopt the opposite default rule, of course, by presuming that negotiating parties *do* intend to assume precontractual commitments, thus forcing any party without such an intent to clearly say so. If most negotiating parties would not intend such a commitment, however, the "no commitment" default rule would at least have the advantage of minimizing the number of parties who were obliged to specify otherwise. Alternatively, the law could adopt a penalty default rule slanted against the more sophisticated party—the franchise company in *Hoffman v. Red Owl,* for example—in order to put the burden of specifying otherwise on the party most likely to know about this burden.

Under any default rule, though, the courts would also have to decide what combinations of words and facts were sufficient to indicate an intention to specify otherwise and overcome the default rule. Moreover, this decision would presumably have to be made on a case-by-case basis on the basis of the factors discussed above in notes 2 and 3. The need for such case-by-case interpretations would detract somewhat from any certainty that might otherwise be created by the default rule.

5

Unconscionability and Other Defenses

There are other requirements for contract formation besides those discussed in Chapter 4. In addition to a valid offer and acceptance, the formation process must be free from fraud, duress, or any other conditions that would taint the validity of all or part of the resulting contract. Defenses to contract enforcement that rest on the absence of free assent are the subject of the readings in Chapter 5.

The first essay, by Richard Epstein, discusses the traditional common-law defenses such as fraud or duress. While these defenses are not very controversial, the Epstein essay also discusses whether these defenses should be pushed beyond their narrow, common-law boundaries. For example, if (through no fault of the seller) a buyer will starve unless he or she agrees to the seller's terms, is this sufficiently like duress to make the seller's terms enforceable? This and like questions test the boundaries between the common-law defenses and more open-ended doctrines.

The remainder of the chapter deals with these more open-ended doctrines such as unconscionability and reasonable expectations. These doctrines are similar in that each sets aside part or all of the written contract and substitutes some other term instead—usually, a term that seems fairer to the court according to some criterion of fairness. The doctrines thus raise questions like the following: (1) When should courts set aside the terms of the written contract? (2) If the terms of the written contract are set aside, how should courts decide what terms to supply instead? (3) Are the courts likely to do a very good job at selecting terms to supply? (4) Are sellers likely to do any better at selecting terms, given the market incentives that sellers face?

The second section of this chapter explores the reasonable expectations doctrine in articles by David Slawson and Kenneth Abraham. The third section focuses on the doctrine of unconscionability (though many of the readings in that section are relevant to the reasonable expectations doctrine as well). For convenience, the readings in the third section have been arranged according to the market failure or other procedural concern that might justify courts in setting aside some or all of the written contract. The third section thus includes two readings on market power, two more on imperfect information, three on standard form contracts, and one on buyer incompetence.

§ 5.1 **Traditional Defenses**

Unconscionability: A Critical Reappraisal

RICHARD A. EPSTEIN

The classical conception of contract at common law had as its first premise the belief that private agreements should be enforced in accordance with their terms. That premise of course was subject to important qualifications. Promises procured by fraud, duress, or undue influence were not generally enforced by the courts; and the same was true with certain exceptions of promises made by infants and incompetents. Again, agreements that had as their object illegal ends were usually not enforced, as, for example, in cases of bribes of public officials or contracts to kill third persons. Yet even after these exceptions are taken into account, there was still one ground on which the initial premise could not be challenged: the terms of private agreements could not be set aside because the court found them to be harsh, unconscionable, or unjust. The reasonableness of the terms of a private agreement was the business of the parties to that agreement. True, there were numerous cases in which the language of the contract stood in need of judicial interpretation, but once that task was done there was no place for a court to impose upon the parties its own views about their rights and duties. "Public policy" was an "unruly horse," to be mounted only in exceptional cases and then only with care.

Richard A. Epstein, Unconscionability: A Critical Reappraisal, 18 Journal of Law and Economics 293, 293-301. Copyright © 1975 by The University of Chicago. Reprinted by permission.

This general regime of freedom of contract can be defended from two points of view. One defense is utilitarian. So long as the tort law protects the interests of strangers to the agreement, its enforcement will tend to maximize the welfare of parties to it, and therefore the good of the society as a whole. The alternative defense is on libertarian grounds. One of the first functions of the law is to guarantee to individuals a sphere of influence in which they will be able to operate, without having to justify themselves to the state or to third parties: if one individual is entitled to do within the confines of the tort law what he pleases with what he owns, then two individuals who operate with those same constraints should have the same right with respect to their mutual affairs against the rest of the world. . . .

Traditional Common Law Defenses

Duress

We begin our examination of contractual limitations with duress. In its simplest form the defense of duress allows A, a promisor, to excuse himself from performance of his part of the bargain because the promisee, B, used force or the threat thereof in order to procure his consent. The defense makes perfectly good sense even in a regime that respects the freedom of contract once it is recognized that the initial distribution of rights under the tort law protects both a person's physical integrity and his private property. Duress is an improper means of obtaining A's consent because it requires him to abandon one of these two initial rights ("your money or your life") in order to protect the other. A's case is crystal clear where for example, he transfer[s] goods, under a threat of force to his person; and it is but an easy extension to the next case where force is used to procure not the transfer of goods, but the promise thereof. The defense of duress allows A, as defendant in a contract action, to vindicate *both* his initial entitlements, even though he has yielded to the force of the moment. A's consent has been given, but there is good reason to set it aside.

The issue of duress is important in another class of cases, those involving the so-called problem of the "duress of goods." Suppose that B has agreed to clean A's clothes for $10. After the work is done, B tells A that he will return the clothes only if A pays, or promises to pay, him $15. If A pays the $15, it is quite clear that he has an action to recover the $5 excess. B has put him to a choice between *his* clothes and *his* money. As in the case of duress by the threat of force, B has required A to abandon one of his rights to protect another, and the action to recover the $5 is designed to make certain that A will be able to protect them both. . . .

The defense of duress, though capable of extension, is also subject to limitations, the most important of which concerns the question of "economic duress." Suppose that B at the outset refuses to clean A's clothes unless A pays him $15, even when B's previous price had been $10. There is no doubt that A is worse off on account of B's decision to make a "take it or leave it" offer, but it would be the gravest mistake to argue that B's conduct constitutes actionable duress because it puts A to an uncomfortable choice. Indeed the case is sharply distinguishable both from the threats or use of force and from the duress of goods. In those two cases of duress, B put A to the choice between two of his entitlements. In this situation he only puts A to the choice between entitlement and desire, between A's money, which he owns, and B's services, which he desires. It is the very kind of choice involved in all exchanges. A could not complain if B decided not to make him any offer at all; why then is he entitled to complain if B decides to make him *better off* by now giving him a choice when before he had none? If A does not like B's offer, he can reject it; but to allow him to first accept the agreement and only thereafter to force B to work at a price which B finds unacceptable is to allow him to resort (with the aid of the state) to the very form of duress that on any theory is prohibited. There is no question of "dictation" of terms where B refuses to accept the terms desired by A. There is every question of dictation where A can repudiate his agreement with B and hold B to one to which B did not consent; and that element of dictation remains even if A is but a poor individual and B is a large and powerful corporation. To allow that to take place is to indeed countenance an "inequality of bargaining power" between A and B, with A having the legal advantage as he is given formal legal rights explicitly denied B. The question of duress is not that of the equality of bargaining power in a loose sense that refers to the wealth of the parties. It is the question of what means are permissible to achieve agreement. Where, as with force, the means themselves are improper, the threat to use them is improper as well; where those means are proper, so too is the threat to use them. It is a mistake to assert that the law of duress is designed to protect "freedom of the will" without specifying those things from which it should be free. "Economic duress" is not a simple generalization of the common law notions of duress; it is their repudiation. The integrity of the law of contract can be preserved only if that notion is flatly and fully rejected, and the role of duress limited to the case where one party puts the other to a choice between two of his entitlements by means, such as force or the breach of contract that in and of themselves are valid.

Fraudulent Misrepresentations

The case against fraudulent misrepresentation is easy to make out. As a moral matter, a person should not profit by his own deceit at the expense of his victim; and as a general matter, no social good can derive from the systematic production of misinformation. It is quite true that a person is in a better position to defend himself against fraud than against force, if only because he can check out the representations from his own sources or walk away from their maker without adverse legal consequences. But the carelessness of a victim does not excuse, much less justify, the perpetration of fraud. Where a promise induced by fraud is yet to be performed, the fraud is a good defense against an action for breach. Where the promise has been performed, the fraud is good reason to give the promisor the remedy of rescission, or where that is inappropriate, money damages. As with duress, the agreement is not respected because of the process of its formation. The reasonableness of the terms are of no concern to the court, and the same is true of the market position of the parties. (A monopolist can be a plaintiff in a fraud action.) The conduct of the promisee alone is sufficient to allow the promisor to repudiate the agreement.

There are strong common law limitations on the reach of fraud doctrines, similar to those applied to duress. True, it has been always possible at common law to maintain actions for concealment (as with the man who papers over cracks in the walls of a house to hide evidence of termites from a prospective purchaser). Yet, by the same token, a contract cannot be set aside on account of the simple *nondisclosure* of facts, which if known might have put the other party off the agreement that was in fact reached. This position has, of course, come under attack, as many have advocated the imposition of affirmative duties to disclose in a wide variety of contexts. Yet that course is fraught with difficulties of its own. First, it is difficult to determine as a matter of public policy what information must be disclosed because it is "material." It is most likely much cheaper and more effective to allow the parties in question to ask for the information that *they* regard as material, after which the general rules governing fraud and misrepresentation may be applied to the responses that are given. Second, disclosure requirements are always awkward because the party subject to them will have to act as a fiduciary toward someone with whom he wishes to deal at arm's length. Why must A be required to reveal to B at no cost information that he possesses no matter what its cost to him? The common law has been reluctant to impose affirmative duties to speak, just as it has been reluctant to impose affirmative duties to act. The undistinguished record of legislative disclosure laws, be it in truth-in-lending or in securities regulation, suggests that the traditional common law response to the problem was indeed a sound one, which insured that the prohibition against fraud was not by artifice allowed to swallow the basic premise in favor of freedom of contract.

Defenses of Incompetence: Infancy, Insanity, and Drunkenness

. . . [Q]uestions of infancy, insanity, drunkenness, and the like . . . [all go] to the competence of the contracting parties. With the competence thereby called into question, it becomes difficult to argue that the consent, even if given, is in the best interests of the party who has given it, or that the punctual enforcement of the agreement is likely to advance the public good. The important question is, how can we minimize the cost associated with the rules governing incompetence? These costs are of two sorts: first, enforcing contracts that should not be enforced and, second, not enforcing those that should be enforced. The rules fashioned to minimize them should, I believe, take into account three considerations. First, they should attempt to identify broad classes of individuals who in general are not able to protect their own interests in negotiation. A case-by-case analysis of incompetence is for the most part too costly to administer, and it generates too much uncertainty in all transactions. Second, the rules should be designed to allow third parties to identify persons in the protected class in order that they may steer clear of contractual arrangements with them. It is one thing to prohibit exploitation of incompetents; it is quite another to say that people must deal with them, even to their own disadvantage. Third, the rules in question should not create artificial incentives for parties to lower the level of competence they bring into the marketplace. It is dangerous to allow people to plead their own incompetence in any transaction that they wish, with the benefit of hindsight, to repudiate. . . .

Unconscionability Applied

The merits of these exceptions to the general rule in favor of contractual enforcement is one question; their proof in particular cases is quite another. The courts could place upon the defendant a burden of proving fraud, duress or incompetence, say by the preponderance of the evidence. That approach would tend to insure that each of these defenses will be established only where the facts of the case so warrant. There is, however, a cost created by putting this burden upon the defendant. Solely because he cannot meet the appropriate standard of proof, he may not be able to establish a contractual defense in a case where it in fact applies. If this last form of error results in substantial costs, then it should be appropriate to modify the rules of evidence in a manner that makes it easier for the defendant to establish fraud, duress or incompetence. It may well be that the relaxation of the standard of proof required to make out any defense will increase the number of instances where the undeserving defendant is able to defeat the plaintiff's legitimate contractual expectations. But if the costs thereby created are low, then the change in the rules of proof is justified on the grounds that it reduces the total error in enforcement, even though all error is not thereby eliminated.

The legal system has long used this rationale to justify some restrictions on the freedom of contract. The Statute of Frauds, which requires that certain kinds of agreements be put in writing, has the prevention of fraud as one of its chief objects. Yet its application necessarily insures the nonenforcement of certain untainted oral transactions. The parol evidence rule, which prohibits the use of oral evidence to vary or contradict the provisions of an "integrated" written contract, is also designed to control fraud; and it, too, frustrates the enforcement of legitimate consensual arrangements. One can attack either of these two rules on the ground that the control of fraud comes at too high a price (measured by the number of proper transactions nullified), given the alternative means of its control and prevention. But neither rule can be attacked on the ground that it is directed toward an illegitimate end.

The doctrine of unconscionability, properly conceived and applied, serves the same general end as the Statute of Frauds and the parol evidence rule. Ideally, the unconscionability doctrine protects against fraud, duress, and incompetence, without demanding specific proof of any of them. Instead of looking to a writing requirement to control against these abuses, it looks both to the subject matter of the agreements and to the social positions of the persons who enter into them. The difficult question with unconscionability is not whether it works towards a legitimate end, but whether its application comes at too great a price.

Notes on Traditional Defenses

1. The circumstances under which fraud and nondisclosure violate a buyer's rights, or otherwise invalidate a contract, are obviously related to the earlier readings on mistake and nondisclosure (supra at pages 160–74). Many of the same issues are also relevant to the remainder of this chapter.

2. Somewhat different issues are posed by cases involving "necessity" or "economic duress"—for example, the shipwrecked sailor whose only alternative to signing the rescuer's contract is to go down with his ship. In these cases, the problem is not the quality of the complaining party's information, but rather the quality of that party's alternatives to signing the offered contract. This issue is closely related to the contract modification issues discussed in the excerpt by Aivazian, Trebilcock, and Penny (supra at pages 211–19) and to the readings on market power (infra at pages 306–14).

Influential earlier discussions of duress, and its possible extension to cases of "economic duress" or "necessity," include Robert L. Hale, Bargaining, Duress, and Economic Liberty, 43 Columbia Law Review 603 (1943); John P. Dawson, Economic Duress—An Essay in Perspective, 45 Michigan Law Review 253 (1947); John Dalzell, Duress by Economic Pressure, 20 North Carolina Law Review 237 (1941) (Part I), 341 (1942) (Part 2). More recent discussions include Anthony Kronman, Contract Law and Distributive Justice, 89 Yale Law Journal 472 (1980); Peter Benson, Abstract Right and the Possibility of a Nondistributive Conception of Contract: Hegel and Contemporary Contract Theory, 10 Cardozo Law Review 1077 (1989); and Richard Craswell, Property Rules and Liability Rules in Unconscionability and Related Doctrines, 60 University of Chicago Law Review 1, 41–50 (1993).

3. The Epstein excerpt endorses a libertarian position: duress (and other forms of improper bargaining) invalidate the resulting contract if and only if they involve an actual or threatened violation of the plaintiff's *rights*. To give this position any content, however, we have to specify the rights that contracting parties possess. For example, to decide whether duress or nondisclosure invalidates a contract under this theory, we need to know whether buyers have a right to be free from physical violence, and whether they have a right to have all material facts disclosed to them.

Does the Epstein excerpt provide an adequate theory of buyers' rights? For example, does it explain why buyers have a right to be protected against fraud, but not against mere nondisclosure? Or is Epstein's objection to protecting buyers against nondisclosure based more on administrative concerns ("it is difficult to determine as a matter of public policy what information must be disclosed," supra at page 291) than on any philosophical theory of buyers' rights?

4. Is there anything inconsistent about using administrative concerns (and any other relevant costs and benefits) to define the rights buyers are deemed to possess? That is, is it possible to marry a utilitarian theory of buyers' rights to a libertarian's insistence that any contract induced by violating buyers' rights is unenforceable?

Libertarianism is usually thought of as an individualistic theory, which should resist allowing some individuals' interests to be sacrificed just to increase the total happiness in society or to serve any other aggregate utilitarian goal. Is this objection overcome if utilitarian concerns are used only to define individuals' initial rights, so long as individual rights (once defined) must then be strictly respected? But what if the individuals' rights are themselves defined in utilitarian or cost-benefit terms—for example, "a buyer has a right to the disclosure of all information whose value exceeds the cost of disclosure"? A right defined in this way could be strictly respected by invalidating any contract in which cost-effective information was not disclosed. Moreover, once buyers' rights were defined as including only the right to a cost-effective level of information, no buyer's *rights* would ever have to be sacrificed just to improve overall social utility. But if buyers' rights are defined in so obviously a utilitarian fashion, is this marriage of utilitarianism and libertarianism any different from unadulterated utilitarianism? Compare Epstein's suggestion (supra at page 289) that his theory can be defended on either libertarian or utilitarian grounds.

5. There are some contracts that will not be enforced no matter how voluntary and how informed the parties' choice may have been. A contract to become another's slave is unenforceable; so is a contract to sell one's babies, or (in some states) a contract to serve as a surrogate mother. These absolute prohibitions are difficult to reconcile with a general principle of freedom of contract. For discussions of the issues raised by such prohibitions, see Randy E. Barnett, Contract Remedies and Inalienable Rights, 4 Social Philosophy and Policy 179 (1986); Richard Epstein, Why Restrain Alienation? 85 Columbia Law Review 970 (1985); Margaret Jane Radin, Market-Inalienability, 100 Harvard Law Review 1849 (1987); Susan Rose-Ackerman, Inalienability and the Theory of Property Rights, 85 Columbia Law Review 931 (1985); Michael J. Trebilcock, The Limits of Freedom of Contract (Cambridge, Massachusetts: Harvard University Press, 1993). There is also an extensive philosophical literature on "victimless crimes," or (more generally) on protecting people from their own decisions. Much of this literature is surveyed in Joel Feinberg, Harm to Self (New York: Oxford University Press, 1984).

§ 5.2 Reasonable Expectations

Mass Contracts: Lawful Fraud in California
W. DAVID SLAWSON

Every prominent authority on contract law, from the treatises of Williston and Corbin to the leading cases of our own state, sets forth principles from which the conclusion follows that a written instrument is a contract only if it is the parties' mutual manifestation of agreement and only if it means what the parties should reasonably have expected it to mean. The standard form in the typical consumer transaction today meets neither of these requirements. It cannot possibly be given the consumer's manifestation of agreement unless the consumer is given a reasonable opportunity to read it understandingly before he chooses to buy. In fact, this opportunity is rarely given, and under the circumstances in which mass contracting occurs, it rarely could be, because normally neither the mass contractor nor the consumer is willing to spend the time. The standard form also does not usually mean what the consumer expects it to mean, because the consumer's expectations about the transaction are gained from the advertising and other sales representations of the mass contractor, not from the standard form. Any information contained in the form is usually provided, when and if the consumer has occasion to read it, *after* the transaction has been made.

In the typical insurance situation it is so easy to see that the standard-form policy is ordinarily not the contract, that what requires explanation is not this, but why the courts of this and every other state ever assumed it was. The typical purchaser of insurance is not even given a policy until weeks or months after he has agreed to pay for his insurance and the insurance has gone into effect. Imagine if, after one businessman had concluded an agreement with another, the first businessman were to receive from the other some weeks or months thereafter a statement containing terms inconsistent with the prior agreement, politely informing him that the new statement was henceforth to be the source of his rights and obligations. Would any lawyer, or for that matter, any layman, expect that the law would so permit one party unilaterally to dictate to the other the terms of their supposed "agreement"? The law of California and of every other state defines the contract as the *parties'* (not just one party's) *mutual* manifestations of agreement. Something which one party states after the agreement has been

W. David Slawson, Mass Contracts: Lawful Fraud in California, 48 Southern California Law Review 1, 11–14, 21–22 (1974). Copyright © 1974 by the University of Southern California. Reprinted with the permission of the Southern California Law Review.

reached is not mutual; it is not the manifestation of both parties; and it is not necessarily part of their agreement.

The same conclusion follows even if the standard form is delivered to the consumer before or concurrently with his making the transaction in other respects, if the seller should reasonably expect that the buyer will not read it understandingly, if at all, before he enters into the agreement. A typical insurance policy can *never* be part of the contract in a consumer situation, no matter when it is delivered, because it is written in language that a typical consumer cannot reasonably be expected to read understandingly, no matter how long he has to read it before the closing of the transaction. Likewise, even simple "warranty forms" on consumer goods such as washing machines, which sometimes could be read understandingly in the time allowed, are still not parts of the contract of purchase if the seller does not reasonably expect that the typical buyer will in fact have his attention called to the warranties and will in fact read them understandingly before he buys. Most often, such "warranty forms" are simply tucked in the package, which is not opened until the buyer reaches his home.

The conclusion that standard forms as they are ordinarily written and used today are not contracts is compelled by more than the application of principles of contract; it follows from the principles of a free society. In a free society no one is held to duties to which he has not, in some way, had the opportunity to give or withhold his consent. He has had this opportunity as to the duties which the laws of his governments impose upon him, because his governments are democratic. The traditional law of contract also provides this opportunity as to the duties which his contracts impose upon him, because the traditional law of contract characterizes as a contract only that to which both parties have given their "assent." "Assent" in a contractual context carries the same meaning as "consent." Thus, a standard form to which the buyer has not had the opportunity to give or withhold his consent cannot be treated as his contract without violating the fundamental principle that a person shall not be held to duties to which he has not had the opportunity to give or withhold his consent.

Of course, a standard form *can* be a part of a contract or can even constitute the whole contract. For example, the simple one-page forms which insurers commonly use for the purpose of listing the kinds, amounts, and prices of insurance that an insured agrees to buy, and which are designed to be filled in and signed by the insured, normally should be considered as contracts. These forms *are* understandable to the average person, and the insurer *can* reasonably assume that the average buyer will read them before he buys. But, merely by describing how a standard form *could* be a contract in a consumer situation shows how rarely, under present consumer contracting practices, it *is* a contract. . . .

Determining the Contract when the Contract is Not the Standard Form

What is the contract if it is not automatically assumed to be the standard form? It is the parties' mutual manifestations of assent to the transaction which has occurred. Sometimes, even in a modern consumer transaction, their mutual manifestations of assent will be written down. For example, the short order form by which a buyer of insurance typically states the kinds and amounts of insurance he is ordering is usually a contract. The buyer has filled it in himself and presumably has read its short, simple contents understandingly before he agreed to buy. More often, however, the contract in a consumer transaction will have to be implied from the context. The question in each case is: what was the buyer led to expect? The seller's advertising and sales methods are normally a source of expectations. The general expectations of almost all consumers as to the kind of product concerned are another important source. This is how the old law of sales and presently the Uniform Commercial Code determine the implied warranties that accompany the sale of a tangible product.

Under this approach, a case like *Stevens v. Fidelity & Casualty Co.*, where a widow recovered on her husband's airplane crash insurance, could be decided simply. Since the buyer was led to expect by the signs on the insurer's policy-dispensing machine that he was getting life insurance for his "airline trip," this expectation would be a part of his contract. The attempt by the insurance company to detract from this by saying in the policy that the coverage did not extend to certain kinds of airlines even if they were part of the trip would conflict with the contract and so would not be enforceable.

If the counsel for a seller were to argue that this approach ought to be rejected because it would lead to uncertainty, the answer should be that if it did, the seller would have only himself to blame. A seller of a product which his accompanied by a standard form can make a buyer's expectations on any point as clear as the seller wants them to be. The seller could clearly advertise what a buyer should expect, for example, or he could clearly state what the buyer was getting in a standard form which the seller took steps to insure was actually read and understood by the buyer before the buyer entered into the purchase contract.

Some may object at this point that it would be unfair to place upon the mass contractor the burden of clearing up any uncertainties in the buyer's mind, without permitting the mass contractor to clear them up by what he says in his standard form. This objection misses the point. If the standard form really were effective for clearing up the uncertainties about a transaction in the mind of the buyer before he decided to buy, the form *would be* a contract. In this event, the standard form would be the mass contractor's *and* the buyer's mutual manifestation of assent. But this is not the situation with which I am here concerned. When the standard form is *not* the contract, the reason it is not is because the buyer has *not* had a reasonable opportunity to read it understandingly before he decided to buy. To allow the mass contractor to "clear up uncertainties" by language he inserts in a standard form in this situation is not to allow him to clear them up in a potential buyer's mind before the decision to buy, but simply to allow him to dictate to the buyer how they will be resolved after the fact. It allows a mass contractor to exploit the uncertainties in a potential buyer's mind before the decision to buy by leading the buyer to think, however vaguely, that what he would be buying is something quite valuable, and then to "clear up" this expectation by telling him afterwards that he has in fact bought something worth much less.

Judge-Made Law and Judge-Made Insurance

KENNETH S. ABRAHAM

Although there were rumblings earlier, it was not until the 1960's that courts began to accept the argument that they should honor the reasonable expectations of the insured, despite the presence of policy provisions negating coverage. The argument first succeeded in connection with accident, life, and liability insurance, but was soon successfully invoked in other situations as well. By 1970, there were enough cases for Professor Robert E. Keeton to express the principle in formal terms:

> The objectively reasonable expectations of applicants and intended beneficiaries regarding the terms of insurance contracts will be honored even though painstaking study of the policy provisions would have negated those expectations.

. . . The expectations principle is best understood by examining the opinions that have developed it. Several themes common to those opinions are worth noting in advance. First, although literal application of the principle seems to require that the insured in question actually expected coverage and that his expectation was reasonable, courts have generally focused instead on whether any reasonable insured might have expected coverage. Second, the courts often have not explained the source of the insured's "reasonable" expectation of coverage. Third, almost all the cases involve ordinary consumers without a sophisticated understanding of insurance. Finally, by sometimes finding ambiguities where none apparently existed, courts have avoided explaining why an insured's expectations, even though reasonable, should override the language of the policy. An effort to understand why courts have honored an insured's expectations, therefore, must focus not only on the language of the opinions, but also on the underlying facts of each case and on the patterns that emerge from application of the principle in a variety of disputes. . . .

Although classification risks oversimplifying the interplay of factors affecting each decision, identifying patterns both organizes and enhances the value of any analysis. In what follows, the decisions in which the insurer misleads

Kenneth S. Abraham, Judge-Made Law and Judge-Made Insurance: Honoring the Reasonable Expectations of the Insured, 67 Virginia Law Review 1151, 1152–55, 1170–75, 1185–89 (1981). Copyright © 1981 by Virginia Law Review Association. Reprinted by permission.

the insured about the scope of coverage (misleading impression cases) are therefore distinguished from those in which the insurer does not act misleadingly (mandated coverage cases).

In the first group of cases, the insured, or a reasonable person in his position, reasonably could have expected coverage. Significantly, in most of these cases it turns out that such an expectation is reasonable largely because the insurer's words, conduct, or a situation for which the insurer is responsible have created that expectation. In the second group of cases, although the courts speak of the insured's "reasonable expectations," it is much less likely that ordinary insureds would have expected coverage in the situations in question. In those cases, the courts have gone beyond honoring actual or reasonable expectations. They have relied on the expectations principle when in truth they are mandating coverage where it seems desirable. The courts have engaged most frequently in this kind of activity where the omitted coverage is otherwise unavailable from other sources. In cases where such coverage is available, the courts do not find a fictional expectation in order to create coverage where in fact it does not exist. . . .

Why Honor an Insured's Expectations?

Economic Efficiency

. . . Modern economic theory postulates that because people know best what is good for them, they will allocate resources to their most efficient uses by pursuing their self interests. If people do not have adequate information about available goods and services, however, they will not be able to make the choices that are best for them.

By imposing liability on the insurer where the insured possesses inaccurate coverage information, the expectations principle theoretically should encourage the production of more accurate information and thus create a more efficient market. With such information, the insured may better assess the value of coverage offered him and may shop for coverage more completely meeting his expectations. If a particular form of coverage is unavailable but desired, promoting disclosure of its absence from standard coverage may encourage insurers to offer it.

Although the generation of increased market information is a valid goal, there may be limits on the extent to which increased information will promote efficiency, and there are certainly limits on the extent to which the expectations principle will promote increased information. First, it is unclear whether requiring the generation of increased information will, in fact, promote efficiency. The expectations decisions are efficient only if their economic benefits outweigh the costs they impose. The difficulty of evaluating these costs and benefits has resulted in sharp disagreement among economists over which approach is actually more efficient—government intervention or a "hands-off" attitude toward an imperfect market.

Second, even if requiring increased information were guaranteed to promote efficiency, it is questionable whether reliance on the expectations principle would produce a great deal more coverage information. The principle does not mandate disclosure; it merely imposes liability for failure to comply with judicial guidelines. Thus, insurers will disseminate additional information only when the risk of increased liability outweighs the costs of doing so. In deciding whether to provide additional information, an insurer will consider both the cost of anticipating the insured's expectation of coverage and the cost of dispelling it. It is important to recognize these cost restraints in analyzing the probable effectiveness of the expectations principle in generating increased information.

Finally, because estimates of these costs are likely to be very imprecise, the expectations principle may encourage the production of information generally, but it cannot dictate whether the production of that information will be efficient in a particular case. In theory, a competitive market should produce optimal amounts of information and a desirable variety of coverage alternatives, without legal intervention. If for various reasons, however, the insurance market is not fully competitive—because of a history of price regulation or the high capital requirements for entry, for example—encouraging the production of more information through an expectations principle may help make the market more competitive.

In light of these limits, several generalizations can be made about use of the expectations principle to promote the disclosure of coverage information. First, there is an important connection between the insurer's responsibility for the insured's expectations and the feasibility of dispelling inaccurate expectations. The greater the insurer's responsibility for creating the insured's expectations, the better will be its position to anticipate them. Even if the insurer is not directly responsible for the expectations, they may be so common that an insurer could easily anticipate them. . . .

Second, the fact that the insurer is responsible for the insured's expectation provides strong evidence that the conduct creating the expectation is within the insurer's control and that it has the capacity to dispel it. This does not mean, however, that such conduct can be altered inexpensively, or that accurate information can be distributed inexpensively. It may be relatively easy to change the location of a vending machine or the wording of a printed solicitation, but if the expectations are created or furthered by the entire marketing situation, the insurer may be able to dispel them only at the cost of a complete revision of its marketing practices. Even if the coverage is sold through an agent, disclaimers may be only partially effective in dispelling false expectations. . . . Third, expectations may be difficult and costly to dispel when they involve minor or highly technical components of coverage. Finally, the difficulties of a case-by-case application may render the expectations principle less effective in practice than in theory.

In sum, the expectations decisions promise only limited success in promoting the dissemination of coverage information. Although the principle may encourage alterations in insurer conduct that can be achieved easily and inexpensively, it is unlikely to effect more fundamental features of insurer behavior where such changes would be contrary to the economic interests of insurers.

Informed Assent

The additional disclosure of coverage information that the expectations principle encourages may promote informed assent as well as economic efficiency. Encouraging fully informed choice is not only a means to an economic end; it is in a sense an end in itself. Choices made in comparative ignorance are more like guesses than choices. The production of information facilitates genuine choice and thereby serves the interests of autonomy and freedom of contract, as well as economic efficiency. Legal doctrines such as informed consent and the duty to warn about the dangers of unsafe products encourage the dissemination of information to protect those interests. The expectations principle is analogous to these doctrines and serves similar interests.

As noted earlier, the effectiveness of the expectations principle in promoting disclosure is likely to be limited. The sole justification for the principle, however, does not lie in its success in promoting disclosure of additional information. Regardless of whether a decision to honor the insured's reasonable expectations affects the insurer's future behavior, such a decision acknowledges the value of informed assent by treating its absence as a circumstances of special legal concern. The decision to honor an insured's expectations thus recognizes an ideal. Despite this principled justification, however, and because the effectiveness of the expectations principle in promoting the disclosure of additional information is questionable, the principle must find justification in considerations that reach beyond economic efficiency and informed assent. . . .

Risk Distribution

Insurance is a tool for distributing the risk and the cost of various kinds of losses among groups of risk bearers. A court's decision to honor the expectations of an insured may thus be justified as furthering this cost-spreading feature of insurance. The ultimate effect of any particular decision or decisions will depend on the kind of expectations being honored and the kind of coverage involved. The results of the cases may therefore be evaluated by tracing these distributive effects and by assessing the costs and benefits of the increased risk sharing achieved by the expectations principle in particular situations.

In this regard, there is little basis for drawing an initial distinction between the misleading impression and the mandated coverage cases. Whether the absence of coverage is unexpected, both unexpected and unfair, or merely undesirable has little relevance to the desirability of mitigating the effects of an otherwise uninsured loss. But the expectations decisions may have beneficial effects in addition to the immediate cost spreading that results from invoking the expectations principle. An increase in the flow of information and, consequently, in market efficiency may result in a greater variety of coverage options or in coverage for previously uninsured activities.

On the other hand, the expectations decisions may have other, perhaps undesirable, distributive effects. By upsetting private risk-distribution decisions, the expectations principle may reallocate wealth. Because purchasers of insurance have entered a risk-sharing arrangement voluntarily, it is perhaps more legitimate to ask that they share additional risks than almost any other contracting parties. Even given this tacit consent to a limited amount of wealth redistribution, however, questions remain whether the final distribution is equitable and whether courts are well suited to fashion it.

The increased possibility of adverse selection is the first long-term distributive effect of the expectations principle that should be noted. For example, the flight insurance cases may increase the number of charter fliers who purchase coverage, the temporary life insurance cases may increase the number of applicants for insurance who have a better-than-average chance of dying during the application period, and so forth. If this pressures insurers to raise premiums, then other insureds will have to pay higher premiums than their own loss potential would otherwise require. Whether this is appropriate depends on many factors, including the nature of the characteristic that distinguishes the two groups.

Other things being equal, it would seem fairer to pool risks where insureds have no control over risk-increasing characteristics than where they do. For example, some people prefer not to fly on unscheduled or charter airlines because of their safety records; others willingly take the increased risk of flying on such carriers. Decisions like . . . *Stevens [v. Fidelity & Casualty Co.]* tend to force these two groups, perhaps unfairly, to share the risk of mishaps on such flights. Similarly, the duty-to-defend cases pool two distinct groups: those who are sued for committing intentional torts and those who are not. To the extent that the former groups has some control over the activities that give rise to this litigation, it might well be thought unfair to ask the latter group to help "subsidize" these activities.

In contrast, many of the decisions could have distributive effects that seem intuitively acceptable. In the accident insurance cases, the extension of coverage to all losses generally understood as accidental spreads the risk of having a natural handicap or physical infirmity to all those in the insurance pool. This appears to be fair, because the odds are good that everyone has infirmities that could combine with external forces to cause "accidental" injury. Even in connection with these cases, however, questions may be raised about the distributive effects of honoring expectations of coverage. These questions stem not from the fact that these effects are necessarily inequitable, but from the improbability of achieving a consensus as to exactly what constitutes risk-sharing fairness in a given situation. When this is so, the most that can be said for the rearrangement achieved by judicial intervention is that it is "not inequitable." In such cases, a satisfactory justification for the judicial use of the expectations principle to promote risk spreading alone does not seem available.

Moreover, a second possible byproduct of the risk spreading entailed in the expectations decisions is that they will limit the insured's freedom of choice by involuntarily increasing the scope of the coverage he purchases. Any increase in a policy's package of insurance protection will often increase its price. Those who would prefer the narrower but cheaper coverage will then be forced to accept more insurance than they want in order to obtain the coverage they need. Where having insurance coverage is optional, some people will choose not to buy it at all—the increase in cost caused by the expectations principle will have priced them out of the market. Where the coverage is effectively mandatory—automobile liability or fire insurance on mortgaged real estate—insureds will have to give up other noninsurance goods in order to buy the mandated but overly broad coverage.

This analysis demonstrates the importance of tracing the distributive effects of the expectations decisions where distributive concerns are a significant factor in the decision. It is deceptively easy to find an insured's expectations "reasonable" and require that they be honored in order to mitigate the consequences of the insured's unfortunate loss. It is more difficult, but very important, to realize that the award of compensation in such cases may have more profound effects than a simple reduction of the insurer's surplus. Saving one insured from catastrophe may not simply spread a neutral risk to all insureds; it may spread it in ways that a court sensitive to the consequences of its actions would find disturbing. Because of the potential wealth redistribution effects of the expectations principle, the degree of fault in an insurer's behavior may be a much more reliable ground for application of the principle than the desire to achieve an appropriate distribution of risks.

Notes on Reasonable Expectations

1. Although the Abraham excerpt deals entirely with insurance contracts and the Slawson excerpt uses insurance contracts as its primary example, the doctrine of "reasonable expectations" has been applied to other contracts besides contracts of insurance. For the recent history of the doctrine, tracing its growth beyond insurance contracts, see W. David Slawson, The New Meaning of Contract: The Transformation of Contract Law by Standard Forms, 46 University of Pittsburgh Law Review 21 (1984).

2. Doctrinally speaking, the doctrine of reasonable expectations assigns no legal effect to the contested portion of the seller's printed form and treats the contract between the parties as consisting of the buyer's reasonable expectations rather than the language of the printed form. As the Abraham excerpt notes, similar results have often been reached by accepting the seller's printed form as the legal contract between the parties, but finding an "ambiguity" in that form and interpreting it so that it happens to coincide with the buyer's reasonable expectations.

Similar results can also be reached under the doctrine of unconscionability, which accepts the seller's printed form as the legal contract between the

parties, but then strikes down as unconscionable those portions of the form
that conflict with the buyer's reasonable expectations. Indeed, very similar
questions about the enforceability of standard form contracts are discussed
below in the excerpts by Todd Rakoff (infra at pages 324–28) and Arthur Leff
(infra at pages 329–32). Many of the other readings in this chapter are also
relevant here.

3. How should courts determine what it is reasonable for buyers to expect? One possible approach would emphasize the normative component of
the phrase "reasonable expectations" and use the expectations that buyers
ought to have. However, this approach requires some normative theory to tell
us what expectations buyers ought to hold.

The Abraham excerpts suggests an efficiency theory: courts could decide
how much insurance coverage buyers ought to receive and hold that this is the
level of coverage that *reasonable* buyers should expect. However, are courts
well-suited to decide what kinds of insurance buyers ought to receive? A
similar question was raised in the readings on impracticability, implied warranty, and the other doctrines discussed in Chapter 3, where several readings
asked whether courts were well-suited to decide which party ought to bear the
risk of various contingencies. See, in particular, the excerpts by Kull (supra at
pages 149–54) and Schwartz (supra at pages 187–93).

As insurance contracts are concerned entirely with allocating responsibility
for bearing risks, it is not surprising that similar issues should arise here. For
example, Abraham's concern with adverse selection and cross-subsidization
(discussed supra at pages 302–03) echoes similar concerns expressed in earlier
excerpts t Quillen (supra at pages 72–76) and Priest (supra at pages 174–79),
among o ers. Whenever the courts must decide how risks ought to be
allocated whether because the contract was silent and the contingency was
unforeseen, or because the contract was a standard form that did not match the
buyer's expectations—such issues will inevitably arise.

4. A second approach to defining "reasonable expectations" is to look to
what coverage buyers *actually* expect sellers to provide—perhaps by taking
marketing surveys, or by allowing expert testimony. This approach is subject
to three possible objections, though the objections are not necessarily fatal.
The first objection is a practical one: such surveys are not always very reliable.
The second objection is that buyers' actual expectations often simply do not
exist—that is, there may be no "fact" for a survey to discover—because the
risk in question is one that buyers had never thought about. Third, if the
expectations that buyers actually hold are unreasonable ones—for example, if
buyers expect sellers to pay for injuries buyers inflict on themselves—it may
not be justifiable to hold sellers accountable for those expectations.

5. Still another approach is to look initially to sellers' conduct rather than
to buyers' expectations and hold sellers liable if and only if the sellers should
have taken more pains to disabuse buyers of any potentially inaccurate expectations. For example, in the cases that Abraham describes as "misleading
impression" cases, the insurance company could have taken steps to avoid

misleading their policyholders about the policy's coverage. Even in cases where the insurance company was silent, it might still have been possible for the company to take affirmative steps to point out the unexpected limits in the policy. If the court imposes liability on this basis, it is not requiring the insurance company to offer any particular level of coverage, but merely requiring it to be more explicit about whatever level of coverage it has chosen to provide. This might seem to eliminate the concerns discussed above about whether courts are very good at choosing an appropriate level of coverage. Under this approach, the company can offer whatever level of coverage it likes, as long as it is completely open about it.

On the other hand, the concept of "completely open" is not very easy to define. Information is often costly to communicate, especially in mass-market situations, and few would argue that customers ought to always be told *everything* about a contract before they buy. After all, most insurance contracts are complex documents, which could not be *completely* explained to lay consumers except by requiring every customer to attend a three-day seminar prior to their purchase. As a result, the goal of trying to encourage a seller to be more explicit about all the limits of its contract may be practically unworkable. As Abraham points out (in his discussion of the "mandated coverage" cases), this means that the effect of liability will be to alter the amount of coverage insurance companies are effectively required to provide, rather than altering the amount of information given to purchasers of insurance. In these cases, then, the courts must take much more care to make sure they have chosen an appropriate level of coverage. In particular, in these cases, all of Abraham's concerns regarding adverse selection, cross-subsidization, and the other distributional effects of increased insurance coverage become relevant once again—as does the question of whether courts are well-suited to decide how much coverage ought to be provided.

6. The Slawson excerpt makes an additional argument for the reasonable expectations doctrine: "In a free society no one is held to duties to which he has not, in some way, had the opportunity to give or withhold his consent" (supra at page 296). See also W. David Slawson, Standard Form Contracts and Democratic Control of Law-Making Power, 84 Harvard Law Review 529 (1971).

How does this argument apply to the case of a buyer who believes, perhaps unreasonably, that his automobile warranty does cover even damage caused by driving the car without oil? If the warranty is enforced according to its terms, thus denying this buyer compensation, is the resulting obligation being enforced without the buyer's consent? Or is it enough that the buyer *appeared* to consent by signing the seller's form in a context where most buyers (or most "reasonable" buyers) would not have expected coverage for damage due to owner negligence. For an argument that this sort of consent usually ought to be sufficient, by analogy to the objective theory of contract interpretation, see Randy E. Barnett, The Sound of Silence: Default Rules and Contractual Consent, 78 Virginia Law Review 821 (1992).

Notice, though, that this notion of "consent" requires courts to be able to decide what most buyers (or most "reasonable" buyers) would expect. As discussed in the preceding notes, it is not entirely clear how courts decide what it is reasonable to expect in any given case, especially in cases where the seller could not practicably have corrected the buyer's actual expectations. Moreover, courts' beliefs about the appropriate level of coverage seem to play an important role in determining what level of coverage it is reasonable for buyers to expect. If courts are not very good at deciding what level of coverage is appropriate, is it still fair to say that a buyer has "consented" to whatever level the court decides it is reasonable for buyers to expect?

§ 5.3 Unconscionability

§ 5.3.1 Market Power

A Reexamination of Nonsubstantive Unconscionability
ALAN SCHWARTZ

Courts and commentators have suggested that a seller's ability to exercise market power should militate against the enforcement of an agreement. This suggestion apparently rests on the premise, which empirical evidence fails to confirm, that a monopolist or an oligopolist is less responsive than a competitive seller to consumer preferences regarding the content of contracts. This section, however, demonstrates the existence of plausible conditions under which market power is unrelated to the satisfaction of consumer wants. Although resting on untested behavioral assumptions, this demonstration, when considered together with empirical evidence contradicting the conventional wisdom, strongly suggests that resources devoted to proving the existence of market power would be wasted. Therefore, courts should not recognize market power as an aspect of nonsubstantive unconscionability.

In the landmark case of *Henningsen v. Bloomfield Motors, Inc.,* the court, refusing to enforce the disclaimer in the standard automobile warranty, set forth a classic statement of the view that market power should militate against enforcing an agreement:

The gross inequality of bargaining position occupied by the consumer in the automobile industry is thus apparent. There is no competition among the car makers in the area of the express warranty. Where can the buyer go to negotiate for better protection? Such control and limitation of his remedies are inimical to the public welfare and, at the very least, call for great care by the courts to avoid injustice through application of strict common-law principles of freedom of contract.

The *Henningsen* court implied that sellers exercising market power tend to ignore buyer preferences respecting contractual terms. But this view is intuitively implausible. If a monopolist's customers prefer to have warranties rather than disclaimers, and if these customers will pay the premium for additional warranty protection, the monopolist would be irrational not to offer a warranty. Offering only a disclaimer would cost him potential profits.

A more plausible version of the *Henningsen* claim is that a monopolist or an oligopolist tends to be less responsive than a competitive seller to consumer preferences. If so, a monopolist, for example, would make fewer or narrower warranties than a competitive seller. Even as reformulated, however, the argument is suspect because of an absence of empirical support. In fact, some empirical evidence seems to support the opposite conclusion. Automobile warranties, made by sellers thought to have substantial market power, seem qualitatively and quantitatively similar to warranties issued in more competitive markets. Neither the content of warranties nor the content of other contract clauses has been shown to vary with market structure.

Consistent with the empirical evidence contradicting the second *Henningsen* rationale, a theoretical argument, outlined below, indicates plausible conditions under which market structure does not affect the extent to which a buyer's preferences for contract content are satisfied. The argument assumes that, ceteris paribus, the extent to which a buyer's preferences are satisfied varies with the quality of whatever he purchases. Contracts, like products, can be ranked by quality: a "high quality" contract has more characteristics that a consumer prefers (for example, a comprehensive warranty) or less that he dislikes (for example, an acceleration clause) than a "low quality" contract. The argument made below demonstrates that contract "quality" is not a function of market structure. Although developed, for the sake of convenience, with reference to a consumer product, the argument is equally applicable to a contract, because contracts also vary in quality.

The argument that product quality is unrelated to market structure rests on three assumptions. First, it assumes that consumer demand for quality does not vary with the amount of physical product consumed. Second, the argument assumes that all firms within a competitive industry use the same technology regardless of the level of industrywide output. Third, it assumes that the production function for a monopolist is "similar" to that of a competitive industry in the sense that the monopolist and competitive industry face the same cost-minimizing factor combinations at any level of output. The second and third assumptions, taken together, assure that considerations of

cost-minimization will lead a monopolist and a competitive industry to produce identical products, albeit at possibly different levels of output.

Given these three assumptions, a firm will produce the same level of
product quality regardless of whether the firm is a monopolist or a perfect
competitor. In light of the first assumption, a monopolist and a competitor
will face the same demand for quality even though a monopolist will tend to
produce less physical product than a perfect competitor. In light of the second
and third assumptions, a firm, regardless of market structure, will face the
same cost constraints when producing a given level of product quality. To the
extent that neither demand nor cost vary with market structure, a monopolist
and a competitor can be expected to produce goods of the same quality. And,
to the extent that a contract is analogous to a product, a monopolist and a
competitor will offer contracts of the same quality.

Although none of the three assumptions discussed above has been rigorously tested, the first assumption—that demand for quality is unrelated to
physical output—gains support in our everyday behavior. A consumer who
pays a premium to purchase a high quality product typically will pay the same
premium for the extra quality when he purchases another unit of the product.
Assuming, for example, that high-test gasoline is five cents per gallon more
expensive than regular gasoline, a consumer will pay the same five cent per
gallon premium regardless of whether he purchases one gallon or twenty
gallons. The premium consumers pay for quality apparently is independent of
the level of output.

Further study is required to prove the foregoing argument that market
structure does not affect contract quality and, therefore, does not influence
the satisfaction of consumer preferences. As previously noted, the key assumptions underlying the arguments are untested. Moreover, economists
have only recently begun to explore the relationship between market structure and quality. Their efforts lend some support to the conclusion reached
above, and they certainly have not reached a theoretical consensus supporting
the contrary conclusion that a monopolist tends to underproduce quality.

Distributive and Paternalist Motives in Contract and Tort Law

DUNCAN KENNEDY

There are two disparate strands to the rhetoric of unequal bargaining power.
First, because the parties were not equal in power there was no "real"

Duncan Kennedy, Distributive and Paternalist Motives in Contract and Tort Law, With Special
Reference to Compulsory Terms and Unequal Bargaining Power, 41 Maryland Law Review 563,
614–21 (1982). Copyright © 1982 by Duncan Kennedy. Reprinted by permission.

assent—the terms of the contract were dictated by the stronger party—and it is therefore not legitimate to sanctify the bargain by appealing to the idea of free contract. This point is purely negative: now we know there was no real assent, but what follows from that? The decision maker could respond by throwing out the agreement, leaving the parties to their non-contractual remedies, as may happen when a contract is invalidated because of fraud or duress or illegality. But the second element to this body of thought is that what the decision maker does is to rectify the balance not by throwing out the contract as a whole, but by throwing out the offending term, and reading in a term that is more favorable to the weaker party.

What this means is that there is an ambivalence in the way people assert unequal bargaining power to justify compulsory terms. On the one hand, they usually sound as though they were committed to the system of freedom of contract, and to the market system in general. If the objection to this contract is lack of "real assent," it seems to be implied that there is nothing wrong with contracts between people who are on an equal footing. On the other hand, the advocate of compulsory terms will almost always indicate a clear desire to help the weaker party at the expense of the stronger. The rhetoric of unequal bargaining power is distributionist in that it assets the desirability of intervention in favor of the weaker party in situations where there is nothing like common law fraud, duress or incapacity. . . .

Inequality of Bargaining Power

There are a number of different things people seem to be referring to when they identify a situation as involving unequal bargaining power. What I will do here is to show in very summary fashion that none of these subtests is likely to help us pick out situations in which compulsory terms will help the weak at the expense of the strong. In other words, I want to show that if you just went about finding all the situations that, according to these subtests, represent unequal bargaining power, and in each case imposed on the stronger party the duty the weaker party is asking for in the lawsuit, you would act more or less at random from the point of view of the distributive interests of the beneficiary class ("buyers"). I will take up the following tests: the industry is "public"; the terms were drafted by the seller and offered on a take-it-or-leave-it basis; the seller is a bigger entity than the buyer; the sellers have monopoly power in the relevant market; the commodity in question is a necessity; and there is a shortage which permits sellers to exploit buyers.

The public interest category seems to me of little use, simply because we no longer have a viable conception of what distinguishes the public from the private. Everything is public from some points of view; everything is private from other points of view. The label "public" is, these days, more likely to represent another name for the conclusion that it is OK to impose non-disclaimable duties than it is to represent an actually operative element in the analysis. "Publicness" seems only randomly correlated with the factors, such as the shapes of supply and demand curves, that will determine the appropriateness of using compulsory terms for distributive purposes.

Neither the drafting of the terms by the seller, nor the seller's offering them on a take-it-or-leave-it basis, nor the absolute size of the seller affects the buyer's power in any sense we should care about it. If there is competition among sellers, and good information about buyer preferences, sellers will offer whatever terms they think buyers will pay for. We cannot test the ability of buyers to influence the content of the bargain by the ability of an individual buyer to dicker with an individual seller. There may be no bargaining because bargaining is expensive, and buyers as a group are unwilling to pay the increased cost of individualized transactions. Further, in a truly competitive market, no one gets to negotiate terms with anyone else. You can't argue that market power skews bargains and then object in those very situations where, because of competition, no one gets any individualized say at all.

The notion that size, the knowledge necessary to draft the contract, and the practice of imposing take-it-or-leave-it terms give the sellers the power to dictate to buyers is belied by recent experience all over American industry, from automobiles to typewriters. It is ironic that pathbreaking cases like *Henningsen v. Bloomfield Motors* justified compulsory terms for auto warranties by emphasizing that the customer was helpless in the face of gigantic bargaining opponents. Those helpless buyers have somehow induced a proliferation of seller warranty experiments, and then more or less destroyed the auto industry by their preference for foreign cars. Detroit can no longer serve as the textbook case of seller omnipotence. While there is a powerful subjective experience of impotence for most buyers in the market for most goods, it is irrational to translate that experience, *without more,* into a preference for intervention. . . .

The concept of bargaining power is most obviously useful in understanding markets in which there are only a small number of buyers or sellers. It makes sense to say that the monopolistic seller has more bargaining power than the seller who is one of many, and that there is inequality of power when a single seller faces many buyers (or vice versa). In these cases, the test of equality is that there should be about the same degree of competition on each side of the transaction. Unfortunately, the case of market power on the side of sellers is one of those where it is least likely that compulsory terms will have a redistributive effect.

This is intuitively obvious in the case of a monopolist facing many buyers. The monopolist may refrain from passing along the whole cost of a nondisclaimable duty, since at some point he loses more on volume than he gains on price. But the mere existence of marginal sellers who flee the market will have little influence on his behavior. If there are lots of buyers who value the product highly and also think the compulsory term would be a good deal at just a little less than it costs, the monopolist will raise his price substantially. By contrast, competitive sellers may find themselves absorbing the whole cost of the term because the departure of a relatively small number of marginal buyers keeps the price down.

This is not to say that compulsory terms can never redistribute from a monopolistic seller to his buyers. The point is just that there is no guarantee that this will be the outcome. And there will be markets where the strategy of redistribution is particularly hopeful but there is no substantial market power on the seller side. If our concern is redistribution, this test seems as random as that of size/dictation. If our concern is with the other consequences of monopoly power, then the remedy of compulsory terms seems curiously inapt. Whatever effect they may have on buyer income, they have no effect at all on the other aspects of concentration.

I'm all in favor of splitting up concentrated economic powers, and in favor of public control of those it seems inappropriate to split up. If the choice is between having sellers (monopolistic or not) dictate the terms of contracts and having the courts do it, I don't have the slightest preference for sellers. If in a particular case it looks like the official decision maker will look after the interests of buyers better than the seller, then the decision maker should go ahead and do what he thinks will work. My point is that the situations in which it is desirable to impose compulsory terms can't be identified by looking for unequal bargaining power, if that term is defined either in terms of size/dictation or in terms of market power. These factors invalidate the seller's free contract claim without indicating what the appropriate response should be.

Neither the notion of a necessity nor that of a shortage is any more useful. Both are likely to be second-string answers to the question, what do you mean by unequal bargaining power? Answers that point to dictation of contract language or to seller market power undermine the legitimacy of almost all the important bargains people make in a modern economy. If we took them seriously as tests, we would impose compulsory terms just about everywhere. By contrast, the beauty of necessities and shortages as triggers for official action is that they leave most of economic life intact, that is, in the hands of sellers.

But the fact that a good is a necessity—say, food or shelter—does not mean that sellers have more power to dictate price or terms than sellers of other goods, such as luxury yachts. If there are many sellers of a necessity, none of them will be able to charge more than the going package of price and terms without losing all his buyers. If there are few sellers of a luxury item, they will have substantial power to set price and terms, even though there is not a single person who would be materially worse off if the industry went out of business altogether. Moreover, that a good is a necessity does not mean that buyers will necessarily tolerate larger price hikes without reducing demand than they would tolerate in the case of a luxury. One can go on consuming most necessities, and particularly food and housing, long after one has passed the stage at which they are "necessary." The price of vegetables may be determined at the margin not by buyers who must buy at any price or starve, but by those who are deciding between a surfeit and absolute gorging. They will go for a mere surfeit if the price gets too high. Where the marginal buyers *are* the poor, it may be particularly difficult for a monopolist to pass costs along, because poor marginal buyers may be quick to reduce their consumption when prices rise.

In other words, sellers of necessities may have more, less, or the same ability to pass on costs as sellers of other commodities. Ironically, a regime of compulsory terms will be most likely to redistribute wealth from sellers to buyers in those cases where sellers have *least,* not most, of this ability. The rhetoric of necessities will therefore give the decision maker the wrong signal if it indicates he should impose compulsory terms where seller power looks greatest to him.

References to a shortage usually mean that a change in supply or demand is causing a rise in the price of a commodity, so that people who had come to expect to buy it are threatened with having either to do without it or to reduce their consumption of other things. Shortages often give rise to demands for price control, either to limit the ability of sellers to make windfall profits on short term fluctuations, or to guarantee the position of those who were consuming the good at the old price against those who now threaten to bid the good away from them. But people also often invoke the idea of a shortage to explain why the buyer's consent to a contract term reflected unequal bargaining power, so that the court should disallow the bargain and impose a regime of nondisclaimable duties instead. It should be clear from the foregoing analysis that *without price control,* the compulsory terms may or may not work to the advantage of the buyer class. That buyers are invoking the idea of a shortage suggests that they feel pressed to accept terms they would earlier have resisted. If this is true, sellers will pass along all or close to all the cost of nondisclaimable duties, rather than absorbing some part themselves. If this happens, intervention defeats its own purpose, just as conservatives argue it will.

Notes on Market Power

1. Neither of these excerpts challenges the proposition that monopoly sellers—especially those whose monopoly is protected by some form of barrier to competition—will charge a higher price than would be charged in a competitive market. The issue addressed by the Schwartz excerpt is a slightly different one: Whether monopolists' *nonprice* terms (such as warranty provisions) will be less favorable than those that would be offered in a competitive market. The Schwartz excerpt argues that even when monopolists have the ability to charge too high a price, they will not necessarily have any incentive to offer less favorable nonprice terms. Instead, the monopolist may offer favorable nonprice terms to make the product more attractive to its customers, thereby enabling it to charge those customers an even higher price. More precisely, monopolists will have an incentive to offer more favorable terms whenever the cost of offering the more favorable term is less than the extra amount buyers are willing to pay for the more favorable term (over and above the amount they would be willing to pay for the same product without the more favorable term).

For earlier discussions of unequal bargaining power in connection with unconscionability, see M. J. Trebilcock, The Doctrine of Inequality of Bargaining Power: Post-Benthamite Economics in the House of Lords, 26 University of Toronto Law Journal 359 (1976); and Alan Schwartz, Seller Unequal Bargaining Power and the Judicial Process, 49 Indiana Law Journal 367 (1974).

2. The Schwartz excerpt in this section lists several conditions (or "assumptions") necessary to guarantee that the terms offered by a monopolist will be the same as those offered by a competitor. It then acknowledges that, in markets where these conditions do not hold, monopolists may indeed have an incentive to offer less desirable terms than would be offered if the market were perfectly competitive.

Even when the monopolist's nonprice terms would be worse than those offered in a competitive market, however, it is a separate question whether buyers would be made better off if the law required the monopolist to improve its terms. As the Kennedy excerpt points out (supra at page 312), if the law requires the monopolist to improve its terms *without controlling the monopolist's price,* or without doing anything to eliminate the monopoly, this allows the monopolist to respond by offering the required terms and charging buyers an even higher price. While some buyers might consider themselves better off paying a higher price for more favorable terms, others will regard this as a worse deal. At least as a matter of economic theory, there is no way of predicting whether buyers as a class will end up better off or worse off on balance. For a recent review of the economic theory on this point—though presented quite technically—see David Besanko, Shabtai Donnenfeld, and Lawrence J. White, Monopoly and Quality Distortion: Effects and Remedies, 102 Quarterly Journal of Economics 743 (1987).

3. If courts do refuse to enforce a monopolist's contract terms, what terms should they enforce in its place? As the Kennedy excerpt notes (supra at page 309), courts could bar the monopolist from enforcing any contractual claims at all, just as if the monopolist had been guilty of fraud or duress. However, courts typically strike down only those portions of the contract whose substance they deem unreasonable, perhaps replacing them with terms the courts deem to be more reasonable. (If courts were to take the more stringent approach, no monopolist could ever write an enforceable contract.) In doctrinal terms, if monopoly is deemed to represent a form of procedural unconscionability, courts also require some showing of substantive unconscionability before they will grant the buyer relief.

Are courts well-suited to decide what "reasonable" terms ought to be enforced in place of the monopolist's terms? For example, can courts easily decide how long a monopolist's warranty ought to last, or what components ought to be excluded from the warranty? Can they (or should they) take into account the possible effect of different terms on the price the monopolist will charge? The difficulty of identifying "reasonable" terms was raised earlier in the notes concerning the reasonable expectations doctrine (supra at pages 304–05). The same issue arises in several of the remaining readings in this chapter, including the excerpt by Arthur Leff (infra at pages 329–32). See also

Richard Craswell, Property Rules and Liability Rules in Unconscionability and Related Doctrines, 60 University of Chicago Law Review 1 (1993).

4. Both of the excerpts in this section (and all of the preceding notes) focused on the *economic* effects of monopoly power and of judicial regulation of monopolists' terms. Can it be argued that, regardless of the economic effects, the monopolist's contract still should not be enforceable because the buyer did not really have any *choice?* This argument might resurrect the distinction between necessities and other products, criticized in the Kennedy excerpt (supra at page 311).

But should a buyer's necessity be sufficient to justify judicial review of the monopolist's contract? One philosophical counterargument was presented in the discussion of "economic duress" in the Epstein excerpt (supra at page 290). That is, if the monopolist had no legal duty to sell the product to buyers on any terms, what right do the buyers have to complain if the monopolist's terms are unattractive?

A second, more practical counterargument might be based on the administrative considerations discussed in the Kennedy excerpt and in the preceding two notes. That is, if striking down the contract means that the courts themselves must decide which terms are "reasonable" and if the economic effect of the terms selected by courts is as likely to be hurt buyers as to help them, why should the courts intervene? On this view, the merits of this "freedom of choice" argument depend entirely on the answer to the question: are courts better suited than sellers to select reasonable terms on behalf of consumers? The Kennedy excerpt can be read as arguing that the presence of a seller monopoly, or the fact that the product is a necessity, is largely irrelevant to the question of whether courts are likely to be better then sellers are at selecting reasonable nonprice terms.

5. What if buyers were either unwilling or unable to pay any more money for a more favorable term? For example, even relatively wealthy consumers might be unwilling to pay for a more favorable term if they lacked sufficient information to realize how important that term might be. Poor consumers, even if they knew how important a better term was, might simply lack the money to pay enough to cover the cost of offering the term. In either of these cases, a monopolist would not necessarily have any incentive to offer the more favorable term.

On the other hand, in either of the situations just described, firms in a competitive market would also lack any incentive to offer the more favorable term. Even perfectly competitive markets only produce whatever products and contract terms buyers are willing and able to pay for. Thus, if buyers' poverty or lack of information is what prevents the market from offering the more favorable term, it makes no difference whether the market is competitive or monopolistic, so the presence of monopoly is still irrelevant. (Problems of imperfect information are discussed in more detail in the following subsection, infra at pages 315–24. Problems of poverty are discussed in another excerpt from this Schwartz article, infra at page 338–40.)

§ 5.3.2 **Imperfect Information**

Distributive and Paternalist Motives in Contract and Tort Law

DUNCAN KENNEDY

The following examples illustrate how positing the right attitudinal background can help to produce formally correct efficiency arguments that could justify almost anything. In each situation, assume that a few sellers face a large number of middle-class, well-educated consumers. Sellers are specialists in providing the commodity in question, and engage in many transactions. Consumers engage in few transactions of this kind. The deal is a complicated one: there are lots of things that can go wrong with the product, and it is being sold on credit. It pays sellers to invest a little in lawyers who master this complexity, but a *rational* consumer might conclude that it is just not worth it to do likewise.

The term that we are concerned with would make sellers liable if a particular event occurred with respect to the product. This event will not happen to all consumers; consumers have only limited knowledge of the probabilities that apply to them at the time of making the contract. There are many other unexpected disasters that might also afflict them, and they may *rationally* decide that spending even a little time on the terms of legal protection from each would be a waste of effort. Assume that buyers also suspect that sellers in general tend to lie about the contract terms they offer, and that even when they have legally assumed an obligation to buyers, they tend to resist honoring it if it falls due, so that the consumer may have to pay more in legal fees than the value of the injury if he wants to enforce a contract clause covering anything less than a major catastrophe.

Given all of the above, well-educated middle-class consumers in this market have decided that the risk of being cheated, injured or abused by sellers is one of the inevitable risks of life in our economy, and that it is not worth it to invest time or money in the obviously futile enterprise of fighting over contract language. It is more rational simply to ignore the terms and hope that you have happened on an honest seller who is more interested in building a reputation for fair dealing than in extracting the maximum possible gain from each individual transaction.

Here are three ways in which, under these circumstances, we may end up

Duncan Kennedy, Distributive and Paternalist Motives in Contract and Tort Law, With Special Reference to Compulsory Terms and Unequal Bargaining Power, 41 Maryland Law Review 563, 599–604 (1982). Copyright © 1982 by Duncan Kennedy. Reprinted by permission.

with an outcome that might be made more efficient by the imposition of a compulsory term:

The Simple Information Cost Case

All sellers are honest. Moreover, they know that if buyers fully understood the benefits to them of the term in question, they would willingly pay a price premium that would make it worth the seller's while to write the term into all contracts. Sellers also know that if they band together to carry out an educational drive, they can convince buyers to pay the premium. *But* this campaign will be expensive, given the consumer attitudes just described, with consumer distrust likely to be intensified by outrage that the sellers have all along been selling the product without a "necessary" protection. Once buyers are persuaded, the payoff on providing the term will be more than its cost, but that payoff is not great enough to give a reasonable return on the capital investment for a joint seller campaign of information. So no one provides the term. The occasional buyer who asks for it is told to take or leave the standard contract that omits it, since it is uneconomical to settle these matters case by case.

Simple Freeloader Case

This situation is like the "simple information cost" case, except that the cost of the educational campaign is small enough so that it would be well worth undertaking *if* the firm that did it could get all the new business it would generate. But the seller who does the educating finds that other sellers, who have invested nothing, are able to jump on the bandwagon, offer the new contract term, and retain their old customers plus their proportional share of the new business. Although undertaking the campaign makes sense from the point of view of the industry as a whole, it makes sense from the point of view of each seller to sit tight and let someone else do it, hoping to move in to reap unearned benefits when they do. As a result, no one undertakes the campaign.

The Case of "Competitive Pathology"

Thus far, we have two cases in which sellers do not assume added obligations in spite of their belief that consumers would pay more than they cost, if consumers were better informed. Now suppose that the practice in the industry has been to sell under a contract that does offer the term in question, although buyers know nothing about it (until disaster strikes) because they make no attempt to understand their contracts. One seller discovers that he can disclaim this liability (or simply eliminate the term) without losing any customers. He also discovers that when the disaster at which the term is directed occurs, he can stand on his new right not to compensate the particular affected buyer, without buyers as a class finding out about it or changing their attitude toward him as a seller of the product. Since he can charge the old price and sell the old quantity of a newly degraded product (degraded by subtraction of the contract term), he makes more money.

Whether he decides to reduce prices and increase his market share, or just increase salaries or dividends, his course of conduct will put pressure on his competitors. They, too, have market shares, employees and stockholders to protect, and he is threatening all three. They may believe that his disclaimer policy is immoral, or that in the long run it will reduce industry revenue, but it may well be the case that there is no way for them to persuade buyers that there is any difference between the contract they offer and the one the "chiseler" offers at a lower price. To preserve their market share, employees and stock prices, they may find themselves "forced" to disclaim as well. If the industry is competitive, there will then be a flurry of price cutting until all the gains from disclaimer have been passed on to consumers.

Perhaps the seller who began the cycle will try to preserve his ill-gotten gains by further devaluing the contract, eliminating what duties remain. Maybe his competitors try briefly to expose him through an advertising campaign, but no one believes them, and soon all sellers are forced to offer the worst contract permissible under the law. Consumers congratulate themselves on across the board cuts in the price of the commodity; the sellers as a group are slightly worse off than before, since they are selling less product. The seller who started it all may be *much* better off, if he has held on to his initial gains from each devaluation of the contract, and his success is a lesson to knowing businessmen in other fields.

False Case of Competitive Pathology

When people make a case to the decision maker for compulsory terms, using the efficiency analysis of competitive pathology, the supposed chiseler or bad apple sometimes appears to argue against the change. He will typically admit that if things were as the proponents say, there would be a case for non-disclaimable duties. But, he will protest, his ability to cut prices and increase his market share had nothing at all to do in fact with the degradation of the consumer contract. He will claim either that the consumer waiver is fully knowledgeable or that the term was not cost-justified—it raised prices by an amount greater than the amount it saved its eventual beneficiaries, supposing one accurately valued the supposed benefits. He will argue that he charges less than his rivals because he is a more efficient producer and operates on a lower profit margin per unit than they. He will point out that those arguing for compulsory terms have argued in the past for occupational licensing, that they are the least efficient, highest-priced sellers, and that their real goal is to drive him and others like him out of business by driving up their costs.

Who is right? It all depends on empirical data that no one ever seems to
have ready to hand.

Reflections on the Appeal of Efficiency Arguments

Once they have at least somewhat mastered the technical apparatus, people
just love to argue for their favorite proposals on efficiency grounds. For years,
it was mainly a liberal fad, then it fell into favor with the conservatives, and
the liberals are now trying to reappropriate it. Given a choice, almost everyone seems to prefer to cast a difficult rule change proposal in these terms
rather than in those of paternalism or redistribution. The paradox is that the
standard objection to paternalism and distribution as motives is that they are
intrinsically "subjective," "uncertain," and therefore political and controversial. What this means is that they evoke the unresolved conflicts between
groups within civil society about who deserves how much and what is the
nature of true consciousness. Regimes of compulsory terms are part of that
battle, no matter how carefully we refer to efficiency as the only motive for
imposing them, and efficiency arguments are, if anything, even more subjective, uncertain, and therefore potentially controversial than the other kinds.
Why is it that the patent manipulability of efficiency arguments does not
impair their attractiveness, while distributive and paternalist arguments,
which are actually easier to grasp and to apply, seem excessively fuzzy?

At least part of the answer, I think, is that the move to efficiency trans-
poses a conflict between groups in a civil society from the level of a dispute
about justice and truth to a dispute about *facts*—about probably unknowable
social science data that no one will ever actually try to collect but which
provides ample room for fanciful hypotheses.

Such a transposition from one level to another makes everyone, just
about, feel better about the dispute. The move from a conflict of interests or
consciousness to a conflict about facts makes it seem—quite falsely—that the
whole thing is less intense and less explosive. That it is *imaginable* that someone could one day actually produce the factual data makes it seem irrelevant
that no one is practically engaged in that task, or ever will be. In this sense,
the transposition to the cognitive level allows efficiency to act as a mediator of
the intensely contradictory feelings aroused by disputes about the shares of
groups and the validity of their choices—a mediator that defuses rather than
resolves conflict.

It seems obvious to me, but maybe I'm just wrong, that efficiency is also
attractive because it legitimates the pretensions to power of a particular subset
of the ruling class—the liberal and conservative policy analysts, most of whom
are lawyers, economists, or "planners" by profession. Efficiency analysis, like
many another mode of professional discourse, is an obscure mix of the normative and the merely descriptive; it requires training to master; it provides a
basis for an internal hierarchy of the profession that crosscuts political align-
ments. Its high value in legitimating the outcomes of group conflict in "non-

ideological" terms is the basis for the professional group's claim to special rewards and a secure niche in the good graces of the ruling class as a whole.

Proposals for Products Liability Reform
ALAN SCHWARTZ

The risk that a product will cause harm is the likelihood of a defect multiplied by the costs that the defect would create. Let p^* equal the true probability of a defect and C^* equal the true cost of a defect. Then the true expected value of the harm is $R^*=p^*C^*$. If R, p, and C are the consumers' subjective estimates of these parameters, then a consumer is "optimistic" if she believes that risks are lower than they actually are—that is, when for her $R < R^*$. Consumer optimism thus could occur when: (1) $p < p^*$ and $C < C^*$; or (2) $C > C^*$ but $p < p^*$ by enough to dominate any overestimation of defect costs; or (3) $p > p^*$ but $C < C^*$ by enough to dominate any overestimation of loss probabilities. . . .

The Decision To Insure

Consumers will underinsure if they underestimate the risk of harm. For example, let $R^*=\$50$ and $R=\$30$. A consumer would insure against the "perceived risk" at a price of $30, but no one would sell insurance at this price. Hence, the consumer will be underinsured—he will agree to too narrow a warranty. Whether consumers routinely underestimate risk levels is a complex question . . . [because of] "cross effects." A cross effect occurs when the consumer makes offsetting mistakes in p and C. For example, a consumer believes that the probability of harm is lower than it is in fact—"such an accident never could happen to me"—but also believes that accident costs are higher than they really are—"If I do have such an accident, I'll surely be crippled for life." . . . [C]onsumers could [also] overestimate probabilities but underestimate costs—"Everything I buy breaks, but with this product at least there is little danger of physical harm." Because pessimistic mistakes—"everything happens to me"—could dominate optimistic mistakes—"at least I'll never get seriously injured"—it seems impossible to say a priori that optimism in risk assessment is routine. . . . Whether consumers are systematically optimistic thus seems to pose a factual, not a theoretical, question.

Alan Schwartz, Proposals for Products Liability Reform: A Theoretical Synthesis. Copyright © 1988 by The Yale Law Journal Company. Reprinted by permission of The Yale Law Journal Company and Fred B. Rothman & Company from The Yale Law Journal, volume 97, pages 374–82.

Imperfect Information and the Product Safety Goal

Markets could produce insufficient safety for two information-related reasons. First, consumers may buy too many unsafe products. This is a "quantity effect." Second, each product that is purchased may be less safe than it should be. This is a "quality effect." Optimism of the kind just discussed could produce quantity effects. To see how, suppose that a firm has invested in accident reduction to the point where the cost of further investment would exceed the gain in increased safety. The firm's product is then optimally safe and poses a defect risk of R^*. If the firm disclaims liability and its customers perceive $R < R^*$, customers will suppose the total cost of the product to them—price plus risk—to be lower than it is in fact. As a consequence, they will buy more of the product than their better-informed selves would want. There also will be too many accidents, since the number of accidents usually varies directly with the number of products purchased. . . .

Whether unwanted quality effects will occur turns not on whether consumers are optimistic as to the current level of risk, but on how consumers' risk perceptions alter when firms alter risk levels by making (or deleting) safety improvements. As will be seen, products could be less or more safe than informed consumers would want, even if consumers perceive current risk levels correctly. . . .

In each of the following cases, suppose that consumers are optimistic as to the current risk level and firms then make a safety improvement. In the first case, consumers correctly perceive the effect of this change. Optimism respecting the risk level will then have no effect on product safety. Consumers who can correctly perceive changes in risks will buy any cost-justified change that the seller offers; optimal outcomes thus arise because consumers are perfectly informed. In the second case, consumers become more optimistic as the product becomes safer. For example, the product in the illustration poses a .1 probability of harm in its unmodified state, while consumers believe the probability is .08. A firm then reduces the probability of harm to .08, but consumers believe the probability has fallen to .05. If consumers react to safety improvements in this way, firms will overproduce safety: People who overestimate the benefits of improvements will pay for too many improvements. For people who believe it is better to have fewer accidents than many, this second case has two desirable features. First, although too many products are purchased, because consumers perceive $R < R^*$ at all levels of risk, these products are excessively safe, so one bad effect—too many sales—is at least partly offset by the less bad effect—too much safety. Further, though consumers underinsure, because $R < R^*$, they buy excessively safe products; again, one bad effect works against the other. Matters are less clear from an efficiency viewpoint since the more errors there are, the further from the "Pareto frontier" society may be.

In the third case, consumers become more pessimistic as the product becomes safer. This could occur if consumers inferred from the making of improvements that the product is dangerous. Suppose that in the example in the previous paragraph, the firm reduces the probability of harm from .1 to .08, but the consumer believes that the risk has been reduced from .08 only to .07; she is still optimistic, but by less than she previously was. Consumers who undervalue safety improvements—that is, who perceive risk reductions as being smaller than they are in fact—will underpay for safety, and firms will respond by underproducing it. Thus, if consumer optimism decreases as safety increases, not only will too many accident-causing products be purchased, but these products will be less safe than informed consumers would want.

The same three results would occur if firms could reduce defect risks by reducing the costs rather than the odds that defects will occur and if consumers could correctly perceive the odds but may misperceive the effect of reductions in costs. If consumers correctly perceive the effect of a seller's reduction in defect costs, firms will optimally reduce them; if consumers overestimate the effect of cost reductions, firms will reduce accident costs by too much; and if consumers respond pessimistically to changes in a product's ability to harm, firms will make suboptimal reductions in this ability.

When the assumption that consumers make only one mistake at a time is relaxed, matters become murkier. For example, in the second case above, consumers knew the costs of defects but overestimated the effect of a firm's reduction in the probability of harm. Firms overproduced safety as a result. This outcome would often occur when consumers mistake the probability and cost variables simultaneously, but not always. When consumers greatly underestimate the costs that defects could cause, reductions in defect probabilities that do not affect consumers' cost perceptions may cause consumers to react pessimistically. This is because consumers who believe the risk is very small will think there is little room for improvement and so may undervalue improvements that the seller makes. If many consumers perceive risks in this way, firms will underproduce safety. On the other hand, if consumers overreact to safety improvements that reduce both the odds and the costs of defects, the optimism effect is exacerbated; more safety will be produced than in the second case above, in which safety already was excessive. Given the possibility of cross effects and the ability of firms to affect odds and costs together, it becomes very difficult to say as a theoretical matter just when consumer optimism respecting changes in either variable would result in too little safety.

Notes on Imperfect Information

1. The "competitive pathology" referred to by Kennedy (supra at page 316) has also been labeled the "lemons equilibrium." For formal economic models of this phenomenon, see George A. Akerlof, The Market for "Lemons": Quality Uncertainty and the Market Mechanism, 84 Quarterly Journal of Economics 488 (1970); Michael Spence, Consumer Misperceptions, Product Failure, and Producer Liability, 44 Review of Economic Studies 561 (1977). These economic models confirm one of the central points of the

discussion of "quality effects" in the Schwartz excerpt (supra at page 320). That is, manufacturer's incentives to produce too low a quality does not depend on whether consumers over- or underestimate the level of quality, but rather on the rate at which consumer perceptions *change* in response to changes in the level of quality produced by any single manufacturer. If consumers underestimate the effect of any single manufacturer's change, such a manufacturer will have too little incentive to improve its quality. Similarly, if consumers overestimate the effect of a manufacturer's change, the manufacturer will have too much incentive to improve quality. Moreover, either of these effects will occur even if consumers have a perfectly accurate assessment of the *average* level of quality offered by sellers.

2. When standard forms are drafted on a mass basis, rather than customized for individual transactions, sellers' incentives depend on the information possessed by buyers in general, not on the information possessed by any individual buyer. Moreover, even if many individual buyers do not have very good information about the contract terms offered by sellers, other forces might still give sellers an incentive to offer reasonable terms. For example, even if most buyers never notice a seller's unreasonable terms, if those buyers who get hurt by those terms complain to their friends, and if they and their friends then stop doing business with the seller, sellers who use unreasonable terms will then be at a disadvantage vis-à-vis their competitors. The same will happen if sellers who use unreasonable terms get a reputation for having undesirable contracts.

This sort of reputation can be viewed as a form of "indirect" or "qualitative" information about sellers' practices. Generally speaking, as long as some minimum number of buyers respond in one of these ways, it will not pay for a seller to degrade its contract terms. For a further discussion of these issues, see Alan Schwartz and Louis L. Wilde, Intervening in Markets on the Basis of Imperfect Information: A Legal and Economic Analysis, 127 University of Pennsylvania Law Review 630 (1979).

On the other hand, in some markets even these indirect or reputational forces may be lacking. When sellers do not expect to stay in business very long (e.g., fly-by-night firms who can unincorporate at a moment's notice), or when sellers are not worried about retaining their customers' repeat business ("tourist traps" whose customers are all people passing through on their way to somewhere else), reputational forces may be especially weak. For further discussions of factors that may retard or accelerate the "competitive pathology" or "lemons equilibrium," see Victor P. Goldberg, Institutional Change and the Quasi-Invisible Hand, 17 Journal of Law and Economics 461 (1974); Ronald H. Coase, The Choice of the Institutional Framework: A Comment, 17 Journal of Law and Economics 493 (1974); Lewis A. Kornhauser, Unconscionability in Standard Forms, 64 California Law Review 1151 (1976). Portions of the Kornhauser article are excerpted infra at pages 332–35.

3. If courts treat the existence of a "lemons" problem as grounds for refusing to enforce a seller's contract, what should they enforce instead? For

example, should they bar the seller from enforcing any contractual claims at all, just as if the seller had been guilty of duress or fraud? If so, it is hard to see how parties in a market characterized by serious "lemons" problems could ever enter a binding contract. A more plausible response might be to only strike down "unreasonable" terms, or terms that in fact had been degraded because of the distortion of sellers' incentives.

However, one of the main points of the Kennedy excerpt is that it is not very easy to tell whether a particular term has actually been degraded. A seller can argue that its terms are in fact more efficient than its rival's, and that its customers are better off getting the product at a lower price than could be offered if the contract terms were more favorable. Are courts very well-suited to decide such questions? This is the same issue discussed earlier in connection with monopoly sellers (see page 313 supra)—and even earlier in connection with the reasonable expectations doctrine, at least in cases where the buyers' inaccurate expectations could not practicably be cured (see page 304 supra).

4. Another possible solution is to try to improve buyers' information so that the "lemons" problem no longer exists. Many regulatory statutes take this approach—especially regulations that try to provide buyers with *comparative* information (automobile miles-per-gallon ratings, cigarette tar and nicotine levels, etc.) to help buyers see just how much better or worse different sellers' offerings are. Devising a similar rating system to represent the quality of contract terms is not very easy. In any event, since comparative ratings systems work best when they are standardized across an entire industry, such disclosure remedies have not been ordered by courts engaged in case-by-case adjudication of contract disputes.

5. The Kennedy and Schwartz excerpts discuss the effect of inadequate information about contract terms or product quality. In some markets, consumers may also be inadequately informed about the prices sellers charge. For example, comparative price advertising may be subject to legal restrictions (though many of these restrictions have been invalidated as violating the first amendment's protection of commercial speech). In some markets, consumers may simply not bother to shop around to find a better price.

Inadequate shopping behavior by consumers may allow sellers to charge supracompetitive prices, but it is unlikely to distort the sellers' nonprice terms. The reason for this is much the same as that discussed earlier for sellers with monopoly power (supra at pages 312–13): if sellers are not constrained in the prices they can charge, they usually have an incentive to make their nonprice terms as attractive as possible in order to raise their prices even higher. As a result, inadequate shopping behavior usually plays a role only in those unconscionability cases that involve a challenge to the seller's price. For a further discussion of the economics of inadequate shopping by buyers, see Alan Schwartz and Louis L. Wilde, Intervening in Markets on the Basis of Imperfect Information: A Legal and Economic Analysis, 127 University of Pennsylvania Law Review 630 (1979); and Alan Schwartz and Louis L. Wilde,

Imperfect Information in Markets for Contract Terms: The Examples of Warranties and Security Interests, 69 Virginia Law Review 1387 (1983).

§ 5.3.3 Form Contracts

Contracts of Adhesion: An Essay in Reconstruction
TODD D. RAKOFF

Some contend, and many may silently believe, that a practical legal system must start with the presumption that form terms are enforceable, because otherwise the task of constructing rules to govern myriad transactions will prove so difficult as to make all judgment chaotic or unworkable. According to this argument, the most that can be allowed is some expansion in the grounds of excuse.

The assumption that ordinary contract law must form the framework for considering contracts of adhesion ultimately seems to derive from the proposition that all the terms of a transaction must be developed in the same way. Once a scheme of complete legal control of all the terms is rejected as unworkable, and once the parties provide any term, even the price term, then "of necessity" every term so provided must be given similar deference. Either the law traces the stated agreement of the parties or it substitutes its own rules, but it may not do both.

Regardless whether it would be preferable to follow the drafting party's lead on the matters covered by the form terms, it is clear that the legal system need not do so. The argument from necessity completely misconceives what might be called the "shape"—certainly the present shape—of contract law. It may be that courts will not often invent a schedule of duties from start to finish and call it a "contract," but courts in fact require the parties to provide very few terms. Those who intend to be bound can and do make enforceable agreements specifying no more than core business terms, such as price and quantity. It is when a term of this sort is omitted that the judge trots out the slogan "courts do not make contracts for the parties."

All the other obligations that arise out of a simple agreement will routinely be specified by a court. Some of these terms may be most easily thought of as tort obligations (the duty owed by a landlord to a tenant to maintain the common ways), some as incidents of special relationships (the extra care owed

Todd D. Rakoff, Contracts of Adhesion: An Essay in Reconstruction, 96 Harvard Law Review 1174, 1180–83, 1231–38 (1983). Copyright © 1983 by the Harvard Law Review Association. Reprinted by permission.

by a common carrier to a shipper), some as matters of procedure (the rule that, if recourse to law becomes necessary, each party will bear its own legal expenses). Others, such as constructive conditions of performance or frustration, or the specification of available remedies, are more commonly thought of as directly part of the law of contracts. The significant point is that, however they are labelled, these implied terms are standardized. They spring not from an indication of the parties' particular intentions, but rather from a set of background rules.

The standardization of implied terms is most visible in regulatory statutes that provide terms, or whole contracts, applicable automatically or at least in the absence of alternative specification by the parties. The phenomenon, however, is much broader. Article 2 of the Uniform Commercial Code, although not ostensibly regulatory, is in large part a catalogue of the implied terms of contracts of sale. Terms derived from the case law have also become standardized. The rule of *Hadley v. Baxendale,* for example, is no longer regarded as one based on the "tacit agreement" of the parties. Similarly, courts now frankly treat judicially provided conditions of performance as broad rules constructed from elements of trade usage, custom, and policy rather than from what can be gleaned about individual party intent. Parties are, of course, still at liberty to refrain from entering into such standardized transactions at all; but beyond that freedom, the parties' contractual power is now exercised primarily in specifying deviation from the standardized plan rather than in defining the obligation ab initio.

The existence of this pattern in "normal" contract law allows one to question the necessity of enforcing contracts of adhesion. The great majority of form terms merely furnish alternatives to terms that the legal system will provide to flesh out simply stated bargains. Indeed, many of the terms in typical form documents are specifically designed to displace clear rules of law that would otherwise govern the transaction in question. Examples include matters as disparate as clauses requiring suit to be brought within a brief period (displacing otherwise applicable statutes of limitations) and due-on-sale clauses in mortgages (displacing background law permitting the sale of real estate subject to a mortgage). Another large group of form terms attempt to give specificity to rules of law whose application to particular situations is indeterminate. Good examples are clauses setting time limits on matters that the law would require to be done in "a reasonable time," and the various forms of force majeure clauses that attempt to specify available excuses against the background of a flexible legal test of "impossibility" or "frustration." Indeed, there are clauses stipulating the entire body of substantive or procedural law that will apply; the usual choice-of-law clauses and arbitration covenants are examples. Certainly all of the provisions commonly thought to raise problems—limitations of warranties, of consequential damages, of liabilities for negligence, and of times for inspecting goods or filing proofs of loss—fall into one or another of the foregoing classes.

No one supposes that a simple agreement is fatally indefinite if it fails to specify the warranty term, or the governing law, or the allocation of the risk of negligence. Accordingly, there is no reason to think that such terms must be considered enforceable, or even prima facie valid, when they turn up in contracts of adhesion. The judicial solution may be better or worse than the drafter's formula, but it clearly is an alternative. . . .

Evaluating the Practice

. . . Llewellyn's argument—that contracts of adhesion improve upon the legal system's allocation of risks and responsibilities by embodying the judgment of commercial experts—founders on the realization that forms are not drafted from such a professional and interest-free point of view. And the assertion that form contracts yield tremendous savings in transaction costs founders on the realization that the alternative is not the bargaining-out of every deal, but rather the use of standardized yet legally implied terms. In the end, the argument seems to be based on whatever can be teased out of Williston's contention that enforcement of signed documents "rests upon the fundamental principle of the security of business transactions." This contention would appear to contain one or both of two propositions.

The first is that form documents serve a useful economic purpose in making specific, for the circumstances of a particular transaction, general rules of law. It simply saves trouble and expense for the parties to know from the start that "a reasonable time" is "ten days." This contention is, however, much less plausible than it initially seems. The assumed certainty must be discounted in two respects. First, it may not be the practice of the organization itself to adhere to the set line that its form stipulates, and thus an adherent who consults the form in time of trouble may be misled. If actual practice conforms to an unspecified "reasonable time," the possibility that "ten days" will be enforced may in reality serve to create uncertainty. Second, if the dispute goes to law, the intermittent yet time-honored practice of the judiciary to ignore apparently clear form terms or to read them against their sense greatly undercuts the position. An argument for certainty has an inherent all-or-nothing quality. Since "all" is unlikely, not much remains.

Alternatively, the claim based on "the security of business transactions" proceeds from the premise that people who need to know the terms of a transaction will discover them from inspecting the written document, to the conclusion that it is desirable to protect the reliance of such people. Of course, enforcement of the form terms merely because the document is signed does not make the salesperson and the customer more certain that their actual, face-to-face understandings will be supported by the law; the operation of the traditional doctrines very often thrusts on events an interpretation held by neither of these individuals. But if we recall that one of the juridical parties is a segmented and differentiated business firm, we see that the traditional rules do give assurance to members of the organization other than the salesperson that the document states the transaction—or at least what they need to know of it—regardless of what the salesperson actually knew, or said, or reasonably believed. The reliance that is being protected is a reliance internal to the firm; the function of the doctrine is to validate the form as a mode of internal communication. . . .

The fair scope of this argument is limited from the outset. Not all form terms are explicable primarily by reference to the structure of the firm; the drafting party, capitalizing on the perverse nature of the market involved, often tries to insulate itself from legal responsibilities for reasons unconnected to internal efficiency. Even to the extent that standard forms do aid in maintaining internal structure, we cannot blandly assume that the structures exist only when they contribute to overall efficiency. To take just the most obvious point, considerations unrelated to productivity—such as status or simple power—may dictate the structure of organizational hierarchies.

More significantly, even if we assume that the use of form contracts enhances the efficient operation of the organization, and even if we assume that this gain is returned to customers in the form of lower prices, we still have no assurance that there has been an overall gain in social welfare. The risks and responsibilities thrown on the adhering party have their own processing costs—archetypically, the transaction costs of arranging for insurance or handling an uninsured risk that eventuates. Indeed, the existence and size of these costs may be one of the reasons the legal system initially places the liability on the drafting party. There is no reason to think that the form draftsman will ever stop to consider how large these costs might be, or whether they exceed the costs entailed in having firms cover the same risks; for whatever the costs to adherents, disowning the obligation saves the draftsman's client, the firm, money. The argument may thus suggest that the legal system, in constructing legally implied rules, should consider, among other things, the institutional costs to firms of adapting to various responsibilities, but it does not justify general deference to the terms stipulated by the drafting party.

In sum, mass distribution is not dependent on the enforceability of contracts of adhesion, because standardized terms can be legally implied. Contracts of adhesion employed in a competitive market may not result in a large redistribution of wealth to drafting parties, but any gain in social welfare resulting from their use and enforcement will at best be relatively modest. The drafting party cannot be expected to assess accurately whether the value of any potential gain outweighs the costs of achieving it. The analysis of the use of adhesive contracts as a matter of the production and distribution of wealth offers no sure guide to the proper legal response; it certainly does not provide an argument for their presumed enforceability. Persuasive bases of judgment will have to be found elsewhere. . . .

"Freedom of Contract"

. . . Refusal to enforce a contract of adhesion, the courts say, trenches on freedom of contract. Implicit in the argument is an equating of the drafting organization with a live individual. For what gives value to uncoerced choice—the type of freedom that the courts have in mind—is its connection to the human being, to his growth and development, his individuation, his fulfillment by doing. But enforcement of the organization's form does not further these fundamental human values; the standard document grows out of and expresses the needs and dynamics of the organization. To see a contract of adhesion as the extension and fulfillment of the will of an individual entrepreneur, entitled to do business as he sees fit, is incongruous. To argue that such a contract represents the cooperative expression of the freedom of all or most of the individuals who comprise the organization is unpersuasive. Most commercial hierarchies are far from being sufficiently participatory to make that claim more than a reified abstraction; form documents are designed by a very few hands, often those of lawyers. Once it is recognized that contracts of adhesion arise from the matrix of organizational hierarchy, the argument for enforcement of form terms as a recognition of "freedom of contract in its usual sense is unsupportable.

Emphasis on the standard analysis also obscures the manner in which individual freedom really is at stake. A conception of contractual freedom modeled on the opposition between individual and state is inadequate in industrialized, organized, and institutionalized society. Institutions other than the state can and do dominate the individual within the framework of private law as ordinarily conceived. We are accustomed to seeing this danger in certain facets of economic life, such as the labor relationship in modern industry, and to recognizing that elimination of such domination, where it exists, is as much a fulfillment of liberty as is the limitation of governmental control. In considering the types of disputes that typically arise concerning contracts of adhesion, however, we often miss the point.

What the courts should say is that enforcing boilerplate terms trenches on the freedom of the adhering party. Form terms are imposed on the transaction in a way no individual adherent can prevent, and a major purpose and effect of such terms is to ensure that the drafting party will prevail if the dispute goes to court. The adhering party is remitted to such justice as the organization on the other side will provide. As Professor Kessler well said, the use of contracts of adhesion enables firms "to legislate in a substantially authoritarian manner without using the appearance of authoritarian forms."

Of course, the realistic alternative to the drafter's term is not a term chosen by the individual adherent, or even by adherents as a class; a solution will be imposed by the law if not by the drafter. Compared, however, to the drafters of forms, judges, legislators, and administrative officials are impartial. They fill roles that encourage them to take a broader view of the common good. Legislators, at least, are subject to popular political control—and the decisions of administrators and judges, ultimately, to legislators. If government is at all legitimate, it is legitimate for the purpose of framing generally applicable legal rules. That cannot be said of the form draftsman.

Contract as Thing

ARTHUR ALAN LEFF

Contract as Contract

What belongs to the class "contract"? The "official" definition gives some clue of the difficulty of that question by answering it with breathtaking circularity. A contract is "a promise or set of promises for the breach of which the law gives a remedy . . . ." In fact, asked generally that way, the question defies less begging answers. A lot of things are contracts, or at least a lot of them have from time to time and presently ended up being somewhat subject to contract principles. But like all classifications, "contract" has a purposive element. It is, as always, extremely doubtful that one can come up with *a* purpose, but a tight cluster is not too much to hope for. To follow Professor Kessler's lead in this (and to implode his admirable and already admirably tight summary) the common law's category "contract" was developed as a method of segregating, for a particular and predictable treatment, contemplated trading transactions between free-willed persons in an assumedly free enterprise, free market economic system. Given this relatively shadowy "aim," there arose a few class-identifying criteria which, when present in sufficient confluence, made things look more contracty than not.

These identifying criteria are, of course, in the form of gross distinctions between alternative possibilities. First of all, contracts seem to be some species of interpersonal behavior (as opposed to person-thing interactions). Second, the interpersonal behavior demanded for a contract seems to be more or less communicative (rather than directly effective, like a punch in the mouth). Moreover, the communication, to look contracty, ought to have a lot of future tense in it, and bear somewhat on the speakers' expected role in that future. Next, bargain and trade seems to smell more of contract than beneficence; there is the continual pressure to separate deals from gifts. Next is the limitedness of contract. There seems to be something significant to contract in the bordered relationship, "the deal," as opposed to more long-term, non-limit-bound interpersonal relationship like husband-wife and father-son. Last (and keep a close and critical eye on this move, for it will become very important to the argument anon), closely allied to the trade-bargain idea, is the process aura of contract. Contract seems to presuppose not only a deal, but dealing. It is the product of a joint creative effort. At least classically, the idea seems to have been that the parties combine their impulses and desires into a resulting product which is a harmonization of their initial positions. What results is

Arthur Allen Leff, Contract as Thing, 19 American University Law Review 131, 137–38, 143–44, 148–50, 155-57 (1970). Copyright © 1970 by The American University, Washington College of Law. Reprinted with the permission of The American University Law Review.

neither's will; it is somehow a combination of their desires, the product of an ad hoc vector diagram the resulting arrow of which is "the contract". . . .

Contract as Adhesion Contract

. . . The intellectual process which led to "contract of adhesion" is, at least with hindsight, and in terms of my earlier classification discussion, relatively easy to describe. Prior to creation of the adhesion-contract category, "the law" had been guilty of a common misclassification technique. Seeing that consumer transactions were communicative (rather than, say, physically coercive), mercantile (rather than charitable or donative), and bounded (rather than status-relational), and so on, "the law" continued to class them as contractual. This overlooked the fact that *these* "contracts" were not the product of a cooperative process, but the creation (essentially) of only one of the parties. In other words, "the law" was classing consumer transactions as contracts on the basis of less than all the criteria which actually shaped that particular class. The adhesion contract theorists corrected that particular error. They detected the non-process nature of some "contracts" (including consumer transactions) and thus created, so they thought, a new category, roughly speaking "that which would be a contract except that no bargaining process really shapes it." For describing such a beast the phrase "contract of adhesion" is not half bad. Its picture, such as it is, is one not of haggle or cooperative process but rather of a fly and flypaper. Thus the phrase focussed on the fact that whatever protection against overreaching and unfairness a process of cooperative creation might be supposed to give, an adhered-to document gave no such protection. Assuming that contract law is in some complex way tied up with belief in the good-maximizing powers of the market, itself based on the assumption that people know better what is of value to them than any state or other guardian, unbargainable deals were critically different from dickered deals.

But that insight got the commentators only half way. Once they had isolated the fact that there was no contracting-process protection in adhesion deals, their metaphor was exhausted. For if their new legal griffin was not an ordinary barnyard type of animal, it was still some kind of "contract," that is, an undertaken obligation which the law would ordinarily enforce. It might be an exotic animal indeed, but it was still an animal—to be fed, watered, and treated more or less like a cow. But *when* enforce these adhesion contracts? Certainly the phrase "*contract* of adhesion" doesn't push the answer "never" briskly forward. And the economics of the mass distribution of goods make that a commercially absurd answer anyway. Certainly not "always;" the whole purpose of the addition of "of adhesion" was to negative that possibility (and in effect reserve it for plain old unmodified "contract"). But if clearly not "never" or "always," then when? The new metaphor gave (and gives) not the slightest color of any answer to that question. The new metaphoric category, since carved from the older category of "contract," and left with all the smell of freedom-of-contract enforceability clinging to it, was naked of any substantive criteria for evaluating the central problem: what is a "bad" contract.

Contract as Thing

. . . There are, it seems to me, only three general strategies for regulating consumer transactions; you can focus your attention on the parties, the dealing, or the product. For current purposes, the critical strategic decision seems to be between deal control and goods control. (Let me interject here, lest you all think me mad, that I am aware that it is not writ large in the heavens that one cannot mix techniques.) Now, keeping in mind the nature, factual and legal, of the usual consumer-goods (or services) transaction, deal control is ordinarily a stupid option; it is silly to seek to shape and control the contours of a process that does not take place. That does not mean that fraud, duress (both classical and "business duress") or even "procedural unconscionability" are totally irrelevant to the consumer-goods field. It's just that such approaches are beside the point most of the time; it's like bandaging a cut on a broken leg. The normal, non-pathological consumer sale, is one in which a form is proffered by a non-monopolist selling non-life-essential goods pursuant to no huge quantum of lying, and in which that form is signed because no one is particularly interested in the sub-set of the whole package which is the contract. In such a context, how does one go about regulating the contract as a process? By facilitating more bargaining? But that is absurd (1) because, as Llewellyn saw as early as anybody, the mass product contract is complementary to, and has the same economic utility of, the mass-produced product; (2) most "objectionable" clauses of a consumer contract have only contingent, often highly contingent, importance, and no buyer not represented by a lawyer (whose job it is almost wholly to think about things which are not going to happen) is going to think much about them; and (3) even if the unrepresented consumer were interested, it is unlikely that he would have the necessary sophistication for such consideration. . . .

What would happen, however, if one pushed the straddle that "contract of adhesion" represents off the wall in the other direction and made a quantum leap to the other regulatory focus, product control? What would happen, that is, of one did actually think about consumer contracts as things?

First of all, that would open up the law's long tradition, accelerating of late, of direct, explicit governmental control of the quality and safety of products. Autos now have mandatory seatbelts, milk is bereft of its tubercles, and outright poisonous substances are barred from the marketplace. Even less reified "things," *when seen as products,* have been regulated as to quality for quite a long time. Life insurance contracts, for instance, have been in effect written by deputy insurance commissioners for years. Nor has this quality-control device always required legislative action. The initial peculiar treatment of injurious foods in sealed packages has almost ripened into the general "dangerous instrumentality" rule of the Restatement (Second) of Torts. . . .

But the limitations of any quality-control approach must be considered, however briefly. Even if one does approach a contract as a thing, it is not so terribly easy to tell when even things should be regulated. One does not solve the evil-defining problem merely by approaching it as a species of goods-quality control. Bluntly, what kinds of clauses ought to be illegal? Certainly the answer is not "harsh ones." Almost all deal breakdowns are going to have harsh effects on one of the parties, and it is certainly not clear that all harshness should be absorbed by the party able to spread the impact over all purchasers. It is not irrefutably clear that stable contract makers ought *necessarily* pick up the tab for those whose contracts break up. For among other things, this quality-control technique has important economic (and, therefore, social) costs. If carried forward with any vigor, no lawful contract could descend in "fairness" or "safety" below a certain qualitative minimum. In certain situations that would have the same effect as some building codes: the cheapest one can get is more expensive than one can afford. One is less able to trade off quality or risk for price even if one wants to and is aware of what one is doing. In brief, the government-quality-control technique is a replacement for market-set quality and price levels and mixes. Whether the fundamental market-economy postulate, that individuals most often know what they want better than a government does, is empirically sound or not, there is certainly evidence that governments tend frequently to make an unacknowledged conflation of the two wildly different concepts "what they want" and "what is good for them," with numerous repressive effects. Strict quality control is, after all, a regulatory approach the basis of which is "I don't care if you want it; you can't have it." This is highly justified when what you want is a poison, when a slip in making your own decision as to what is "good for you" is fatal. It may be justified for lesser risks when the market is, in fact, highly imperfect. As the price of an error becomes more and more merely economic (rather than organic), and the market becomes more subject to normal competitive correctness, this intrusion becomes more and more noxious. While this new metaphor may thus suggest a newly applicable frame of reference—governmental goods-quality-control—it does not without more make any of the hard decisions about when and how to use it.

Unconscionability in Standard Forms

LEWIS A. KORNHAUSER

The notion of contract as a bargaining process arose concomitantly with the doctrines of laissez-faire and of the free enterprise system. Responsibility for

Lewis A. Kornhauser, Unconscionability in Standard Forms. Copyright © 1976 by the California Law Review, Inc. Reprinted from California Law Review, volume 64, number 5 (September, 1976), pages 1152–55, 1167–69, 1179, by permission.

these doctrines has commonly been foisted upon the nineteenth century economists. In the laissez-faire market place, economic agents, unfettered by government restrictions, interacted freely. Agents met, conferred, and, acting in their own self-interest, contracted. The attribution of laissez-faire to the economists may have resulted from the coincidence of the vocabularies of the nineteenth century economist and his contemporary, the social darwinist. The language of vigorous competition and of survival of the fittest masked the economists' concern for static equilibria and for the conditions which marked them; the predominant economic question was what price equated supply and demand in an unchanging world, not how that price was reached or how it changed over time.

In any case, freedom of contract, "the inevitable counterpart of a free enterprise system," arose from this laissez-faire context. Under the influence of Langdell, Holmes, and Williston, contract law evolved into a law of formalities which, if correctly followed, resulted in a legally binding agreement. Contract law did not, and largely does not, examine the contents of the agreement but merely the manner of formation. When the state wished to regulate the substantive contents of the agreement, it looked to a substantive area of law: the law merchant, insurance law, or labor law.

Actual market behavior increasingly has exhibited deviations from the presuppositions of contract law. This has not gone unrecognized by the legal profession. From Karl Llewellyn, Friedrich Kessler, and more recent commentators, lawyers have understood that a growing proportion of agreements— one commentator has said 99 percent of all agreements—do not conform to the patterns of interaction, dickering and other bargaining required by contract law. The deviant transactions are those concluded upon standard forms. Their flaw has generally been the failure to dicker over terms and the parties' lack of mutual assent to the printed clauses. While the parties have agreed upon some things, they have not upon others. The courts must decide how much, if any, of the standard form to enforce. Presumably, the agreement would be enforceable in its entirety if the parties complied with the formalities of contract law. Consequently, legal solutions have concentrated upon means for determining what reasonable parties would have agreed to or accepted had dickering occurred. Formulations vary from a focus on unconscionability to considerations of the parties' reasonable expectations to a test based on whether any reasonable person would agree to a given clause.

Thus, the laissez-faire metaphor remains central to the way the law approaches standard form contracts. The courts continue to investigate the actions and the agreement of the parties before them, even though the Uniform Commercial Code admonishes the judiciary to examine the "commercial setting, purpose, and effect" of the agreement. This judicial emphasis on the bargaining process between the immediate parties to a contract may be misdirected. More significantly, legislative reliance on judicial resolutions of the problems prompting such inquiries may be unwarranted.

Contemporary economic models have generally abandoned the personality and exuberance of the laissez-faire world. Today both the ideal markets of economic theory and the markets actually operating in the U.S. economy are ones where most agents, though still self-interested, interact only with prices; they do not dicker over terms with other agents. The impersonality of the standard form contract conforms with this impersonal world in which the agent of contemporary theory lives. More importantly, however, failures of the imperfect markets of the real world to supply the "best" contract terms or to charge the competitive price may result from flaws which extend beyond the bargaining process between the two contracting parties. . . .

Recent Models of Economic Behavior

The neoclassical world consists of two classes of nameless agents: firms and households. Households supply factors—varying types of capital and labor— to firms and consume final output. The sale of these factors at prevailing prices yields each household an income which constrains its consumption of final goods sold at given market prices. Similarly, a firm's use of capital and labor is determined by the prevailing prices of these factors, its desire to maximize profits, and the market price of its output. Within this environment each house-hold decides the type and amount of factors it will supply and goods it will consume, while each firm must choose the amount and mix of factors to use in production and the quantity of output desired. The question naturally arises whether a set of prices exists such that all agents' actions are compatible. Such a set of prices and the resultant distribution of output among households is called an equilibrium. Equilibrium exists under ideal circumstances—production con-ditions that make it equally or more costly to produce larger outputs, consum-ers with rational preferences, and many agents all of whom act impersonally and without regard to the actions of other agents. When an equilibrium does exist, the announcement of a set of prices to all economic agents will call forth a supply of goods in each market equal to the amount demanded; all firms maximize their profits while all households, given their initial wealth, buy their most desired combinations of available goods. This notion of a competitive equilibrium, then, provides a standard against which market performance may be measured.

    In this model, the most abstract and formal of modern economics, the agents, unlike those in the laissez-faire model, do not bargain with each other over prices or over the physical characteristics of the good. Rather, agents react to prices posted upon detailed specifications of commodities. The eco-nomic "commodity" differs from that of ordinary language. It is a very well-defined entity. It includes not only a physical description but also a location and a date. Thus, to an economist, No. 2 Red Winter Wheat available in Chicago on 1 October 1976 is a different commodity than No. 2 Red Winter Wheat available in Chicago on 1 January 1977. Purchases consummated under standard form contracts better fit this paradigm than the negotiated contract model of the traditionalists. The myriad clauses in fine print represent a detailed description of the commodity to which the price, the decision vari-able of the agent, attaches. One buys not just a Chevrolet of a given model year, with various options, deliverable in two weeks from a specified dealer, but also a complex of rights and liabilities, risks and insurance outlined by the purchase order form. Certain warranties attach to the physical good; a financing plan often accompanies the automobile. The price refers to this complex of characteristics which constitute the commodity or "thing." The consumer in this abstract model of economic equilibrium does not bargain over clauses or dicker over terms. He purchases or fails to purchase.

The standard form contract, therefore, rather than being inimical to this competitive model, fits it remarkably well. If flaws exist in the form—if oppressive terms have been written in—the cause must be other than the failure to bargain or the lack of dickering over terms. To elucidate the causes of these "unconscionable" results requires an investigation of how real markets differ from ideal, competitive ones. From an understanding of these causes may develop curative measures different from those ordered by courts focusing on the bargaining process of the parties before them. . . .

The economic models developed briefly above have one major lesson: that standard form contracts peculiarly suit the type of market and competition which characterize the modern world. No model depicts the contracting event as one of active negotiation. Consequently, the existence of oppressive terms in the form or the charging of an exorbitant price may reflect not the imperfections of the bargaining process nor the absence of negotiations over terms between buyer and seller but the imperfections of the market mechanism. Economic agents have been supplied with inadequate or incorrect information; rational responses in this environment, whether intended to maximize profits or to optimize utility within a budget constraint, do not lead to optimal results. Even in the presence of bargaining or the absence of unfair surprise and advantage the market would not supply the right goods at the ideal (competitive) price to all consumers because the malfunctions do not lie in the two-person interactions but in the more complex movements of the market as a whole.

Notes on Form Contracts

1. The introduction of the phrase "contract of adhesion" into American law is usually credited to Edwin Patterson, The Delivery of a Life-Insurance Policy, 33 Harvard Law Review 1982 (1919). Other significant articles in the development of this analysis include Karl Llewellyn, What Price Contract?—An Essay in Perspective, 40 Yale Law Journal 704 (1931); and Friedrich Kessler, Contracts of Adhesion—Some Thoughts About Freedom of Contract, 43 Columbia Law Review 629 (1943). For a formal economic analysis, see Avery Katz, Your Terms or Mine? The Duty to Read the Fine Print in Contracts, 21 Rand Journal of Economics 518 (1990).

2. Unfortunately, the phrase "contract of adhesion" is not easy to define. A portion of the Rakoff article not reprinted here lists several possible defining factors, including (a) whether the contract was printed as a standard form,

(b) whether the contract was offered on a "take it or leave it" basis, (c) whether the contract was in fact bargained over, (d) whether the seller had a monopoly, (e) whether the product being sold was a necessity, (f) whether the seller was commercially sophisticated or the buyer was commercially unsophisticated, and (g) whether the buyer read and/or understood the significance of the contract's terms. Rakoff, Contracts of Adhesion: An Essay in Reconstruction, 96 Harvard Law Review at 1176–80. Factors (a) and (b) in the list above seem most central to the definition of the phrase "contract of adhesion"—but, over the years, all of the other factors have also been mentioned by courts.

3. This uncertainty about the best definition of a "contract of adhesion" may reflect a deeper uncertainty about which of these factors, if any, ought to justify courts in scrutinizing the reasonableness of a seller's terms. Thus, the earlier discussions of monopoly power (supra at pages 306–14) and imperfect information (supra at pages 315–24) may be relevant to deciding whether contracts offered by a monopolist, or contracts offered by sellers in markets characterized by "lemons" problems, ought to be treated as contracts of adhesion. If this approach is followed, however, the phrase "contract of adhesion" becomes little more than a conclusion or summary phrase to describe "those contracts, whichever ones they are, whose terms ought to be scrutinized by the courts."

4. Can it be argued that even without monopoly power or imperfect information courts should still scrutinize the reasonableness of any contract terms offered in a standard form on a take-it-or-leave-it basis? That is, should factors a and b in the list above be *sufficient* to define a contract of adhesion, and to justify the courts in refusing to enforce any unreasonable terms? The Rakoff article, in a portion not reprinted here, effectively limited its analysis to markets characterized by "lemons" problems. Rakoff, Contracts of Adhesion: An Essay in Reconstruction, 96 Harvard Law Review at 1226–27 (1983). As the Leff excerpt points out, however (supra at page 330), if a contract is defined to require a certain amount of "dealing" or haggling, a form offered on a take-it-or-leave-it basis does not look very much like a contract. Such a document might therefore be viewed suspiciously by the courts even if it were used in a market not plagued by informational problems—for example, even if it were used by a corporation selling to a small number of well-informed industrial buyers.

However, the only reason for singling out take-it-or-leave-it forms as a separate category is if there is some reason to view take-it-or-leave-it forms with more suspicion than other contracts. The Kornhauser excerpt argues against this position on the grounds that, especially in today's mass economy, there is nothing inherently virtuous about individualized bargaining and nothing inherently suspicious about refusing to bargain. Instead, Kornhauser argues that courts should focus on traditional problems, such as fraud or duress, and on broader structural problems, such as monopoly power or imperfect information.

5. In its final paragraph (supra at page 332), the Leff excerpt raises the concern that courts may not be very good at identifying reasonable terms. In particular, he raises the concern that if courts require sellers to offer more favorable terms, this may raise the price buyers ultimately have to pay, which may not always be in buyers' interests. On the other hand, the Rakoff excerpt points out that courts pick reasonable terms all the time in cases where the parties' contract was silent. For example, courts pick damage rules for parties who have not specified what will happen if one party breaches, and they pick rules allocating the risk of unexpected contingencies under the doctrines of impracticability and mistake. Thus, Rakoff seems less concerned about the possibility that courts might not be very good at selecting terms to replace those found in standard form contracts.

One response to the Rakoff position questions whether courts are very good at picking reasonable terms in other areas of contract law, even when the parties' contracts are silent. For example, the Kull excerpt (supra at pages 149–54) on impracticability and frustration questioned whether courts were very good at allocating the risk of unexpected contingencies. The Schwartz excerpt on relational contracts (supra at pages 187–93) questioned whether courts were very good at filling the gaps in long-term, complex relationships. Your views on how well courts fill these gaps may be relevant in deciding how frequently courts should set aside the terms of standard forms and replace them with the law's usual default rules.

A further response is to *compare* the ability of courts to select reasonable terms, on the one hand, against the quality of the terms likely to be produced by sellers responding to the prevailing market incentives, on the other. This, in essence, is the approach of the Kornhauser excerpt, which advocated a search for imperfections in the prevailing market incentives as a prerequisite for judicial intervention. This approach was also implicit in the earlier readings on monopoly power (supra at pages 306–14) and imperfect information (supra at pages 315–24). See also Richard Craswell, Property Rules and Liability Rules in Unconscionability and Related Doctrines, 60 University of Chicago Law Review 1 (1993).

6. The analysis of contracts of adhesion is closely related to the "reasonable expectations" principle discussed in the readings supra at pages 295–306. Indeed, the two principles overlap to a considerable degree, as both argue that (at least some) standard forms should not be treated as enforceable contracts. If there is a difference between the two principles, the "reasonable expectations" principle argues that the standard form fails to qualify as a contract only when (and to the extent that) it conflicts with buyers' reasonable expectations. Moreover, to the extent that the standard form fails to qualify as a contract, the "reasonable expectations" principle argues for enforcing a contract based on buyers' reasonable expectations instead of enforcing the standard form.

By contrast, more general critics of contracts of adhesion (such as Rakoff) argue that at least some standard forms should not be regarded as enforceable contracts *regardless* of whether the forms do or do not conflict with buyers' reasonable expectations—or, indeed, even if reasonable buyers would not have had any expectations one way or the other. As a consequence, these critics are not committed to replacing the standard form with a contract based on buyers' reasonable expectations, but instead are more willing to let the legal system decide what terms would make a reasonable replacement for the seller's standard form. (Of course, to the extent that identifying buyers' *reasonable* expectations requires courts to decide what terms would be reasonable, the difference between these two approaches begins to vanish.)

§ 5.3.4 **Buyer Incompetence**

A Reexamination of Nonsubstantive Unconscionability

ALAN SCHWARTZ

[T]raditional contract law . . . assumed that people act competently to maximize their personal utility. A contrary assumption, however, apparently underlies much of the recent legislation governing consumer contracts. It is now commonly assumed that many people, in particular the poor, cannot competently maximize their utility, because these people are ignorant, inexperienced, or simply bad at making their preferences and purchases congruent. Even though some people are poor at maximizing their own utility, the issue is whether the law should presume that, as a general rule, consumers act competently. The traditional premise of consumer competence, it is argued below, should continue to govern, because the evidence of widespread incompetence is too unreliable to justify the costs of assuming that some consumers cannot act in their own best interests.

In unconscionability cases, courts often infer from the terms of the agreement itself (such as a broad security interest or an apparently extravagant purchase) that a consumer is incompetent. This approach, however, is circular: the party is ruled incompetent because the deal is bad, while the deal is ruled bad because the party is incompetent. Although the content of a contract is often relevant to whether a consumer has acted competently, a court or legislature that inferred incompetence only from the terms of a

Alan Schwartz, A Reexamination of Nonsubstantive Unconscionability, 63 Virginia Law Review 1053, 1076–82 (1977). Copyright © 1977 by the Virginia Law Review Association. Reprinted by permission.

contract would in fact be banning a clause for substantive reasons alone. The question of whether, in a given case, the substantive objections to a clause, standing alone, are sufficiently compelling to justify nonenforcement should be faced openly, not concealed in the apparently neutral language of behavioral judgments.

Inferences of incompetence, however, are sometimes based on evidence extrinsic to the terms of an agreement. But these inferences rarely are drawn by experts—psychiatrists or psychologists; they reflect judgments about large groups rather than the results of individual examinations or controlled experiments; and they are often based on anecdotal evidence or subjective impressions. For example, a lawyer-scholar, influential in the consumer protection movement, once argued:

> [T]he consumer is often so preoccupied with achieving his primary objective of obtaining credit that he does not evaluate any creditor's offer of exchange in terms of whether the benefit sought is to be paid for at a higher price, in terms of dollars and relinquishment of rights, than he perhaps should pay. The lack of sensitivity to contract terms tends to prevent the consumer from considering alternative arrangements that may be available to him in the market.

Also, a frequently cited note asserted:

> Low income consumers are frequently concerned to satisfy non-material needs by their purchases: status seeking and escapism heavily influence their buying patterns. . . . Many of the poor are shy and unwilling to deal with strangers. . . . Low income consumers . . . are . . . generally less able to make rational choices among products than their middle income counterparts. . . . A low income consumer is profoundly different from a middle income consumer . . . . Underlying his problems . . . is a crucial lack of motivation. Many of the poor . . . no longer regard the attempt [to get more for their money] as worthwhile.

Inferring incompetence in this manner is objectionable, not because such inferences have been proven false, but rather because the evidence is in fact inconclusive. With regard to the evidence, the previously quoted note reflects the common assumption that the poor are much less competent than the middle class at maximizing their personal utility. Empirical studies of the behavior of poor consumers, however, often contradict this assumption. A recent case study of rural poor families concluded:

> [P]oor people do perceive and act in accordance with marginal costs and returns. . . . They are . . . close to their optimum given their circumstances, which is the most we can say of anybody. This means it is appropriate to assume . . . that the poor behave rationally and in their best interests.

Another study examining the consumer credit behavior of 650 "representative" households and 150 black families observed: "In general low-income and minority buyers appear to have realistic perceptions of the structure of the credit market, at least by comparison with other groups in the population." Although some studies do suggest that poor people are incompetent consumers, the empirical evidence is inadequate to justify an assumption that poor people in general are too ignorant, too inexperienced, or too inept to maximize their own utility. Before intervening to strike a contract clause, a court or legislature, therefore, should require strong evidence of such incompetence, because people who are wrongly classified as incompetent are significantly disadvantaged. . . .

The costs of an overinclusive classification of incompetence are twofold. First, an overinclusive classification based on poverty stigmatizes the poor, as being unable to do what "ordinary" people can do. Second, overinclusive classifications unnecessarily restrict a competent consumer's freedom to contract. For example, the Uniform Consumer Sales Practices Act lists, as a factor relevant to a finding of unconscionability, the fact that a seller "knew or had reason to know" that he had taken advantage of a consumer's inability reasonably to protect his interests due to the consumer's incompetence. In light of the recent trend toward increased consumer protection, courts are likely to interpret the phrase "had reason to know" as imposing upon sellers a substantial duty to investigate the competence of consumers. Because such individualized investigations are apt to be too costly to conduct in mass transactions, a seller operating in such a market might refuse to deal with, or raise prices to, persons belonging to groups in which a substantial percentage of members are likely to be judged incompetent. These groups would probably be defined in terms of race, wealth, and other sociologically obvious characteristics. As a result, some members of social groups, such as poor blacks and Chicanos, would face unnecessary restrictions upon their freedom to contract.

In sum, an overinclusive classification of incompetence creates both unfair stigmatization and undue restrictions upon freedom of contract. Moreover, the rigorous evidence that now exists does not strongly support—indeed, it tends to contradict—the assumption that many consumers are incompetent. Therefore, a court or legislature should not allow the possibility of consumer incompetence to militate against enforcing a contract.

Notes on Buyer Incompetence

1. In cases involving one of the traditional forms of incompetence—for example, if the contracting party is under age—the court can usually refuse to enforce the contract without having to decide whether the substance of the contract was reasonable or not. In theory, then, legal incompetence releases the incompetent party from *all* contract terms, not merely those terms the court deems "unreasonable."

In some cases, though, courts may evaluate the reasonableness of a contract's terms in an effort to evaluate the party's competence. While some forms of incompetence (infancy) are easy to recognize, others (mental illness or undue influence) are hard to verify. In borderline cases, courts' judgment about the competence of the contracting party is likely to be influenced by the court's assessment of the reasonableness or unreasonableness of the contract terms. The classic analysis of this judicial tendency is Milton D. Green, Proof of Mental Incompetency and the Unexpressed Major Premise, 53 Yale Law Journal 271 (1944). More recent discussions include George J. Alexander and Thomas S. Szasz, From Contract to Status via Psychiatry, 13 Santa Clara Lawyer 537 (1973); Duncan Kennedy, Distributive and Paternalistic Motives in Contract and Tort Law, With Special Reference to Compulsory Terms and Unequal Bargaining Power, 41 Maryland Law Review 563, 603 (1982); and Alexander M. Meiklejohn, Contractual and Donative Capacity, 39 Case Western Reserve Law Review 307 (1988–89).

2. By contrast, if poverty is treated as a form of legal incompetence, it could hardly justify excusing poor consumers from *all* contracts, regardless of the contracts' terms. Such a rule would leave poor consumers legally unable to sign an enforceable contract—even one whose terms were perfectly reasonable—and thus would discourage sellers or other contracting parties from dealing with poor people. The most that can be argued, then, is that poor customers should perhaps be released from *unreasonable* terms, not that they should be released from all contracts whatsoever.

Are courts well suited to decide which terms are reasonable for poor consumers? In particular, can courts decide what effect more generous terms would have on the prices consumers would have to pay—and whether the more generous terms would be worth the higher prices? This is the analog of the question asked in earlier notes, about the ability of courts to choose reasonable levels of insurance coverage (see the readings on the reasonable expectations doctrine, supra at pages 295–306) or to choose reasonable terms to replace the terms of monopoly sellers (see the readings on market power, supra at pages 306–14). Is it easier or harder for courts to choose appropriate terms for poor consumers?